SEARCHING FOR THE CATASTROPHE SIGNAL
The Origins of The Intergovernmental Panel on Climate Change

SEARCHING FOR THE CATASTROPHE SIGNAL
The Origins of The Intergovernmental Panel on Climate Change

Bernie Lewin

GWPF

ISBN 978-0-9931189-9-9

Set in Minion.
Printed by Createspace.
Published by The Global Warming Policy Foundation.

Earlier versions of some sections of this book appear on the blog *Enthusiasm, Scepticism and Science.*

Contents

Timeline

Acronyms

AEC   US Atomic Energy Commission; replaced by the ERDA in 1974.

AGGG Advisory Group on Greenhouse Gases, 1985–7

CFCs   Chlorofluorocarbons; synthetic chemicals used as propellants in aerosol spray cans

CIAP   Climatic Impact Assessment Program, assessing the potential impact of SSTs, 1971–1974

CO_2 Carbon dioxide

CoP   Conference of Parties to a UN Convention, hence CoP1, CoP2, CoP3, for successive sessions

CRU   Climatic Research Unit at the University of East Anglia, UK, established by Hubert Lamb in 1972

DoE   US Department of Energy; replaced the ERDA in 1976

DoT   US Department of Transport

EDF   Environmental Defense Fund; founded in 1967 for a legal campaign to ban DDT

EPA   US Environment Protection Agency

ERDA US Energy Research and Development Administration; founded 1974 and replaced by the DoE in 1977

FCCC   UN Framework Convention on Climate Change

FDA   US Food and Drug Administration

GARP Global Atmospheric Research Program; WMO/ICSU 1967–; transitioned to WCRP 1980–1983

ICSU   International Council of Scientific Unions; with membership including national academies of science

IEA   Institute for Energy Analysis, Oak Ridge, Tennessee; directed by Alvin Weinberg, 1974–1984

IFIAS   International Federation of Institutes for Advanced Study; private charity established in 1972

IGY   International Geophysical Year, 1957–8

IMOS   US federal inter-agency taskforce on the Inadvertent Modification of the Stratosphere, 1975–

INC   Intergovernmental Negotiating Committee for the FCCC, reporting to the UN General Assembly, 1990–2.

IPCC   Intergovernmental Panel on Climate Change, WMO/UNEP 1988–

NAS   US National Academy of Science

NASA   US National Aeronautics and Space Administration, 1958–

NCAR   US National Center for Atmospheric Research, in Boulder, Colorado, 1960–

NGO   Non-governmental organisation

NOAA   US National Oceanic and Atmospheric Administration, 1970–

NO_x Oxides of nitrogen, NO and NO_2

SBSTA   Subsidiary Body for Scientific and Technological Advice, UN FCCC, 1995–

SCEP   Study of Critical Environmental Problems conference, Williamstown, Massachusetts, 1970

SCOPE   Scientific Committee on Problems of the Environment, ICSU, 1969–

SMIC   Study of Man's Impact on Climate; conference following on from SCEP; Stockholm, 1971

SST   The US Supersonic Transportation program, 1963–1971

UNEP   United Nations Environment Programme, 1972–

WCP   World Climate Programme; begun after the 1979 World Climate Conference

WCRP   World Climate Research Programme, transitioned from GARP as one of WCP's four components, 1980–

WMO   World Meteorological Organization

Madrid, 1995

If we were to fashion a comic strip of a great showdown over the scientific evidence for global warming, we might imagine our evil antagonist as the chief delegate from some fabulous kingdom of Big Oil.

He would arrive in costume from the Arabia of sand dunes, oil and Mohammad. He would be Mohammad, yes, but Dr Mohammad, a scientist with the best education the West could offer, enunciating graciously the *lingua franca* of modern diplomacy. And he would have the most wonderful Big Oil title:

Economic Advisor to the Minister of Oil for the Kingdom of Saudi Arabia.

And so it was Dr Mohammad Al-Sabban from Jeddah who raised his flag once again to speak. Ever polite, but never afraid to re-state his point if it were slightly misconstrued...and persistent... *Boy, is he persistent!* He is legendary at climate conferences for his ability to keep going, tenaciously labouring a point, sometimes solo against the whole room, politely—*And just one more matter if you please Mr Chair*—and, miraculously, all day and into the night if necessary, one time even until dawn, only stopping when the chairman simply said *Enough is enough!* This is diplomacy by exhaustion. Then it becomes consensus by exhaustion.

This sort of negotiation is not for everyone. Those who master it manage to stay calm and hang in through the day and into the night and then rise again the next morning, clear and attentive among delegates drunk with exhaustion. They appear blessed with some superhuman tolerance for what would do in the heads of any of us mortal folk. Mortal folk like Ben Santer for instance. He was the scientist leading the drive to persuade the gathered country delegates to change the intergovernmental report to make way for a weak 'detection' claim.

After so many reports of government and intergovernmental panels, for the first time an official assessment would claim that the balance of evidence was pointing towards a human influence on global climate. However, Santer was not there yet. Indeed, Dr Al-Sabban's resistance was formidable. And he was

1

not working alone, constantly receiving notes for yet another question passed from the Big Business Lobby hovering at the sidelines. Santer could only tolerate so much of this before he snapped. For a moment he lost his cool, barking back at the Saudi: *If YOU are so interested in this topic then why have YOU not joined the Side Group to discuss it?*

It was already mid-morning on the final day when the 'Detection and Attribution' side group reported back to Working Group 1 of the Intergovernmental Panel on Climate Change (IPCC) session in Madrid, November 1995. Santer had undertaken to finalise a proposed new statement on detection, following extensive discussions in that group, which had been established on the first day after the resounding response to his presentation on his latest 'fingerprint' studies. Given this new evidence, Santer had explained, the chapter on detection was out of date and needed changing. The co-chair of the session, Sir John Houghton, agreed. And so did most of the delegates. However, some resistance was expected, and the convening of side groups at these meetings was a strategy to avoid unwieldy debates crippling the plenary discussion, allowing vexing issues of conflict to be thrashed out and resolved in less formal and more intimate exchanges. They were also seen as a way to short-circuit the blocking strategies of those intent on preventing the conference proceeding to some positive resolution. Other side groups had been formed on the first day, but none caused quite the controversy that this one did when reporting back to the plenary.

There are a number of accounts of that final day in Madrid, but the *Australian Delegation Report* is notable for its frankness:

> Dr Al-Sabban suggested that—because the text had been prepared by a special group which he could not attend because his was a small delegation that was unable to cover the various parallel meetings—the Session use instead the exact text of the 'Concluding Summary' of the relevant chapter of the complete report. A brief acrimonious exchange followed with Dr Santer telling Dr Al-Sabban that if the issue was as important as he said, he (Al-Sabban) should and could have given it priority. He declined to accept Dr Al-Sabban's suggestion to use the 'Concluding Summary' on the basis that 'it is not an accurate summary of the science'.[1]

The Concluding Summary to which Al-Sabban referred was in Chapter 8 of the final draft of the IPCC's second scientific assessment, which had been circulated to all the country delegations well before the meeting. Chapter 8 was the critical chapter on 'Detection and Attribution' and its conclusions were thoroughly sceptical of any claim that a warming *detected* in the global climate record could be *attributed* to the human influence. This plenary meeting of country delegations was only supposed to agree on a summary of the entire report; the so-

2

called 'Summary for Policymakers'. It was not supposed to change the under-lying report itself, which had been written by experts and subject to extensive peer review. Santer had led the writing of Chapter 8 but now he was asking for it to be changed.

> Dr Al-Sabban sought a ruling from the Chairman as to whether IPCC procedures did not require that the Summary for Policymakers be made consistent with the underlying scientific chapters rather than have the un-derlying chapter brought into line with a politically negotiated Summary for Policymakers. Sir John Houghton ruled that there is nothing wrong with changing the underlying chapters to bring them into line with a bet-ter formulation of the science agreed by the Working Group.

Bert Bolin, head of the IPCC, was also at the working group session and he gave his support to Houghton's ruling. Dr Al-Sabban then protested that,

> ...through the last six years of his involvement in IPCC, he believed he had several times been prevented from inserting text in the [Summaries for Policymakers] on the basis that it was not based on the underlying chapters. He indicated that he had always accepted that position and be-lieved the ground rules were now being changed.[2]

Indeed they were. The 'Concluding Summary' of Chapter 8 was removed to make way for the detection claim. Many years later, Houghton published an article reflecting on this IPCC working group session under a banner 'Meetings that changed the world'. According to Houghton, without the triumph of science over vested interests at this Madrid meeting, global action on climate change could not have proceeded to the climate treaty protocol agreed in Kyoto two years later.[3] According to Houghton, the IPCC's detection claim saved the treaty process. It also saved the IPCC.

1 Introduction

In 1987, when the World Meteorological Organization (WMO) decided to assess the threat of climate change through an 'ad hoc intergovernmental mechanism', there was a general expectation that this would be a limited affair: experts nominated by the few dozen interested countries would draft a short document to assist the policy development of its members. Three decades later, several hundred international conferences and workshops have been convened, producing nearly 50,000 carefully drafted and extensively reviewed pages of science, including five full assessment reports, with the prospect of another on the way. A science-policy experiment unfolding on the grandest scale, the Intergovernmental Panel on Climate Change (IPCC) remains at the centre of perhaps the greatest global policy adventure since the fall of the Berlin Wall.

Yet despite its importance there are gaping holes in the public record of the IPCC's early days. Moreover, published accounts of its origins in the postwar science-policy arena remain vague and fragmentary at best, with important developments during earlier atmospheric scares obscured by heroic gloss. For those wishing to come to grips with this science-policy monolith, a full account of its emergence from postwar science policy is now required. In this account, the origins of the scientific preoccupation with global warming is traced back to the rise of global environmentalism during the early 1970s, when began the first in a series of global atmospheric scares. From there, we move through these various scares to the beginnings of the global warming scare, the establishment of the IPCC and its first two assessments.

The reader should note right away that this is not primarily a story about politics or policy. Rather, the story of the origins of the IPCC is a story of modern science. In all the sound and fury about future global atmospheric catastrophes, with all the lobbying, the rhetoric, the testimonials, the panels, the reports and the headlines, we sometimes forget that this is not just another wave of scares surging through society, much as they have since time immemorial. These recent forebodings of self-inflicted global catastrophes are different in that they all defer to the authority of modern science. Without the authority of

5

science, they are nothing.

It can be astounding at first to consider that such a premodern phenomenon as apocalyptic foreboding should re-emerge so powerfully upon this so-very-modern authority. Yet perhaps we should not be so surprised by this turn of events when we consider that it was some time in the late 20th century in many Western societies that the institutions of modern science surpassed the churches as the ultimate oracles of truth. And, just as once religious doctrine informed the policy of state, so now science is called on to inform social policy. In the span of a few generations, science and its institutions have been elevated to this great status, with all the funding and privileges that entails. Thus perhaps we should only expect that, as with religion, science could also sometimes be drawn into corruption by the powerful social influence of apocalyptic scares.

In order to find some perspective on the extent of this corruption, we need to recall exactly *what is this thing called science?* This is not so hard to grasp as some commentators imagine. Firstly, there are the formal sciences: mathematics (arithmetic and geometry) and logic. Then there are the sciences of experience. Empirical science is about analysing experience and developing a coherent formal account of it. Natural sciences have always been like this; theories account for nothing without validation in experience. This science of observation is the science that has been promoted and defended by the Royal Society and by national academies of science since the 17th century.

In this understanding of natural science, the science behind the global atmospheric scares that emerged in the 1970s is remarkable for its detachment from any empirical grounding. The science behind efforts to remove smog from cities, to remove industrial pollutants from streams, to preserve and regenerate habitat and many other *local* environmental problems is based on observed environmental effects. Even with problems that some sceptics might question and consider illusory, such as acid rain, the debate is still about the measurable impacts. However, when we consider global warming and the other global atmospheric scares before it, we find that they are all based on predictions about impacts in some far-off future. It is only then that real validation can be achieved. In place of any direct validation by evidence, we have, instead, theoretical models. These models garner no small status due to their ever-increasing sophistication, and they might be important, useful and justifiable in some way. But while they remain theoretical, *they are not science.* Nevertheless, reporters, activists and politicians take the models as supported by the full authority of science, and they then often extrapolate and exaggerate their predictions to conjure up fanciful images of imminent catastrophe. The ensuing scare generates funding to support further scientific investigations of the threat, where the predictions of the modelling distort the approach and bias the findings.

6

Many scientists who work under the considerable funding generated in this way are all too aware of this detachment from the empirical ground, yet they remain committed to the ideals of science. Keen to reconnect their work with observations, they strive to detect the first hints of evidence of the disasters predicted by the models. When first detection is achieved, it legitimises the science in the eyes of the scientists, and also in the eyes of the public, and so such claims have considerable political force. This is why our story of science and its corruption gives much attention to the struggle for 'first detection'. It is also why our story ends with the first, albeit weak, global warming detection claim by an official assessment body. This came with the publication of the IPCC's second assessment in 1996. Widely celebrated as it was, this detection claim ensured the IPCC's future. However, criticism of the process by which the IPCC came to announce this claim threatened great reputational loss, not only to those who based their policy position on the claim, but also to government-funded science generally. As a consequence, many institutions of science closed ranks behind the panel to protect it from criticism and to actively promote its detection claim as underpinning a policy response to the global warming scare.

Part I

Ozone

2 The postwar boom and its discontents

Our story is a story of America. It is true that the most sustained public and political support for the global warming scare would come from Europe. And it is true that a coalition of poor countries would shape the global response to the crisis even before the treaty negotiations began at the United Nations. However, the character of the scare, especially its scientific character, was nurtured in the polity of the United States. It was born there and it matured there. Even today, it parades its Yankee colours more than many of those involved, friend or foe, care to admit. Therefore, it is in the United States that we begin, and specifically in the economic boom that followed victory in the Second World War.

And what a boom it was! The United States gained all the advantages of reconstruction without having first been destroyed. The flood of US dollars into the European renewal shifted market control away from London and into the hands of the bureaucrats in Washington and the traders in New York. Science drew great advantage from the US-led recovery. The enormous federal investment of the war years continued throughout the 1950s and 1960s, with the tax burden justified by the Cold War. For example, aeronautics research continued apace under military command until 1958, only then to give way to a sudden impulse for so much more. Not ten years after its foundation, while trying for the moon, the National Aeronautics and Space Administration (NASA) had become a vast conglomeration of scientific and technological specialities employing 36,000 people across many states. This was not the only civilian outgrowth of military research: out of the Manhattan Project grew the Atomic Energy Commission, which served civilian as well as military purposes and employed thousands of scientists. The massive federal funding extended also to private industry, as we shall see.

11

This postwar expansion of American science would leave behind the British, with their considerable achievements across the previous centuries, but comparison with the earlier prestige of German science is perhaps the most striking. Prior to the war, much of the research in fields such as chemistry, nuclear physics and geology was led by German-speakers and published in the German language by scientists in central, eastern and northern Europe. From the late 19th century and through the First World War, those scientists who could not read German had to rely on translated summaries if they were to keep up with the latest developments. But then the war that devastated the fatherland also devastated this culture. In the 1950s, while Germany was still rising from ruin in divided foreign tutelage, the pick of surviving German chemists and physicists were working in US industries and for the US government. This US domination of science and technology meant that English—the language of the fading empire which it now outshone in the 'post-colonial' world—became the undisputed technological *lingua franca*. The universal language of air traffic control would also become the command language of computer operating systems. Soon enough, any claim to scientific precedent was all but ignored until published in the language of US science.

The triumph of US science in the postwar years had a popular manifestation in the celebration of American 'know-how'. This was an unbridled confidence in scientific successes and demonstrable technological progress. While food rationing continued in Britain, America had the highest living standards in the world, symbolised by the middle-American daddy motoring the wife and kids down the highway of the future in his brand-new car. A cliché, but true, is the prevailing ethos of rapid and interminable progress. Progress became an unquestionable virtue, at least in the rhetoric of the time. There was progress in healthcare; the eradication of diseases that had plagued mankind for millennia. There was progress in wealth, an endless improvement in comfort and convenience. And, in the historical imagination of Americans, this progress was towards and beyond frontiers. In its greatness, the frontiers of America were now beyond America; indeed, theirs had become the frontiers of mankind. At the peak of this aspiration, during the Kennedy presidency, there was no doubt where the new frontiers lay. They were up there in the boundless skies.

The conquest of the atmosphere, of space and of outer-space, realised a science fiction fantasy of a quintessentially American enterprise that was propelled into life by the shame and embarrassment of 1957, the so-called 'International Geophysical Year'. An old-world idea, long in the planning, the summer of 1957 would begin a year of cooperative, coordinated and simultaneous investigations of our shared planet. The Americans boasted much, spent big, and then delivered impressive results. However, this was all for naught when the Soviets—*the communists!*—on time and as promised, sent the first manmade satellite arching across the sky. *Across the American sky.* To surpass this audacious display of technological prowess the Americans would break free of Earth's gravitational control to venture into deep space. *The moon was only the beginning.** The race to become the first space nation so dominated its work that it is easy to forget that NASA was more than just the Space Administration. There was in fact much activity in aeronautics and in atmospheric science, and these lesser conquests are more important to our story, especially where they involved the lower atmosphere—the troposphere—and its turbulent weather.

From the earliest colonial days, North America has had a reputation for weather extremes. Massive snowstorms, floods, droughts, tornadoes and hurricanes featured prominently in local and national folklore.[6] The successive waves of pilgrims and migrants might have arrived in a land free from old-world oppression, but unanticipated environmental disasters—if only the 'dustbowl' of the 1930s—served to remind modern Americans that, economically and socially, they remained subjugated to the vagaries of weather. The postwar conquest of the weather was not only about prediction, it was also about control. President Kennedy was only evoking the *zeitgeist* when, in a 1961 address to the UN General Assembly, he imagined 'further cooperative efforts between all nations in weather prediction and eventually in weather control'.[7],† Even if we take only prediction, this was boundlessly ambitious. It is easy to forget how at this time it was widely believed that advances in the monitoring of weather systems would eventually deliver such precision to forecasting that weather surprises would become a thing of the past. Up until the early 1960s, there was no particular reason to doubt that one-to-one representation of natural phenomena could be achieved, or at least approximated, in a faster, more powerful computer simulation. There was no reason, not until this Laplacian vision was shattered when Edward Lorenz founded 'chaos theory' with a demonstration of the impossibility of precisely predicting the future state of stable systems of the kind that seem to command our weather.

---

*The year the war ended, a report was circulated entitled *Science, the Endless Frontier: A report to the President on the Program for Postwar Scientific Research*.[4] This outlined a plan for continuing the wartime federal science funding in order to push forward science's 'endless frontier' during times of peace. Implementation of the plan included the establishment of the National Science Foundation in 1950. The NSF's 10th anniversary was marked by the report's republication with an introduction by its director Alan Waterman. In summarising the remarkably successful realisation of the original plan, Waterman notes the unplanned and unanticipated emphasis on space research that followed the shock of the Soviets' 'Sputnik' launch three years earlier.[5]

†This suggestion by Kennedy would be followed by a UN General Assembly resolution (Res 1721 (C)) that was the impetus for the WMO and the ICSU to establish the Global Atmospheric Research Programme (GARP). This was begun in the late 1960s (see p. 140).

The mission to make the forecasting of a meteorological event as precise as the forecasting of planetary eclipses was one thing, but, importantly, Kennedy also envisaged 'weather control'. The weather modification movement took off in postwar America with all the intensity of a military campaign. It was *a war on weather*. This should be no surprise, although we often forget the military roots and associations of modern meteorology. In fact, many modern national weather services developed under military control, including, no less, the UK Met Office, which came under air force command in 1919 and remained an agency of the Ministry of Defence until very recently. In the USA, atmospheric science was critical to military aerospace development; after all, it is through the atmosphere that their planes and rockets must fly. But the military also required weather forecasters, and in droves. In his book, *Fixing the Sky*, James Fleming explains:

> During World War II, the US Army Forces and the US Navy trained approximately 8000 weather officers, who were needed for bombing raids, naval task forces, and other special routine operations worldwide. Personnel of the army's Air Weather Service, an agency that was non-existent in 1937, numbered 19,000 in 1945. Even after demobilization, the Air Weather Service averaged approximately 11,000 soldiers during the Cold War and Vietnam eras. In 1954 a National Science Foundation survey of 5273 professional meteorologists in America revealed that 43 percent of them were still in uniform on active duty, 25 percent held Air Force Reserve commissions, and 12 percent were in the Navy Reserve. Thus almost a decade after World War II, 80 percent of American meteorologists still had military ties. Postwar meteorology also benefited from new tools such as radar, electronic computers, and satellites provided by or pioneered by the military.[8]

By the middle of the 20th century, the science of weather modification already had a long history, but this was mostly in attempts to advance the ancient art of rainmaking. Then, during the 1950s and 1960s, all sorts of interventions were envisaged against the raid of hail and the scourge of frost. Hurricanes would be quelled or diverted out of harm's way. Massive geo-engineering projects were also proposed, with some aimed at permanently changing the climate. These included the redirecting of ocean currents and the removal of Arctic ice in all sorts of ways, even by nuclear explosions. Most of this did not advance beyond excited speculation, and then, where it was applied, it did not get very far.[9] But, *who was to know?*

Here we come to another aspect of postwar technological progressivism, where the dark side of this progress is acknowledged in pessimistic resignation to its inevitability. This technological fatalism is encapsulated in a pop-

14

ular essay published only a decade after the war's end, entitled 'Can we survive technology?'[10,11] Its author, the Hungarian immigrant John von Neumann, had become famous for advancing electronic computing during the secret wartime development of the atomic bomb. Less well known is his leadership of a team pioneering the computer modelling of weather systems. In this 1955 essay, von Neumann predicts that techniques of weather and climate modification would unfold within a few decades 'on a scale difficult to imagine at present'. He foresees 'fantastic effects', where 'the climate of specific regions... might be altered' and envisages that these steps would be undertaken without any reservations because of the enormous and undeniable benefits. The main difficulty would be in predicting the full scope and degree of intended and unintended consequences. However, he believed that knowledge of the dynamics and the controlling processes in the atmosphere would soon be sufficient to make such predictions. Lorenz and the chaos theorists might soon disagree with that, but in 1955 von Neumann had little doubt that we will be able to carry out the analysis needed for controlled and predictable intervention on any desired scale.

For von Neumann, the dark side of these inevitable developments in science and technology would be those consequences entirely *intended*. While the underlying science and much of its technology is generally value-neutral, von Neumann predicts that purposeful destructive applications will be unavoidable. Sometimes the destructive will come first. Referring implicitly to his own work—first with the Manhattan Project and then, after the war, with the Atomic Energy Commission—he says that 'even the most formidable tools of nuclear destruction are only extreme members of a genus that includes useful methods of energy release or element transmutation'. In other words, we must take the good with the bad... and hope we survive. Survival through control of the destructive applications would require a level of international cooperation that might be hard to achieve. Any political crisis that would ensue from the development of any 'particularly obnoxious form of technology' could not be resolved by trying to prohibit or inhibit its development because any 'banning of particular technologies would have to be enforced on a worldwide basis, but the only authority that could do this effectively would have to be of such scope and perfection as to signal the *resolution* of international problems rather than the discovery of a *means* to resolve them'. For von Neumann, it seems that such *resolution* was nowhere in sight.

So when it came to the idea of climate and weather control, von Neumann was in no doubt that such knowledge could be used for nefarious purposes: while expecting 'the most constructive schemes for climate... control' we should also recognise that they will be based 'on insights and techniques that would also lend themselves to forms of climatic warfare as yet unimagined'.[12,13] That was before Sputnik. After Sputnik, well, *it was anybody's guess what else the Soviets were up to.*

Behind discussions of weather modification, the prospect of environmental engineering as a technique of war was always lurking. Some of the published accounts of possible military applications were framed defensively in terms of another surprise from the Soviets. But, even there, ambiguity remained about whether the schemes were responses to a genuine threat or, otherwise, thinly disguised grant proposals.

From the late 1950s, when weather modification research projects were being funded, the secrecy surrounding the topic bred speculation of conspiracy. Not long after the Cuban Missile Crisis, Fidel Castro accused the USA of redirecting Hurricane Flora on its destructive path across Cuba. Castro had good reason to theorise conspiracy, and, indeed, conspiracy became reality for the scientists working at that very time on the hurricane control experimentation known as 'Storm Fury'. The eventual realisation of the scientific futility of their efforts was devastating, and doubly so when they also discovered that they really had been working towards the sinister military ends anticipated by Castro. Confirmation of conspiracy also came, although much later, when the Pentagon Papers revealed that the US had not only used defoliant in Vietnam, but they had also attempted to bog down the enemy by triggering deluges.[14]

These were remarkable times in America, and especially so for scientists. They could be forgiven for believing that the protection, prestige and prosperity of the Union depended entirely on their ingenuity. In the strength of this belief, they had never had it so good. But, of course, not every American, and not every scientist, shared this view. Scepticism of sci-tech progressivism was already developing during the brief Kennedy presidency. In the early 1960s, movements for civil rights and against the waging of war were coalescing into an opposition that was deeply principled. It was this 'counter-culture' of the burgeoning postwar generation that nurtured a movement promoting the *natural* over the *manmade*. A new 'environmentalism' arose through a broadening of specific campaigns against environmental destruction and pollution. It began to target more generally the industries and technologies deemed inherently damaging. Two campaigns in particular facilitated this transition, as they came to face-up squarely against the dreams of a fantastic future delivered by unfettered sci-tech progress. One of these challenged the idea that we would all soon be tearing through the sky and crossing vast oceans in just a few hours while riding our new supersonic jets. But even before the 'Supersonic Transportation Program' was announced in 1963, another campaign was already gathering unprecedented support. This brought into question the widely promoted idea that a newly invented class of chemicals could safely bring an end to so much disease

16

and destruction—of agriculture, of forests, and of human health—through the elimination of entire populations of insects.

DDT and the Environmental Protection Agency

At the outbreak of the Second World War, while working for an international chemical company in neutral Switzerland, Paul Müller discovered the marvellous qualities of the organic pesticide DDT and its related compounds. One of the outstanding qualities he noted in these chemicals was their persistence, and it was this persistence that was behind DDT's first extraordinary wartime success, when occasional dustings down shirt-fronts or up skirts was enough to control lice among soldiers, refugees and prisoners. DDT would persist in clothing and on surfaces for weeks, even after washing. Its persistence would also be important in later years in the eradication of malaria and yellow fever through periodic spraying in dwellings. Müller's 1948 Nobel Prize citation claimed that there was no doubt DDT had 'already preserved the life and health of hundreds of thousands'.[15]

Some of the first peacetime successes of DDT were in America, when the new organic pesticides were applied on the home front against the hazards of nature. For one of its missions at least, the results appeared nothing short of miraculous. The objective was to eradicate malaria completely from the United States. Operating in an arc of 13 states across the southeast, a key aspect of the Malaria Eradication Program was to spray DDT on the interior surfaces of rural homes. Within three years, millions of homes had been treated, and by 1951 malaria had all but disappeared.[16] Another mission was to exterminate the insects attacking US crops, and aerial spraying became a popular mode of application. Then, it seemed, organic pesticides were everywhere. Swamps, parks, roadsides and other public lands were treated. Various formulations at high concentrations were promoted for unrestricted use on private lawns and gardens. Confidence in the power of these manmade poisons led to even more ambitious plans. Some pests, especially introduced insects, could be eradicated entirely by ambitious programs involving blanket spraying of vast areas.

One of the most ambitious campaigns was to eliminate the Gipsy Moth. In 1957, a plan was hatched to blanket spray across four states, including heavily populated areas such as Long Island. The following year a similar campaign of eradication was mounted against the fire ant, this time across nine states. And so millions of American homes, gardens, ponds, playgrounds, pets and people were showered with the tiny pellets of this relatively new and unknown poison. As campaigners would not neglect to point out, these silent showers were reminiscent of the mystery fall-out descending on farms and towns following secret nuclear testing in the Nevada desert. As with the nuclear testing, DDT was applied without consultation or redress. Fierce protests ensued and campaigns were organised. It was in and around these campaigns that the new environment movement emerged as a revolution within the old nature conservation establishment.

The North American conservation movement had matured during the first half of the 20th century. It had an affinity with political conservativism and a wealthy white constituency, personified in the Republican president Teddy Roosevelt, with his support for the national parks movement, and one of its leaders, John Muir. Muir had founded the Sierra Club, which had grown rapidly and which, by the 1950s, had become an important promoter of nature preservation in the political mainstream. When the huge DDT spraying programs began, the Sierra Club's immediate concern was the impact on nature reserves. But then, as the movement against DDT developed, and as it became increasingly involved, it began to broaden its interest and transform. By the end of the 1960s it and other similar conservation organisations were leading the new environmentalism in a broader campaign against DDT and other technological threats to the environment.

This transformation was facilitated by the publication of a single book that served to consolidate the case against the widespread and reckless use of organic pesticides: *Silent Spring*.[17] The author, Rachel Carson, had published two popular books on ocean ecology and a number of essays on ecological themes before *Silent Spring* came out in 1962. As with those earlier publications, one of the undoubted contributions of the book was the education of the public in a scientific understanding of nature. In warning about the use of these insecticides, Carson explained the interdependence of species; how the elimination of one pest could cause a plague of another or the starvation of a predator or, more generally, how the killing of one species could have repercussions across entire ecosystems. In order to promote the alternative use of biological controls, she revealed to her readers tiny insect parasites and explained their vital role in the whole ecological system. Carson used DDT's extraordinary stability to show how poisons could accumulate in the flesh of animals and pass up the food chain, from insects to small animals and birds, right up to eagles, and pass out into regions far removed from their source (traces of DDT residue would soon be found as far away as Antarctica[18]). Carson used the emerging problem of pesticide resistance to explain how small populations of rapidly reproducing organisms, such as insects, can adapt relatively quickly to changes in their environment. And in general, with her call for caution in our interventions, she revealed to her readers how little science really understood the interdependencies in the web of life. With Carson, popular 'ecology' was born, but it was born to protest. *Silent Spring* became one of the bestselling natural history books of all time.

We will never know how Carson would have responded to the complete ban on DDT in the USA. She was suffering from cancer while writing *Silent Spring* and died shortly after publication (leaving the royalties from its sale to the Sierra Club), but the ban was not achieved for another decade. What we do know is that a full ban was never her intention. She supported targeted poisoning programs in place of blanket spraying, and she urged the authorities to look for alternative and 'integrated control', along the lines of the 'Integrated Pest Management' approach that is common and accepted today. Indeed, DDT presented undeniable advantages, not only in the eradication of malaria but also in allowing farmers to do away with arsenic. While the poisoning effect identified in DDT is limited to insects, arsenic is much less effective as a pesticide and yet much more poisonous for animals and humans. But DDT became much more than a substitute for arsenic, and today, with the benefit of hindsight, there is little dispute that the introduction of new synthetic chemicals warrants a level of caution that was clearly absent when Carson began researching her book in the late 1950s. There had been some investigations of its potential dangers, but more research and monitoring of its unintended ecological consequences was surely advisable before such extensive application. For Carson, an overriding concern was the lack of vigilance and regulatory oversight. While the US Food and Drug Administration had an eye to drugs and foods, there was no independent body regulating pesticides. The Department of Agriculture, which was promoting farming through the promotion of its protection from pests, could hardly be expected to act as a regulator. Carson argued for an independent body to consider the dangers for wildlife and humans.

Overall, by today's standards at least, Carson's policy position was moderate, and so we should be careful not to attribute to her the excesses of her followers. The trouble with Carson was otherwise: it was in her use and abuse of science to invoke in her readers an overwhelming fear. In *Silent Spring*, scientific claims find dubious grounding in the evidence. Research findings are exaggerated, distorted and then merged with the purely anecdotal and the speculative, to great rhetorical effect.

Historically, the most important area of distortion is in linking organic pesticides with human cancers. The scientific case for DDT as a carcinogen has never been strong and it certainly was not strong when *Silent Spring* was published. Of course, uncertainty remained, but Carson used the authority of science to go beyond uncertainty and present DDT as a dangerous carcinogen. And it was not just DDT; Carson depicts us 'living in a sea of carcinogens', mostly of our own making, and for which there is 'no safe dose'. The title of her chapter on cancer, 'One in every four', refers to the claim that the chance of a person getting cancer had recently increased from 1 in 5 to 1 in 4. She links this increased rate of cancer to our increasing exposure to synthetic chemicals during the same period. The inference of the chapter is clear: the reason there is a cancer epidemic is because the unfettered chemical industry has produced a 'world filled with cancer-producing agents'.[19]

That Carson was carefree in her appeal to scientific authority has been noted by many critics, both at the time and afterwards.[20,21] Refutations of the supposed causal link are numerous,[22] yet the cancer scare promoted by *Silent Spring* would run across three full decades, often doing great damage in the process.

Why cancer scares are so successful in generating community fear is easy to understand when their affinities with pre-scientific scares over demonic influences are considered. Carcinogenic effects are often insensible, insidious. They can also take years, even decades, to manifest. However, once a cancer risk is proposed, the label sticks and (as with the witches of old) the burden of proof shifts to those seeking to remove it. Scientifically this is a slow and difficult process, but there is no time to await unequivocal evidence that might clear up the case, as this could be too late for too many victims. While doubt remains, all sense of proportion is lost in the prescription of precaution. But anyway, in the 1960s there was a recent ominous precedent to support the plausibility of any new cancer scare. This was the emerging understanding of the insidious, delayed and deadly carcinogenic effects of nuclear radiation and the risk from deliberate and inadvertent exposure during postwar atmospheric testing.

Armed with Carson's magnificent contribution, the anti-DDT protestations picked up through the 1960s. It was not only the established conservation organisations like the Sierra Club that stepped up to consolidate the campaign. New organisations also formed. The most successful of the newcomers was undoubtedly the Environmental Defense Fund (EDF). Its substantial financial backing drove some spectacular legal challenges and it quickly rose to dominate the arm of the campaign that eventually succeeded in winning a complete ban.

This was no easy victory. Right up to the very end, even after success in the courts, it looked as though evidence-based science might hold out against widespread popular alarm. The Nixon administration had responded to general pollution concerns, and specifically to the DDT scare, by establishing the Environmental Protection Agency (EPA), just the sort of regulatory body Carson had envisaged. An EDF win in the US Federal Court delivered an order for the EPA to begin deregistration of DDT, but this was met with resistance from the first head of the new agency, William Ruckelshaus. Ruckelshaus had already been involved in the controversy before he took up the leadership of the EPA. Two years earlier, in another case brought by EDF, he had filed a briefing that said that DDT...

> ...is not endangering the public health and has an amazing exemplary
> record of safe use. DDT, when properly used at recommended concen-
> trations, does not cause a toxic response in man or other mammals and
> is not harmful. The carcinogenic claims regarding DDT are unproven
> speculation.[‡]

Now as the head of the new regulator, he cited studies by his own staff to support his rejection of immediate deregistration. Instead, the EPA would conduct its own public hearings. These turned out to be no small affair. Testimony
taken from 125 witnesses over seven months was recorded in a transcript that
ran to nearly 10,000 pages. Cross-examination was used to good effect, weed-
ing out dubious claims that might otherwise have gone unchallenged. In the
end, and in the face of enormous public and political pressure, the hearing examiner, Edmund Sweeney, found only limited endangerment to humans and
wildlife from excessive use of DDT. In his report, he recommended against a
full ban.[24] But just two months later this finding was summarily overruled by
Ruckelshaus himself. In the summer of 1972 he ordered the registration of DDT
to be cancelled.

If we are to understand how the EPA ban came about, it is important to
realise that this action succeeded in breaking a policy stalemate that was becoming increasingly hazardous for the increasingly embattled Nixon administration. On one side of this stalemate were the repeated scientific assessments
pointing to a moderate position, while on the other side were calls for more and
more extreme measures fuelled by more and more outrageous claims. Consider, for example, that in earlier hearings held in Wisconsin in 1968–9, the
EDF had claimed that scientific studies showed that DDT dissolved in seawa-
ter would debilitate the photosynthesis of plankton, trigger massive ecological
disruption and lead to extinctions of marine and bird species.[25,26] Such spectacular claims might not withstand scientific scrutiny (in fact, this one turned
out to be entirely spurious), but they would linger in the public consciousness,
and they would serve to encourage another part of the conservation establishment, the Audubon Society, to take an extreme stand on the subject; the society
campaigned not only for a domestic ban on DDT but for a ban extending to ex-
ports, thereby threatening World Health Organization antimalarial programs
in impoverished regions of the world.[27]

[‡]Ruckelshaus WD, Brief for the Respondents, US Court of Appeals for The District of
Columbia Circuit, No. 23813, on Petition for Review of an Order of the Secretary of Agriculture, 31 August 1970. Quoted in an article by Jukes.[23]

Meanwhile, on the other side of the stalemate was the National Academy of Sciences, the US equivalent of Britain's Royal Society. A wide-ranging report by the Academy's life sciences public policy committee was published in 1970. Where it discusses the difficulties of assessing environmental hazards, DDT is presented as exemplar of the case for moderation. The report acknowledges the great difficulties and the significant time-lag in coming to a full assessment of some identified environmental hazards. Indeed, it cites the public health case of smoking, the carcinogenic dangers of which had only recently become evident in the results of longitudinal studies. The lesson taken with respect to DDT is that it should be 'patently advisable' in such cases to 'minimize exposure' until reliable evidence becomes available.

> But surely a rule of reason should prevail. To only a few chemicals does man owe as great a debt as to DDT. It has contributed to the great increase in agricultural productivity, while sparing countless humanity from a host of diseases, most notably, perhaps, scrub typhus and malaria. Indeed, it is estimated that, in little more than two decades, DDT has prevented 500 million deaths due to malaria that would otherwise have been inevitable. Abandonment of this valuable insecticide should be undertaken only at such time and in such places as it is evident that the prospective gain to humanity exceeds the consequent losses. At this writing, all available substitutes for DDT are both more expensive per crop-year and decidedly more hazardous...[28]

Such sober assessments by scientific panels were futile in the face of the pseudo-scientific catastrophism that was driving the likes of the Audubon Society into a panic over the silencing of the birds. By the early 1970s two things were clear: public anxiety over DDT would not go away, and yet the policy crisis would not be resolved by heeding the recommendations of scientific committees. Instead, resolution came through the EPA, and the special role that it found for itself following the publication of the Sweeney report.

It is easy to be cynical about the DDT decision, with its speed and circumstances giving every indication of a purely political intent; the timing *was* spectacularly political. Nixon, who had made much of his record with pollution controls, announced the DDT ban on the eve of the final ministerial sessions of the 1972 UN Human Environment Conference in Stockholm. But cynicism, and any claims of corruption, rest mostly on the presumption that EPA decisions should remain firmly grounded in the science. But, again, if this one had been, it would not have resolved the political stalemate, and science may never have been a primary driver. Consider that Ruckelshaus's background was as an attorney, and he had a long history in public office. Accordingly, he approached his decisions as an advocate for public environmental concerns. Later, in official correspondence with the president of the Farm Bureau, he famously justified his actions by explaining that, while 'scientists have a role to play', EPA decisions are 'ultimately political'.[29] Whatever the original intention in establishing the EPA, Ruckelshaus's attitude set the pattern for its future leadership. While similar science-policy organisations were often run by career scientists, Ruckelshaus's successors over the next three decades all shared his legal background. From the DDT decision onwards—and so, *from the beginning*—the EPA made little effort to hide the fact that it was a political operation; it would act to address public alarm, even where sober scientific assessments would suggest that no alarm was necessary.

This special role that the US EPA came to play is instructive to our whole story. As a quasi-scientific government agency engaged in environmental advocacy, it could take decisions that would be beyond scientific committees and panels, with their overriding concern to preserve their scientific integrity. Indeed, in such scientific bodies there would be much talk about the science-policy interface and how science could be delivered to policymakers while still protecting the scientific process from political corruption. Some scientists and science administrators would try to defend the integrity of scientific assessments against pressure from the policy machine. But when there is an unrelenting environmental scare abroad, eventually something has to give. There are two possible outcomes. Either there is some EPA-like intervention, or the scientific integrity of the assessment is compromised. Sometimes it would not be easy to tell which of these outcomes had occurred in practice, especially when an executive summary—or even a press release—would spin alarming claims out of the heavily-qualified obscurities of the underlying report.

For now, our story remains quintessentially American. Just how American was the DDT scare can be seen by comparison with how the insecticides controversy played out in Europe. In Britain, there was nothing to compare with the obstinacy and ambition of the fire ant eradication campaign. But nor was there anything to compare with the American public's backlash against DDT. As Kenneth Mellanby puts it in *The DDT Story*, the path to a decision on restriction of organic pesticides in Britain was the result of 'discussions mostly held in private, though they were not secret and the reports of all the committees were published. The law was not brought in'.[30] By contrast, in the USA the EDF litigation was very public, very confrontational and heroic, with scientists brought in to play oracles of doom before judges, congressmen and the press. With DDT (as later with the various scares over atmospheric destruction), scientists summoned for testimony found themselves encouraged to reach beyond their expertise and to advocate against industrial interests, with environmentalists cheerleading in the background. Some of them would warm to the game and were able to draw enough attention to their work to compensate for the ensuing animosity from their colleagues.

The smell of conflict attracted media attention and so fuelled a level of public excitement among the American public that would be unimaginable in other countries. Witnessing the remarkable phenomenon of this first great controversy of US environmentalism, British scientists were among those prominent in raising concerns about its impact, not only on the science-policy interface, and not only in the USA, but its more general and corrupting influence on science itself. Consider the perspective of the very scientist who invented the detector that first picked up the widespread traces of DDT residues. Fifteen years after the publication of *Silent Spring,* in his famous essay, *Gaia: A New Look at Life on Earth,* Jim Lovelock reflects on the emergence of environmentalism and its preoccupation with synthetic chemicals, with new technologies and with battling the industries promoting them. He acknowledges the understandable concern, and the need for caution, when, for example, traces of synthetic pesticide residues were found 'in all creatures of the Earth, from penguins in Antarctica to the milk of nursing mothers in the USA'. But he also notes the long-term impact on science when the authority of science is used and abused in the push for a policy response to such findings. Indeed, so he claims, the life sciences have been in 'turmoil' ever since Carson made her famous intervention, 'particularly where science has been drawn into the processes of power politics'.

> When Rachel Carson made us aware of the dangers arising from the mass application of toxic chemicals, she presented her arguments in the manner of an advocate rather than that of a scientist. In other words, she selected the evidence to prove her case…This may have been a fine way of achieving justice for people in those aspects of the problem affecting the community at large, and perhaps in this instance it was excusable; but it seems to have established a pattern. Since then, a great deal of scientific argument and evidence about the environment is now presented as if in a court-room or at a public enquiry. It cannot be said too often that, although this may be good for the democratic process of public participation in matters of general concern, it is not the best way to discover scientific truth. Truth is the first casualty of war.[31]

Before publishing *Gaia,* Lovelock had also used his electron capture detector to trace the spread of another class of synthetic chemicals across the globe. This time it was chlorofluorocarbon spray-can propellants, and the publication of his results triggered another US environmental scare (which we will come to shortly). Lovelock was quick to publish scepticism of the science supposedly

24

supporting the new scare. When questioned by the British press about his response, he expressed frustration at the alarm he had inadvertently initiated. 'The Americans tend to get in a wonderful state of panic over things like this' he was quoted as saying. He disparaged the main scientist involved in raising the alarm for acting like 'a missionary' and called for 'a bit of British caution'.[32]

Lovelock was not the only British expert calling for calm on the chlorofluorocarbons scare. As we shall see, the difference between the USA and Britain (and Europe) on this issue was even greater than with DDT. But, if we return to the late 1960s, we find that this distinctively American excitability also drove the other formative environmental campaign that was to breakout across the wealthy states of the Union at that time; one that gave rise to a very different type of campaigning organisation: Friends of the Earth.

Supersonic transportation

From some time in the late 1950s, and right through the 1960s, there was no doubt about it: the future of transportation would be supersonic. It seemed that everyone in the know believed this. By 1962, the French and the British had finally come to an agreement on building together a prototype passenger aircraft ('Concorde') and the Soviets announced their own similar plans. This supersonic dream was still alive in 1970 when PanAm introduced the first commercial flying 'jumbo', with room for more than 400 passengers in its elephantine belly. By the mid-1970s only a handful of supersonic passenger aircraft were in service and these were restricted to long-haul intercontinental flights. Yet the supersonic dream remained alive. Despite the commercial success of wide-bodied subsonic jets in the late 1970s, supersonic development programs were still being proposed and implemented in the belief that the technological problems would eventually be resolved and that the slow old jumbos would become obsolete. The complete failure of this vision perhaps explains why the rolling controversy over the environmental consequences of supersonic flight is mostly forgotten today.

As with the space program, it was foreign competition that brought the US supersonic passenger aircraft program into life. Back in the early 1960s, the British and French governments, and the Soviets, were pouring massive funding into supersonic R&D, and this was seen to threaten the dominance of US companies in the aerospace industry: their leading positions might be lost if they did not quickly get in on the new game. However, the investment and risks were too high for any private company. And so, in the summer of 1963, President Kennedy announced that his government had been 'spurred by competition from across the Atlantic' towards 'partnership with private industry to develop…a commercially successful supersonic transport superior to that being built in any other country'.[33] That it had to be 'superior' to any other was a *critical* (and some would later say, a *fatal*) specification. Kennedy envisaged that, by the end of the 1960s, a jet faster than twice the speed of sound would be carrying passengers to all corners of the globe. A huge contract to build just two prototypes designed for speeds much faster than Concorde was awarded to Boeing in 1966. Thus began the so-called Supersonic Transportation (SST) public–private partnership, a program that was beleaguered from the beginning, right through to its collapse in 1971.

As much as there were troubles with the design, there were troubles also in getting the American public to come along for the ride. Boyish dreams of the supersonic age were soon moderated by some of the supersonic realities. The first problem was the very signature of supersonic flight: the magnificent sonic boom. It was found that the shockwave trailing a plane passing through the sound barrier startled folks as much as the report of a rifle at close range. The hope was that familiarity would diminish the distress. In order to test this, from the late 1950s trials were conducted over various cities, where preference was given to those with some civic investment in the industry. The most extensive test would be the last, conducted over Oklahoma City in 1964. This involved eight booms per day, one per hour over a six-month period. The boom impact was found to be uneven, with points of extreme intensity cracking buildings, smashing windows and rocking ornaments off shelves. There were thousands of building damage claims and groups formed to demand an end to the torment. Those evaluating the public response tried to generate positive spin, but eventually it would be decided that flights could not be conducted over populated areas. And even then, after concern was raised about the possible impact on wildlife in remote areas, it was accepted that the only tolerable routes would be over the oceans.[34]

The next concern was with the noise of the jet engines at take-off. Initially, it was hoped that noise-tolerance levels for supersonic craft would be lifted, but in the end the costs and constraints of the various attempts to meet the noise restrictions led the program leadership at Boeing to realise that the production of an entire SST fleet would not be commercially viable. This conclusion was reached late in 1969 and it tore a rift between Boeing and those on the government side of the contract. Still, an expectation remained that all the work on the commissioned prototypes would not be wasted and that these at least would continue to be funded and built. The trouble was that the billion-dollar government funding for the prototypes was justified by the viability of commercially-funded fleet production, and so the view to the contrary was not publicised.[35]

The controversy was thus in a curious situation when environmental groups came together with a coordinated anti-SST campaign early in 1970. Prototype construction was challenged on concerns about the expected air pollution from the exhausts of fleet sizes varying from 400 to 850 craft, which in reality no-one intended building. Congressional opposition developed during 1970, and the controversy came to a head on 18 March 1971 when the House rejected a $290 million funding bill. A vote in the Senate two months later signalled the end, and hopes for the revival of the SST program subsided over the following months. Nevertheless, hopes for a commercial supersonic industry of some sort lingered on to 1976, when Concorde was cleared to land on US tarmac, and for many years after.

The contribution to the demise of the SST of the environmentalists' campaign is sometimes overstated, but that is of less concern to our story than the perception that this was their victory. While the DDT campaign was struggling to make headway, the SST campaign would be seen as an early symbolic triumph over unfettered technological progressivism. It provided an enormous boost to the new movement and helped to shape it. Back in 1967, the Sierra Club had first come out campaigning against the SST for the sonic shockwaves sweeping the (sparsely populated) wilderness over which it was then set to fly. But as they began to win that argument, tension was developing within the organisation, with some members wishing to take a stronger, more general and ethical stand against new and environmentally damaging technologies such as this. When, in 1969, the Sierra Club executive director David Brower broke away and established Friends of the Earth, he wasted no time in commencing campaigns against both nuclear power and the SST. It was not long before the Sierra Club came around, joining Friends of the Earth to form the 'Coalition against the SST' just in time for the first Earth Day in April 1970.

The phenomenal success of the Earth Day celebrations and protests across the USA heralded the environment movement's arrival in the mainstream. In the lead-up to the big day, interest snowballed across the country, with schools, churches, politicians, business groups and large corporations declaring their support. The press coverage of preparations was extensive and mostly positive. Then, on the day, the television networks caught smiling schoolchildren picking up litter on a sunny mid-week afternoon. They also beamed home the grand spectacle of an estimated 100,000 city folk parading down New York's Fifth Avenue and into Central Park. New types of awareness-raising events, including teach-ins, attracted interest too, with attention primarily given to the reduction of local water, land and air pollution. The SST campaign gained something of a profile, if not for Congressmen taking the opportunity to air their views on the subject, then for the only arrests of the day, at an airport 'die-in' to protest prospective supersonic pollution.[36,37]

With popular support for environmental causes already blooming across the country, and with the SST program already in jeopardy, scientists finally gained their own position of prominence in the controversy when they introduced some new pollution concerns. This began when the anti-SST coalition worked with the leading anti-SST campaigner in the Senate, William Proxmire, to capitalise on the Earth Day publicity by organising a series of congressional hearings a few weeks later. In the first hearings, questioning of those involved in the program revealed enough of the findings of a suppressed report to present a strong case against the SST on economic grounds alone. The unresolved problem of noise suppression was emphasised by the sensational admission that, on the current engine design, the 'sideline' noise on take-off would be similar to that of 50 Boeing 747s leaving at once.[38] If that wasn't enough, environmental concerns were also raised in the most general and cursory terms about the aircraft's exhaust emissions. These first expressions of pollution concerns would soon be followed by others, from scientists who were brought into the debate to air speculation about various atmospheric catastrophes that would ensue if these supersonic birds were ever allowed to fly.

SST, water vapour and climatic change

The first of these catastrophes concerned the effect of supersonic aircraft on the climate. As air traffic increased during the mid-1960s, attention was drawn to the visible white lines of condensed and frozen water vapour that were sometimes seen trailing behind ordinary subsonic jets. These 'contrails' would occasionally linger for hours and spread into bands of manmade cirrus clouds. In 1963, a *New York Times* front-page report, opening on the question of deliberate weather modification, soon switches to inadvertent effects. It describes the director of the National Center for Atmospheric Research, Walter Orr Roberts, pointing to streaks of contrails crisscrossing a cloudless blue Colorado sky and explaining that 'by this afternoon they will have spread into a thin sheet and you won't be able to tell them from natural cirrus cover formations'.[39] The speculation was that this cloud might retain reflected heat and cause warming. In 1965, another front-page story repeated this speculation with particular reference to the effects of *supersonic* jets. Subsonic aircraft fly in the upper troposphere, but the planned supersonic jets would be flying higher, in the much more stable air of the lower stratosphere. Philip Abelson, the editor of *Science*, the journal of the American Association for the Advancement of Science, had speculated in a speech that at those altitudes the contrails might linger even longer. A whole fleet of SSTs might produce a permanent pale haze.[40]

28

Until this time, most of the interest in affecting the atmosphere was in relation to deliberate attempts to modify the weather. In the late 1950s, President Eisenhower had established an Advisory Committee on Weather Control, and then in 1964 'public interest in weather modification' caused the National Science Foundation to establish a Commission on Weather Modification. Around the same time, the National Academy of Sciences set up the Panel on Weather and Climate Modifications to review the field and to consider its future 'potential and limitations'. However, one section of its report of 1966 revealed a shift of interest around this time, addressing possible *inadvertent* atmospheric effects. Consideration was given to industrial carbon dioxide emissions as well as the emissions of water vapour by supersonic aircraft. On carbon dioxide, it recommended further monitoring and study, although it noted that 'dire predictions of drastic climatic change may well be unjustified'.[41] The report was similarly subdued on SSTs, finding that persistent contrails were unlikely to be problematic. The extreme dryness of the stratosphere meant that clouds would not be expected to form. The only time that clouds had been observed at that height was in the extreme cold of the polar winters. The panel found that whole fleets of SSTs might raise the humidity a little on popular routes, but this would not be sufficient for cloud formation. In other words, the panel found no basis for the climatic concerns about stratospheric flights.[42]

This sceptical view had already been outlined in the 1965 *New York Times* story run prior to the report's release, which nevertheless led with the more newsworthy speculation otherwise. Nor was the report's release enough to settle the matter, and scientists continued to publicise their concerns about the effects of supersonic contrails. The issue received sporadic press attention as the anti-SST campaign was building. The trouble was that there was very little evidence to support or contradict these concerns. What evidence was available did not help the case. Contrails were not known to linger at that height. Humidity seemed to be naturally variable, but always way below saturation point. This fact had been established during the war, when British meteorologists considered the problem of how best to prevent tell-tale contrails forming behind their bombers.[43] Developing instruments to reliably measure the low levels of humidity in such cold and thin air proved difficult, and by 1970 there was still very little data. The best evidence available was a newly-published time series from 1964 of measurements taken in the sky over Washington. This showed significant variability, but when smoothed there was an upward trend across the record. Indeed, there had been a rise of around 50% in just six years.[44,45] These widely-publicised results suggested that the estimated effect of an entire supersonic fleet would be entirely overwhelmed by natural fluctuations.

Nevertheless, the speculation continued. The UN Human Environment

Conference in Stockholm was not scheduled to begin until the summer of 1972, but from late in 1969 preparatory work was undertaken to ensure the successful launch of a cooperative global environmentalism at that meeting. Two months after Earth Day, in the summer of 1970, Caroll Wilson, an energy specialist at the Massachusetts Institute of Technology, led a month-long scientific conference designed to inform the policymakers' discussions in Stockholm. This live-in inter-disciplinary meeting of over 50 scientists surveyed a catalogue of concerns around the *global* environmental impacts of mankind's activities. The Study of Critical Environmental Problems (SCEP) conference, as it was called, involved scientists employed by government agencies and private firms as well as academics. They came out with a few strong recommendations, including a call for a drastic reduction in the use of DDT, although not a full ban, and not due to cancer concerns. But the main outcome of the conference was a conclusion that the general lack of data inhibited proper assessment, so most of the recommendations were calls for more coordinated environmental monitoring of one kind or another. This was especially so with respect to atmospheric pollution and its effects on global climate.

For this SCEP conference, a special working group of atmospheric scientists had been assembled to consider the various possible inadvertent global climatic effects. Led by William Kellogg from the National Center for Atmospheric Research, the group considered all sorts of atmospheric changes, including increases in greenhouse gas levels due to industrial emissions. This was assessed to be both uncertain and relatively unimportant, and it was certainly not headline news at a time when there was a global cooling trend. Indeed, there was not much to excite the press reporters in the two days of briefings at the end of the conference. What did make the front page of the *New York Times* on 2 August 1970 was concern about another climatic effect highlighted in the executive summary of the report. The headline trumpeted 'Scientists ask SST delay pending study of pollution' (see Figure 2.1).[46] The conference had analysed the effect of emissions from a fleet of 500 aircraft flying in the stratosphere, and concerns were raised that the emission of water vapour (and to a lesser extent other emissions) might absorb sunlight sufficiently to have a local or even global effect on climate. Later, in congressional testimony, Kellogg downplayed these findings, noting that the increase in humidity would not be great and that SCEP had only recommended study before full fleet deployment; there had been no suggestion that the prototypes under development by Boeing should be delayed.[47] The climatic change argument remained in the arsenal of the anti-SST campaigners through to the end, but it was soon outgunned by much more dramatic claims about possible damage to the ozone layer.

SCEP had in fact considered ozone effects of water vapour and other ex-

Scientists Ask SST Delay Pending Pollution Study

Continued From Page 1, Col. 6

that fine particles from the exhaust of jet engines would tend to double global averages of such particles, with unknown effects. The particles, distributed in the lower level of the stratosphere where the first supersonic jets will fly, reflect sunlight back into the stratosphere, thus tending to warm it.

The group's preliminary report, called the Study of Critical Environmental Problems, also recommended that a monitoring program be set up promptly to measure the lower stratosphere for water vapor quantities and to determine the measurements for sulphur dioxide and nitrogen oxide and hydrocarbons, which make up most of the fine particle matter.

The report followed a recent call by William M. Magruder, in charge of the Federal Government's supersonic transport program, for a study of the plane's potential environmental impact. The supersonic, or SST project, is currently awaiting Senate action on a $290-million appropriation.

The teachers, scientists and professional men who met here also studied and reported on other aspects of ... wide pollution ...

A Concorde flying over Paris. At least one U.S. airline has this type of plane on order.

the oceans as a result of emission and wasteful practices on land.

¶The increasing use of fertilizers and the growing quantity of animal and man waste destructive ...

ment of sophisticated monitoring facilities for determining facts about air and water pollution and the routes the pollutants travel.

global comp... rporatio... models in... eric r...

problems, but rather to seek recommendations for new programs of focused research and action and to obtain more definitive information on the major wide pollution problems. se ... int ... me

Figure 2.1: Concerns over climatic effects of supersonic aircraft.[48]

The report, on the front page of the *New York Times*, was the result of the press conference at the end of the Study of Critical Environmental Problems (SCEP).

haust emissions. After reviewing previous studies, as well as the recent unpublished work of two of the atmospheric scientists present, it found 'the reduction of ozone due to interaction with water vapour or other exhaust gases should be insignificant'.[49] Its dismissal of these effects as unimportant would be the cause of much derision from at least one of the scientists raising the alarm over ozone in the scares that soon followed. And there was not just one alarm sounded over ozone. Throughout the 1970s, scientific speculation drove a series of ozone scares, each attracting significant press attention. These would climax in the mid-1980s, when evidence of ozone-depleting effects of spray-can propellants would be discovered in the most unlikely place. This takes us right up to the start of the global warming scare, presenting along the way many continuities and parallels. Indeed, the push for ozone protection up to the 1980s runs somewhat parallel with the global warming movement until the treaty process to mitigate ozone damage suddenly gained traction and became the very model for the process to mitigate global warming. The ozone story therefore warrants a much closer look.

31

3 Ozone layer jitters

The meteorology of ozone

The idea of an ozone layer high in the atmosphere was first proposed in the late 19th century as an explanation for the blocking of specific bands of the Sun's ultraviolet light. It was suggested that this effect was caused by an unstable molecule of oxygen, ozone (O_3). This idea was supported by further evidence that also challenged the presumption that temperature decreases with pressure all the way up through the atmosphere. Crude instruments aboard early unmanned balloons confirmed that at a certain altitude—more than 10 kilometres up, where the air pressure is almost one-tenth of that on the ground—the air starts to warm again. This gradual warming from the troposphere/stratosphere boundary—the 'tropopause'—and up through the stratosphere was eventually attributed to the energy absorbed in the photochemical reactions that produce ozone from the regular oxygen molecule, O_2.[*] It was proposed that this ultraviolet action not only creates but also destroys ozone in a continuous cycle. This photochemistry and the basic picture of the ozone layer that we have today was mostly established in the inter-war years; the 1920s and early 1930s mark the golden age of ozone science.[50] The top of the stratosphere is where atmospheric oxygen is fully exposed to the Sun's intense ultraviolet radiation and this is mostly where the ultraviolet rays create and destroy their own ozone shield. The more ultraviolet that gets through at any level, the more ozone is created. Thus, although short-lived, at very low concentrations and in very thin air, the interaction of ultraviolet rays with this high-level oxygen delivers a 'steady state' ozone protection of the atmosphere below. Ozone that drifts down into the lower stratosphere is mostly shielded from destruction in this way, creating a denser 'layer' of more stable ozone. When it drifts further down, into the troposphere, it is soon destroyed by contact with weather-driven solids, especially

---

[*]Note that regular oxygen (O_2) makes up 21% of the atmosphere, while the rest is the very stable molecular nitrogen (N_2) and small traces of many other gases including ozone and carbon dioxide (CO_2).

particles of dust, smoke and ice (which themselves scatter the remaining ultra-violet light). Thus, the lower stratosphere is where most of the ozone resides. It is also where its concentration is most variable, and early empirical research was mostly about mapping this variability.

Directly after the First World War, a spectrographic instrument was invented that could measure the total ozone 'column' above it. A few years later, a spectrograph was designed to adequately control for the variable atmosphere scattering of ultraviolet light; it was sufficiently accurate for comparative readings. Interest grew among meteorologists as early measurements picked up variability sympathetic with changing weather patterns and seasons. Already in the 1920s and through the 1930s, distribution patterns were emerging from recordings at various geographic locations.[51]

In the first place, each measuring station picked up great rises and falls from one day to the next, and sometimes from hour to hour as the weather changed. When this variability was smoothed over hundreds of readings, a distinct seasonal cycle was discovered. This cycle is more pronounced at higher-latitude stations where the greatest concentrations of ozone are found, and found to almost double during winter (see Figure 3.1). Given that ozone production would be expected to vary with the intensity of solar radiation, these findings suggested that ozone is transported from its site of production by the circulation of stratospheric air. Indeed, the explanation of this geographic and seasonal distribution was key to the proposal of this so-called 'Brewer–Dobson circulation.'[52]

If ozone concentrations depend on local production, more ozone should be expected where and when the Sun's radiation is most intense. This should be at the equator and in the summer at higher latitudes. In fact, almost the exact opposite was found to be the case. To explain this, it was proposed that warm, ozone-poor tropospheric air rises at the equator and slowly drifts through the upper stratosphere, where it is enriched in ozone content before descending at higher latitudes and especially near the poles, this descent occurring especially when the air becomes very cold (and so heavy) in the winter. Findings of relatively thin ozone over the rising air of high pressure systems (and thickening over upper-level troughs) came in support of this hypothesis, as did balloon data that eventually included recordings of ozone concentrations at various heights.[53] Ozone levels were also found to vary with the 11-year solar sunspot cycle, and it was this and other suggested links between the sun above and the weather below that sustained interest for a small network of meteorologists in these first scientific explorations of the layer of clear thin air above the highest clouds.

One of the leaders of these explorations was Gordon Dobson, the British meteorologist who had designed the spectrometer that became the standard for

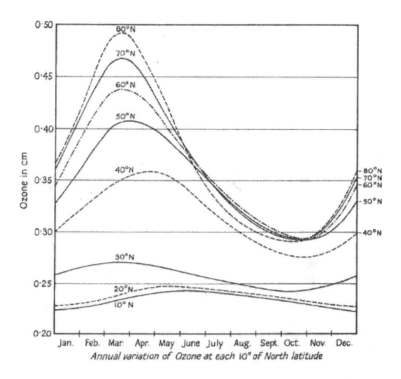

Figure 3.1: Seasonal variations in the ozone layer by latitude.

The variations were demonstrated through extensive data collection during the International Geophysical Year. Dobson summarised these findings in his 1963 textbook *Exploring the Atmosphere*.[54]

measuring the total ozone 'column'. The hut he had built on his property outside Oxford in 1921 became a global centre for the calibration of these instruments, in what became known as 'Dobson units'. When the International Geophysical Year was announced, Dobson had long since retired from his position at Oxford University, yet he was still coordinating much of the data collection and so he took this opportunity to send some of his hefty coffin-like contraptions to new locations of particular interest. The ozone record of the high latitudes of the southern hemisphere remained sparse, and he was especially pleased to hear that one of his instruments had arrived safely on the Antarctic continent for the first time. This was at the British base on Halley Bay during the austral summer of 1956.

Operating the Dobson spectrometer was not easy at the best of times, but the Antarctic presented some extra challenges. As the results arrived back with

Dobson over the following year, he became sure they were in error. The trend was low and flat across the seasons, more like the tropics than equivalent regions of the polar north. The more difficult and dubious readings taken by moonlight in the gloom of winter showed the expected winter peak entirely truncated. But then Dobson noticed that late in the spring the levels returned to expectations. This sudden change around November coincided with a sudden warming of very cold air in the upper stratosphere. When this pattern repeated over the next few years, Dobson proposed a reason for the anomaly:

> Clearly, during November the whole structure of the south polar strato-sphere has undergone a fundamental change. It seems as if in winter the south polar stratosphere is cut off from the general world-wide circula-tion of air by the very intense vortex of strong westerly winds which blow around the Antarctic Continent enclosing very cold air which is rather weak in ozone; neither the ozone nor the temperature rises much until this vortex suddenly breaks down in November.[55]

This explanation of the Antarctic anomaly remains to this day. The Halley Bay ozone station also remains. It has delivered a continuous record from those first readings in 1956, through the start of the satellite era, and into the 1980s, when a new trend was picked up. Beginning in the late 1970s, both the satellite and the Dobson data show a deepening of this Antarctic ozone depletion just before the November breakup of the polar vortex. This brief spring dip, while the sun is emerging low on the horizon, soon became known as the 'hole' in the ozone layer. However, curiously, the term 'ozone hole' has a forerunner.

The idea of modifying the ozone layer

For astronomers, the ozone layer presents an annoying hindrance; it blocks large bands of the radiation arriving from the stars. This problem of atmospheric fil-tering was only recently resolved by having the Hubble space telescope beam stellar radiation data down to Earth. Back in the 1930s another British pioneer of ozone research had a different idea: Sidney Chapman speculated about tem-porarily cutting a hole in the ozone layer. An aeroplane high in the sky would spread a suitable 'deozonizer': a chemical catalyst of ozone destruction.[57,58] Much later, in 1962, after reading Chapman's proposal, Harry Wexler, the di-rector of meteorological research at the US Weather Bureau, asked a chemist to suggest a suitable catalyst, and was told that chlorine or bromine would do the trick. A 100-kiloton bromine bomb would be sufficient to destroy all the ozone over the polar region.[†]

[†]The use of a 'bromine bomb' was raised again in the hearings over CFCs.[59]

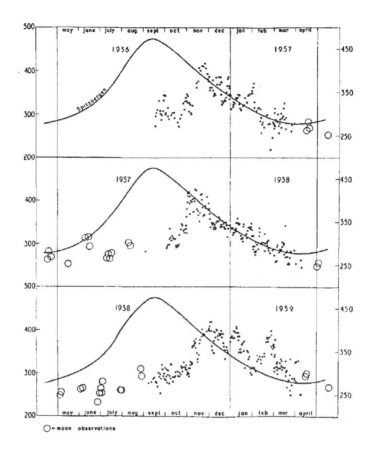

Figure 3.2: The Antarctic seasonal anomaly.

Dobson described and explained the Antarctic seasonal anomaly in publications from 1963. A 1968 retrospective presents the original data that led to the discovery.[56] The ozone layer thickness trend curves for Arctic Spitzbergen have been shifted by six months so they can be compared with the Antarctic Halley Bay data points, including moonlight measurements indicated by circles.

Wexler was one of those who had expended considerable effort speculating about deliberate weather and climate modification. Now the recognition of the catalytic effect of these chemicals led him to consider an *inadvertent* effect of the new rocket age. Chlorine in the exhaust of some rockets might accidentally open a hole in the ozone layer. He raised this possibility in a wide-ranging lecture entitled 'On the possibilities of climate control', which he delivered on a tour to a series of scientific audiences, and it was this rocket pollution speculation that was picked up by the press. This 'new space-age peril' was the headline-grabber in at least three prominent newspapers. 'Space taint could twist the weather' announced the *Boston Globe*. 'Chlorine in the exhaust chemicals', the *Los Angeles Times* explained, 'would dissipate the ozone layer which protects the earth from sterilizing ultra-violet radiation'.[60,61,62,63]

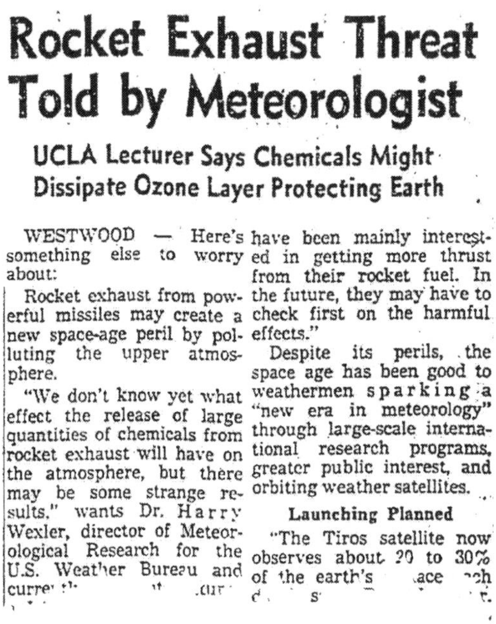

Rocket Exhaust Threat Told by Meteorologist

UCLA Lecturer Says Chemicals Might Dissipate Ozone Layer Protecting Earth

WESTWOOD — Here's something else to worry about:

Rocket exhaust from powerful missiles may create a new space-age peril by polluting the upper atmosphere.

"We don't know yet what effect the release of large quantities of chemicals from rocket exhaust will have on the atmosphere, but there may be some strange results," wants Dr. Harry Wexler, director of Meteorological Research for the U.S. Weather Bureau and curre᠉ ⁱᵗ .cur᠉

Here's have been mainly interested in getting more thrust from their rocket fuel. In the future, they may have to check first on the harmful effects."

Despite its perils, the space age has been good to weathermen sparking a "new era in meteorology" through large-scale international research programs, greater public interest, and orbiting weather satellites.

Launching Planned

"The Tiros satellite now observes about 20 to 30% of the earth's ace ᠉ch

Figure 3.3: Fears of the impact of chlorine on the ozone layer.

A warning about the possible ozone layer impact of chlorine pollution was issued by the director of research at the US Weather Bureau in early 1962.[64] However, this issue attracted little attention until an environmental impact assessment of the Space Shuttle in 1974.

Wexler died shortly after delivering this warning, and interest in chlorine as an ozone destroyer fell dormant for a decade. Meanwhile, in the mid-1960s another scientist working in weapons research in Canada proposed another way stratospheric flights could threaten the ozone layer: from their frozen vapour trails. While other US scientists were speculating that contrail ice crystals would increase cloud formation, he speculated that they might catalyse ozone destruction. This effect was considered by a group of experts at the SCEP conference and dismissed as insignificant. But when the idea was picked up again by a modeller working at Boeing, the concern about SST exhaust began to shift from climate to ozone.

SST, water vapour and ozone

The job that the chemist Halstead Harrison took up at the Boeing research laboratory in Seattle in 1963 was one that many scientists today may only dream about. He was not assigned to any Boeing programs or projects; he needed to be available for occasional in-house consulting, but for the most part he was to dabble in self-directed research. Boeing had whole teams of researchers under much the same instruction. This would allow them, so it was reasoned, to remain abreast of scientific and technical innovations that might affect Boeing's business. This dream model of curiosity-driven research began to break down with Boeing's financial trouble in the late 1960s, but, for Harrison, only slightly. Boeing was competing with NASA for work on missions to Mars and Venus, so Harrison's team was encouraged to investigate the atmospheres of those planets.[65]

Harrison had never been involved in Boeing's SST program, nor was he among its supporters. On the contrary, as he later recalled, he shared 'some half-formed judgments that SSTs would be noisy, expensive, and unprofitable'. He thought that 'enthusiasts who were pushing them were exaggerating their virtues well beyond the reasonable'. What piqued his interest in the subject was the scientific side of the debate, and in particular the work of Bo Lundberg, who had been publishing on the sonic boom and its impacts since 1962, but whose work achieved particular prominence in the USA in 1970 through the efforts of the Coalition Against the SST. Harrison also read some of the speculations about the effects of SST exhaust pollution, including an interview in the local paper with a long-time friend, Bob Charlson, who was quoted with yet more casual scientific speculation on the topic; he had suggested that sulphur in SST exhaust might augment stratospheric aerosol content, which would cool the Earth by veiling the Sun's warmth. Harrison was fairly sure Charlson was wrong—and indeed the effect was later dismissed by SCEP group as entirely insignificant—but the idea nevertheless prompted him to start to look at the SST exhaust question more closely. He modified his Mars–Venus model to the conditions of the Earth and then added into the mix the exhaust of 850 fully operational SSTs, just to see what happened.[66] Harrison's modelling found that the most significant impact of SST exhaust would be when its water vapour content catalysed the destruction of ozone; the fleet would reduce the natural steady-state level by up to 3.8%. He first came to this conclusion early in 1970, just before the SCEP report reignited concerns about the climatic impact of SST exhaust that summer. Harrison alerted the Boeing SST engineers and wrote up his findings in a confidential internal memo. Soon a team was brought together for a general review of any damage the exhaust might cause. In June, this review was leaked to Henry Reuss, an anti-SST campaigner in the House of Representatives, who accused Boeing of suppressing what they knew of the environmental effects of SSTs and demanded to see Harrison's original memo. The next day Harrison was identified as the scientist behind the ozone destruction findings and the press hounded him.[67]

As the controversy developed, Harrison and Boeing tried to downplay the findings in an effort at damage control. Arnold Goldburg, the chief scientist of the SST program, asked Harrison to investigate the empirical evidence on the topic. His model predicted that increasing humidity from *any* source would cause ozone decline. Yet most evidence suggested, if anything, a general *rise* in total ozone during the 1960s, while the best available humidity dataset was that for the sky over Washington,[‡] which also showed a net increase. While dubious about the humidity data, and dubious about its generalisation, Harrison nevertheless charted it against the general ozone trend, thus presenting empirical evidence apparently contradicting his modelling. Much to Harrison's chagrin, Goldburg then used this chart to show that the evidence did not support Harrison's claim, and even held it up for all to see on the local evening news.[68] Nevertheless, Harrison wrote up his findings in a paper published in *Science* later that year. Appearing in November 1970, this would be the only scientific publication supporting the SST ozone scare to come out before the House of Representatives' vote brought the SST program to an abrupt halt the following March.[69]

Meanwhile, another scientist had picked up on the 'Boeing research' and gave it considerably more impact: in early March 1971, just before the vote, his congressional testimony linked ozone depletion with cancer.

SST and cancer

James McDonald was a specialist in the properties of ice crystals, and it had been his calculations that led to the conclusion of the NAS panel back in 1966 that there should be no concern about SST contrails. When the issue arose anew in a highly charged political context during the summer of 1970, the Academy asked McDonald to re-assess the science. This was when he became concerned about the impact on ozone, and the potential increase in skin cancers from the additional ultraviolet light passing through a depleted ozone layer. He first took this concern to the SST program administrators at the Department of Transport, where it was received with less gravity than McDonald would have liked.[70]

[‡] See p. 29.

The department did respond to this and other scientific forebodings that summer, if grudgingly, and not without some underlying scepticism. In July, William Magruder, the head of the SST program on the department side, publicly declared his doubts about both the climatic and ozone effects of water vapour—if only because the SST contribution is likely to be insignificant compared to what thousands of thunderstorms around the world throw up there every day. Nevertheless, in the same speech he announced plans for a huge research program to investigate these new concerns about the exhaust as well as those continuing over take-off noise. According to the *Los Angeles Times*, Magruder justified investigating the exhaust effects by saying that 'while the weight of scientific opinion together with existing evidence argue against the likelihood of any significant environmental effects from SST operations', still 'there is not unanimous assurance that supersonic transport flights will not produce undesirable consequences to life on earth'.[71] Magruder is quoted saying that the research is required because 'it is proper...that we seek such assurance before any [fleet] production commitments are made'. Submission of investigation reports would be timed so that they could inform decisions on such commitments. Meanwhile, on the current commitment to prototype production, the growing opposition to a spending blowout had meant that the previous funding bill only just passed a Senate vote, and so the *Los Angeles Times* interpreted Magruder's announcement as aimed to 'calming Senate opponents'.[72] In that respect it failed.

Two weeks after Magruder's announcement, the SCEP group report reinforced his view that there is as yet little hard evidence upon which to base a case for alarm and so more research is required. For Magruder, as for SCEP, government-funded development of the SST prototypes could continue while the supposed 'undesirable consequences to life on earth' were investigated. But this did nothing to mollify the SST opponents who continued to call for an immediate stop to prototype development. The main opposition in Congress was driven by fiscal conservatives, for whom belated emissions alarm was only convenient to their main objective of cutting government funding.[73] But this was a convenience they were quick to exploit. After first drawing attention to the Boeing research, and then hearing of McDonald's further work exploring the cancer threat, Representative Reuss invited McDonald to present his findings at the Capitol.

McDonald's testimony on 2 March 1971, was long, complex, technical and emotive. He first laid out the case for vapour-catalysed ozone depletion, in corroboration of the leaked Boeing modelling results. On top of this, he added his own calculations of the skin cancer impact. The testimony went decidedly apocalyptical in its portrayal of ultraviolet light as a peril, not only for humanity,

41

but also for 'life on earth'. Life did not emerge from the oceans, he explained, because it could not, at least not until the ultraviolet shield was in place. He also made the remarkable claim that 'the whole history of evolution...has been a battle with ultraviolet light', although his take on evolutionary history was not entirely relevant to his case. What truly made an impact though was his claim that an average depletion of 1% would imply up to 10,000 new cases of skin cancer every year in the USA alone.[74]

White House and Proxmire In Dispute on SST Hazard

Senator Asserts Some Experts Support Theory Linking Plane to Skin Cancer —Nixon Aide Sees Move to Stir Fear

By CHRISTOPHER LYDON
Special to The New York Times

WASHINGTON, March 17 — On the eve of a new House test on the supersonic transport plane, the White House and Senator William Proxmire got into a dispute today over the plane's potential health hazard.

Later, however, Ronald L. Ziegler, the White House press secretary, conceded that "some questions" had been raised, and that the Administration had no intention of moving the air-plane into the production stage.

Figure 3.4: Skin cancer fears.

Senator Proxmire's peer review of McDonald's cancer claim made front page news on the morning of the lower house vote that was to signal the end for the Boeing SST prototype program.[75]

The congressional opponents of SSTs capitalised on this new evidence over the following two weeks, arguing that it would be foolhardy to approve further spending until the questions about skin cancer were answered. On the very eve of the House vote, Senator Proxmire presented a press conference with a pile of letters in support of McDonald's cancer claim. As something of a peer review, he had circulated McDonald's work to 40 leading meteorologists and dermatologists. Three had even agreed to appear with Proxmire, including Jule Charney, the Sloan professor of meteorology at the Massachusetts Institute of Technology. A front-page story in the *New York Times* reported Charney declaring his own long-held opposition to SST on social and scientific grounds.[76] He then vouched for McDonald's scientific credibility and said that he was impressed by the work behind the cancer claim. Indeed, the *New York Times* seemed to sup-

42

port Proxmire's claim that, of the 24 responses presented to the press that day, none contradicted McDonald's work. Some reinforced it, including Gio Gori, the scientific director of the National Cancer Institute. Gori had initially agreed to present his own findings in person, but, according to Proxmire, he had been 'gagged' by the Nixon administration. Calculating on a greater level of ozone destruction, it was widely reported that Gori had come up with a much greater rate of skin cancer in America: up to 103,000 additional cases every year.[77] This was scary stuff.

Scary, yes. But not so scary as it might sound. It is true that in 1971 the link between skin cancer and sun exposure was fairly well established in various ways, including by epidemiological studies that found fair-skinned communities in low latitudes tended to record higher rates. However, the link to ultraviolet light exposure (specifically, the UV-B band) is strongest among those cancers that are most common but are also rarely lethal. The link with the rarer and most dangerous cancers, the malignant melanomas, is not so strong, especially because they often appear on skin that is not usually exposed to the sun.[78] Thus, in the various controversies over the dangers of sun exposure, the term 'cancer' has always had a particularly strong polemical effect. While cancers are popularly assumed to be uniformly deadly, those wishing to emphasise the dangers of sunlight can effectively exaggerate the gravity of the risk by failing to distinguish the cosmetic from the fatal. It has also helped the anti-sunlight cause to ignore the benefits of moderate doses of UV, which include the synthesis of Vitamin D and the prevention of rickets. In the 1970s the link between depression and the lack of sunlight was not established scientifically, but the 'wintertime blues' was certainly evident in the vernacular. And this reflected behaviour: many of those who could were migrating to sunnier regions. The clouds, dust and smoke that blighted northern industrial centres blocked or scattered as much ultraviolet as visible light, only adding to the (unhealthy) pallor of their residents. When 'The Mamas and the Papas' released *California Dreamin'* in 1965, many northerners, both from the USA and Europe, were already realising that dream, heading for the (healthy) outdoor lifestyle of sunny southern California. At the same time, Florida was fast becoming a haven for arthritic retirees, all stripping down to enjoy the sun. Thus, sceptics of the fuss over the risk of a few percent thinning of the already variable ozone layer would point out that the anti-SST crowd did not seemed overly worried about the modern preference for sunshine, which was, *on the very same evidence*, already presenting a risk many orders of magnitude greater: a small depletion in the ozone layer would be the equivalent of moving a few miles south. To the dismay of their environmentalist opponents, the bolder among these sceptics would recommend the same mitigation measures recommended to the lifestyle migrants—sunscreen, sunglasses and sunhats.

However, this was not the main problem with McDonald's testimony. In 1954, he had sighted a UFO, an experience that had launched an abiding interest in the subject that lasted his entire scientific career. His campaigning for more research had been partly successful, resulting in a few government studies and reviews. But it was a congressional testimony in 1969 in which he linked an electricity blackout to alien intervention that would come back to haunt him when he made his anti-SST contribution two years later. His pro-SST opponents used his preoccupation with 'flying saucers' to mercilessly attack his credibility, the force of the attacks suggesting just how great a threat was the cancer claim. From the start of his testimony, McDonald spoke defensively, emphasising that his forewarning of a cancer epidemic is based in sober science not ecological paranoia. At one point he said, 'it is not kooky, it is not nutty, it is not ecological extremism'; rather, it is 'physics and chemistry, photochemistry, cell biochemistry, atmospheric physics'. He compared his surprising cancer findings to the environmental problems caused by DDT, which were entirely unexpected when the pesticide first came into use, and for which a forewarning would have been welcome.[79]

Alas, whatever their position on UFOs (or on DDT), and notwithstanding Proxmire's efforts to present things otherwise, atmospheric scientists were not exactly jumping up in defence of the physics and chemistry behind McDonald's alarming cancer claims. At a scientific conference opening on the very day that the SST program was first voted down, McDonald was left fully exposed to the heckling of Boeing's Arnold Goldburg on UFOs and skin cancer alike. McDonald had introduced a new and powerful campaigning angle for the anti-SST movement, and much was made of it in the lead up to the vote in the lower house, but yet he had become isolated in the scientific community. His marriage was also breaking down at this very time. In April 1971, only weeks after the conference heckling, he attempted suicide, succeeding with another attempt in June.

Meanwhile, a new ozone scare was on the rise. This time the stakes were raised substantially. At the same conference where McDonald was tormented by Boeing's Goldburg, a chemist was fuming in the audience, frustrated that the assembled scientists were chasing a red herring. For Harold Johnston of the University of California, the real problem with SST exhaust would not be water vapour but oxides of nitrogen. Working all night, the next morning he presented Xerox copies of handwritten work projecting 10–90% depletion. In high traffic areas, there would be no stopping these voracious catalysts: the ozone layer would all but disappear within a couple of years. Even when Johnston later settled for a quotable reduction by half, there could be no quibbling over the dangers to nature and humanity of such massive environmental destruction.

Oxides of nitrogen and ozone

The conference that introduced Johnston to the controversy came with part of the initial investigations in preparation for the much grander research program announced by the Department of Transport the previous summer.[§] A panel had been established in September for a preliminary assessment of the state of the science, including scientists in various relevant fields and also an industrial representative. As is often the custom with such panels, those directly involved in the research were not selected. Their opportunity to contribute came at our fateful conference. This was organised by one of the panellists, Joe Hirschfelder, a theoretical chemist, and hosted by William Kellogg of the National Center for Atmospheric Research in Boulder, Colorado.[80]

The two-day conference was unusual for its mixing of scientific disciplines and their disparate discourses, and although such meetings are often considered helpful to the progress of science, they are invariably marred by miscommunication, defensiveness and partisan conflicts. Indeed, the main axis of interdisciplinary conflict prominent later in the ozone controversies emerged at this conference: at one extreme were empirically-based meteorologists and atmospheric scientists, while at the other were theoretical and laboratory-based chemists. There were other lines of conflict, including where a professor of medicine on the panel challenged McDonald on the evidence behind his skin cancer calculations. Add an industrial representative who could not have been more involved, namely Boeing's Goldburg, also add the unfortunate coincidence of the Congressional vote, and the potential for conflict was high. On the first day, Johnston, who had incidentally supervised Halstead Harrison's chemistry doctorate back in the 1950s, now witnessed McDonald's attempt to present his cancer case on the back of the Boeing researcher's modelling. Later Johnston recalled that he was 'distressed... by the tenor of Goldburg's interruptions', saying that he 'had never seen any person at any scientific meeting so abused... during the course of presenting his paper'. Arriving fresh to the controversy, unknown to many of the participants, Johnston would find himself subjected to what he saw as ferocious attacks. As the conference proceeded, his feeling of isolation grew.[81]

Hirschfelder had pressed Johnston to attend because of his expertise on ozone. This was not ozone in the stratosphere, but as one of the key components of the photochemical smog that famously hung low and visible over cities from London to Los Angeles. Nitrogen oxide and nitrogen dioxide (collectively known as oxides of nitrogen, or NO_x) arising from motor vehicle exhaust reacts with regular molecular oxygen to *produce* ozone, a component of this smog.

[§] See p. 41.

So far in the debate over stratospheric ozone and the impact of SST pollutants, the NO_x emissions from jet engines had been dismissed as unimportant. This was the conclusion of the SCEP conference the previous summer, a view largely based on the work of meteorologists, and one in particular: Julius London.[82] Johnston had very little knowledge of the stratosphere. However, in preparing for the conference, he was already coming to the view that, in the chemical environment at those altitudes, the NO_x impact would be most significant. While in the lower troposphere these oxides react to *generate* ozone, up in the stratosphere they would catalytically *destroy* it. During London's presentation on the first day in Boulder, Johnston was disturbed to find that London's dismissal of NO_x seemed due to his ignorance of the chemistry. And so this is how it came about that, when London had finished his talk, a stranger—to London and to most in the audience—launched an emotive surprise attack, not on the substance of London's paper, but on his dismissal of NO_x concerns. Alas, Johnston's protestations garnered little support among the assembled experts in Boulder, and thus began what he saw as a lone fight against a consensus of uninformed meteorologists.[83]

Undeterred, Johnston was determined to prove his point. London had invited all conference participants to a party at his home that evening, but, instead of attending, Johnston worked through the night, laying out the relevant chemical reactions and calculating their rates. His results made it all too apparent that the introduction of an SST fleet would rapidly effect a global environmental catastrophe. If he thought that circulating his shocking findings the next morning would be sufficient to raise the alarm and turn this boat around, then he was seriously mistaken. The program of presentations continued as planned, with no chance to debate Johnston's claims. But then an opportunity presented itself. On the matter of scientific debate generally, Hirschfelder had also become dissatisfied with the proceedings and so he surreptitiously called a small breakaway meeting of selected scientists to thrash out the issues. At this meeting, water vapour effects continued to dominate the discussion until, as one of the participants later described, Johnston lashed out at their continuing neglect of the NO_x problem. This vehemence finally won him an opportunity to make his case.

In the discussions that followed, the major objections came back to the dearth of evidence.[84] In hindsight this is no surprise. Consider that at least with water vapour there had been some measurements of stratospheric humidity and there was even one time series. But in 1971 there was no way to directly measure stratospheric NO_x. No one was even sure whether there *was* any up there. Nor was there any way to confirm the presence—and, if so, the concentration—of many of the other possibly relevant reactive trace gases. This left scientists only guessing at natural concentrations, and for NO_x, Johnston and others had done just that. These 'best guesses' were then the basis for modelling of the many possible reactions, the reaction rates, and the relative significance of each in the natural chemistry of the cold thin air miles above. All this speculation would then form the basis of further speculations about how the atmosphere might respond to the impacts of aircraft that had not yet flown; indeed none had even been built. The best empirical evidence available came from atmospheric nuclear testing in the early 1960s, which had blasted as much NO_x into the stratosphere as several hundred SSTs. Yet there was no attributable dent in the ozone record, which, as we have seen, indicated a general rising trend.

Johnston won no support at Boulder and received a great deal of direct criticism. The conference could only conclude that further monitoring technologies needed to be developed and implemented in order to better assess speculations such as those by Johnston. In particular, direct measurements of NO_x (or its reactant, nitric oxide) were needed. Johnston was not satisfied with this outcome and the frustration that built during the conference propelled him into a one-man campaign to avert the looming global catastrophe.[85] His success in turning the debate around is nothing short of astonishing. When he got back to Berkeley, Johnston telephoned the White House, asking to speak to the science adviser to President Nixon.[86] Johnston could not have expected a welcome reception given that two years earlier Nixon had come out in full support of the SST. Moreover, the response to Proxmire banging the cancer drum on the eve of the House vote was scalding: a White House statement said it was 'a shocking attempt to create fear about something that is simply not the fact'.[87,88] Now, in the lead-up to the Senate vote, Johnston was pushing to raise the stakes almost to the extreme of credibility.

Johnston never got the call back from the science adviser, but a week later a staffer asked that he put his concerns in writing. In doing so, Johnston did not hold back in his depiction of the disastrous impacts.

> All animals of the world (except, of course, those that wore protective goggles) would be blinded if they ventured out during the daytime. ‖

This summary was leaked, to sensational coverage in a small local newspaper. Larger newspapers pressed Johnston and the University of California into releasing a statement, and so the story went out on the new wire services, to appear across the country on the eve of the vote on SST funding in the Senate.¶ A

‖ Quoted by Dotto.[89] See also newspaper coverage at the time.[90,91]
¶ See p. 27.

New York Times reporter contacted Johnston to confirm his claims and although the report he delivered was subdued, the story remained alarming.

> It would take less than a year of full-fleet operations, Dr Johnston said in a telephone interview, for SSTs to deplete half of the stratospheric ozone that shields the earth from the sun's ultraviolet radiation. Scientists argued in the SST debate last March that even a 1 percent reduction of ozone would increase radiation enough to cause an additional 10,000 cases of skin cancer a year in the United States.[92]

The next day, 19 May 1971, a strong negative vote demolished the funding bill. All but a few stalwarts agreed that one more vote in the House and it was all over for Boeing's SST. After that final vote, on 30 May, the *New York Times* followed-up on its initial story with a feature on Johnston's claims. This was written by their leading science writer, Walter Sullivan, an influential science communicator important to our story.

Sullivan had become famous through the 1960s for his ability to translate complex science into brief narratives that engaged a broad audience. Respect for his work, and for the *New York Times* leadership of intellectual debate in the USA, meant that he was highly influential in determining what stories in science gained currency. Indeed, it was Sullivan who had written the front-page story back in 1965 that had raised the possibility of climatic effects of SST contrails.**

This time Sullivan's feature gave a simplified explanation of Johnston's work, complete with diagrammatic representations of the two key photochemical reactions that 'protect the earth', as well as the two key reactions by which NO_x emissions would destroy this protection (see Figure 3.5). In the final section, entitled 'End of ozone', Sullivan explains how these latter reactions...

> ...would first wipe out virtually all ozone in the lower stratosphere, where the planes would operate. The nitric oxide then would slowly diffuse up through the rest of the ozone region, continuing to deplete it even if an alarmed world suddenly called a halt to SST flights.[93]

This popular account was influential in Washington and seems to have been instrumental in Johnston's subsequent breakthrough. Certainly, success was achieved long before his paper on the subject appeared in *Science* three months later.[94] Submitted not two weeks after the Boulder conference, it had been held up in peer review, returning to Johnston for two rewrites with requests to tone down the talk of looming catastrophe that had so excited the press.[95] But already the message had got through to where it mattered: to the chair of the Senate

---

** See p. 28.

48

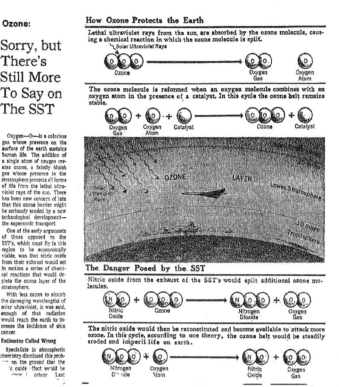

Figure 3.5: Walter Sullivan shows how SST will destroy the ozone layer.

The feature, in the *New York Times* of 30 May 1971, explains how, according to Johnston, NO$_x$ from SST will destroy the ozone layer and 'imperil life on earth'.

Committee on Aeronautical and Space Science, Clinton Anderson. The senator accepted Johnston's theory on the strength of Sullivan's account, which he summarised in a letter to NASA before concluding that 'we either need NO$_x$-free engines or a ban on stratospheric flight'.[96] And so it turned out that directly after the scrapping of the Boeing prototype, the overriding concern about supersonic exhaust pollution switched from water vapour to NO$_x$. Upon the assumption that Johnston was right—that NO$_x$ exhaust would be the main and the prohibitive environmental impact of supersonic aircraft—NASA and its corporate partners would spend tens of millions of dollars over a decade and more. The

49

problem of take-off noise also remained unresolved, and many millions would be spent on that too. In addition, there was the 'Experimental Clean Combustor Program', a huge engineering effort to re-design the jet engine so as to reduce NO_x in the exhaust. By 1979 this had cost NASA $86 million, with corporate partners contributing an equal sum.[97]

As startling as Johnston's success appears, it is all the more extraordinary to consider how all the effort directed at solving the NO_x problem was never distracted by a rising tide of doubt. The more the NO_x effect was investigated, the more complex the chemistry seemed to be and the more doubtful became the original scientific foundations of the scare. In cases of serial uncertainty, the multiplying of best-guess estimates of an effect can shift one way and then the other as the science progresses. But this was never the case with NO_x, nor with the SST-ozone scare generally. After Johnston's initial claim, it only ever went one way. Harold Schiff, who was deeply involved in this research, later reflected that...

> ...every new measurement reduced the predicted impact on ozone. It was like tossing fifty heads in a row.[98]

At first, participating scientists held out against the evidence in support of the theory, and against growing criticism from outsiders. But right around the time when Schiff made his observation of the unlikely odds, the case completely collapsed. By the late 1970s, the evidence was pointing firmly back to the conclusion of SCEP all those years ago, namely, that the impact of NO_x is likely to be insignificant.

The first big blow to the case for alarm over NO_x came in 1972–3, with the first direct measurements of stratospheric nitrogen oxide. They showed that the contribution of a full fleet of SSTs would be relatively minor: *there is so much NO_x up there already that a little extra could not be so destructive after all.*[99,100] These findings helped to bring the projected ozone destruction estimates down into a range under 20%. The picture in 1974–5 was nothing like that graphically presented in the *New York Times* only a few years earlier, but still the message could remain one of environmental disaster, and this would come into the service of a new cause, which was to prevent Concorde from winning landing rights in the USA. It was not until that fight was lost and Concorde was finally permitted to touch down at JFK airport in 1977 that the knockout blow was struck. Around this time, estimates of the rate of nitrogen oxide's reaction with water were revised upwards, to 40 times faster. This meant that the chemical interaction of water vapour and NO_x in the aircraft exhaust would massively moderate the overall effect. When this and other new information was factored in, the modelling started to suggest that SSTs flying in the lower stratosphere might augment ozone production as much as it would reduce it, and so there would only be a small net effect one way or the other (see Figure 3.6).[101,102]

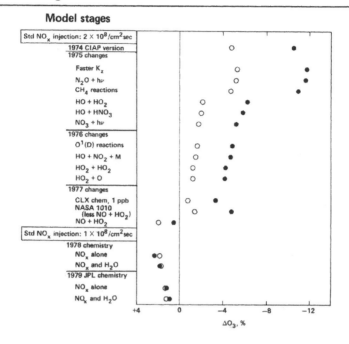

COMPUTED O_3 DEPLETION BY SST's

Figure 3.6: The effect of SSTs on ozone.

This chart, reproduced by Ellsaesser,[103] shows how the modelling results of SST depletion moderated over time until the NO_x and NO_x+H_2O effect was found to be positive. That is, by 1977, calculations suggested that SST exhaust would tend to *increase* stratospheric ozone levels. Not shown here are the results from Johnston that first sparked the scare. His round 50% depletion, as reported in the press, would be way off the page.

Luckily for those scientists who had been promoting the NO_x scare, its embarrassing demise played out after the supersonic dream had faded and under the cover of another ozone scare. Johnston and his speculative modelling could slip quietly into the forgotten recesses of the public record. Not that Johnston was ever ready to entirely back down; more than a decade later, he was asked to review the science behind NASA's huge investment in NO_x emissions-reduction research. In his report, submitted to NASA in 1989, he was still holding to a weaker, more doubtful version of his original claim. However, that same year, new research into the role of stratospheric aerosols estimated the impact of SST exhaust at levels so close to neutral that, even if large flocks of these strange birds ever managed to fly, any impact on ozone would be almost impossible to discern over the random noise of natural variability.[104] In 1993 this view was upheld by a NAS panel, and the case quietly closed.[105] Johnston's theory never held up to scrutiny. Despite this, for more than a decade it had a powerful and expensive influence on the direction of supersonics research and development. It also had a powerful influence on the direction of atmospheric science, at the centre of which was the Climatic Impact Assessment Program (CIAP).

NO_x and the Climatic Impact Assessment Program

The research program into the environmental effects of the SST announced by the Department of Transport in the summer of 1970 was christened in the fall the 'SST Climatological Impact Assessment Program'. Later on, 'climatological' was shortened to 'climate' and even though its attention shifted from climate to ozone, it was ever after known as CIAP (pronounced 'see-app').[106] The program got off to a delayed and faltering start. There was no doubt about the scientific need for it: the SCEP review had made it all too clear that the stratosphere was virtually unknown to science, and so research was urgently required before any proper assessment of pollution impacts could be made. The same call was put out from the Boulder meeting the following spring. But research could not get going until sufficient funding was secured, and no significant funding source was in sight, at least not until Johnston won his breakthrough. And so it was that while the Boeing SST prototypes were being dragged to the scrap heap, the Senate agreed to a $21 million budget for three years of intensive research into the potential effects of their exhausts on the stratosphere. At its completion, CIAP would boast the participation of 1000 scientists from 10 countries, numerous conferences, and a 7200-page final report. When additional investment via government agencies and universities is included, the investigation cost around $40 million.[107,108]

The fortuitous timing of this injection of funds into the atmospheric research community can hardly be overstated. The recession of 1970 ended the longest boom in modern history, and it hit particularly hard on a sector of industry born of that boom. Firstly, producers were in trouble. Boeing's workforce contraction from 103,000 to 38,000 was one of the greatest layoffs ever in corporate history. This rendered almost trivial the sacking of an additional 7000 working on the SST program. Airline operators were also in trouble; when PanAm cancelled its orders for six Concordes early in 1973, one very good reason to escape this extravagance was its four straight years in the red. The NASA bubble was also deflating. Employment levels had peaked a year before the moon landing and took a further dive when the Apollo program was halted in 1972. An estimated 45–50,000 former aerospace employees were out of work at the end of 1970, a figure which only dropped slightly the following year when it was also estimated that there were 40,000 'engineers, scientists and high-level technicians' unable to find any employment.[109] Of those who did find work, stories ran on the television of highly-skilled and formerly highly-paid experts taking on lowly casual jobs, like playing Santa Claus in shopping malls.

In the midst of all this negativity in the aerospace sector, and despite its likely further negative influence, CIAP offered opportunities for scientists and scientific institutions who were anxious to find new ways to apply their expertise. In his history of the SST program, *High Speed Dreams*, Erik Conway explains how, as SST work dried up, some researchers were able to shift into the new field of stratospheric research opening up under CIAP. For NASA, the program marked its movement into stratospheric investigations, and CIAP initiated what Conway calls 'the first "golden age" of stratospheric research'.[110] At least in terms of funding, he is right: CIAP was a far cry from Dobson's tinkering in his back shed. And even if the great advances in monitoring did not arrive until satellites stepped up to the task a few years later, many investigations never before undertaken had their relatively primitive beginnings under CIAP. It was a big house, but it became a crowded house, and there was a rush for shelter under this new roof.

The CIAP research program might have been conceived of the government's SST start-up program, but three years later its report was delivered into a very differently-shaped supersonic controversy. When the report was readied at the end of 1974, negotiations were already underway for the newly-built Concorde to land at US airports. The Concorde program had survived, but only just. With only a few aircraft ready to fly, and with the sonic boom excluding all overland routes, profitability was unlikely. The prospects of anyone, anywhere, developing a full supersonic fleet was not altogether lost, but it was now further away than it had been in the hopeful optimism of Kennedy's announcement of the SST program back in 1963. Still, the 820-page CIAP *Report of Findings* remained focused on the impacts of a full fleet. It concluded that NO_x remained the greatest problem, although the ozone depletion estimate was already down to a range below 20%.[111] Not that this finding was publicised. Indeed, the release of the *Report of Findings* was delayed for weeks after a 27-page executive summary was presented to the press in January 1975. And this summary told a very different story.

This executive summary had been drafted by the Department of Transport administrators of the program without consulting the participating scientists.

SST Is Cleared on Ozone

Associated Press

A three-year study dispels fear that the present fleet of supersonic transports will damage the earth's protective blanket of ozone, the Department of Transportation said yesterday.

Dr. Alan J. Grobecker, who directed the study, said a U.S. fleet of the high-flying planes would not have weakened the ozone shield either. Plans for a U.S. fleet of supersonic transports (SSTs) were scrapped in 1971 during debate about possible health and environmental damage.

The ozone blanket protects the earth from radiation that could cause skin cancer and from excessively high temperatures from the sun.

John W. Barnum, deputy secretary of transportation, said release of the report—ordered by Congress in 1970—doesn't mean the Ford administration is interested in reviving plans for an American SST fleet.

The Transportation Department said it drew on more than 1,000 investigators and 16 U.S. and foreign government agencies to complete the study, entitled "The Effects of Stratospheric Pollution by Aircraft."

The study says the 16 Anglo-French Concordes and 14 Soviet TU-114s now flying or scheduled for service will cause atmospheric changes ʃ⸴ ·ninimal they won't ʼɐ able to be detected.

But the studʼ ɑⱼ ˙   ·· expansion of
ⱼ ·   ʼoⱼʼtorꞏ ·

Figure 3.7: CIAP report clears SST.

The Associated Press story as it appeared in the *Washington Post* following the release of the CIAP executive summary on 21 January 1975. This coverage infuriated scientific proponents of the scare and opened old wounds sustained during the SST controversy four years earlier.

In its principal conclusions there was no mention of ozone, nor of skin cancer. In fact, there was no mention of any findings relating to the impacts of a full SST fleet. Instead, as was soon surmised, the summary was pitched to reassure the public that there should be no concerns about a few Concordes crossing the Atlantic and landing at a couple of US airports. Soon after the release of the executive summary, the Federal Aviation Administration (FAA) and the EPA both produced statements favouring Concorde, with the FAA 'draft' environmental impact statement quoting not from the scientific report itself but, instead, appropriating the conclusions of the executive summary, which reported that a small fleet would have no deleterious effect on 'climate'.[112] This good news story was further emphasised in the press conference. The strange thing was that the

54

impact on *climate* had long been put to bed and was not in dispute. Nor was the impact (whether on climate or on ozone) of a small fleet ever much in contention. A further complication of the summary's release came when the widely-circulated Associated Press story misinterpreted the press conference by saying that a *full* SST fleet would not have weakened the ozone shield. The next morning the story appeared on breakfast tables across the country under headlines like 'SST is cleared on ozone' (see Figure 3.7). According to the story, this huge program of investigation had found that there was, after all, no substance to the claims of the doomsayer scientists who had contributed to the demise of a home-built aircraft a few years earlier. While this conclusion might have been on the money, CIAP had certainly not made that call. Protests from Johnston and some of the other scientists were to no avail, and this journalistic error was not corrected by the Department of Transport until long after the story had died.[††]

The mismatch of CIAP's cautionary conclusions and the positively spun news stories had two predictable and opposing effects. On the one hand the pro-SST lobby directed angry recriminations at the scientists for recklessly falling in with the anti-SST campaign, stirring up a scare that had always had dubious scientific grounds (more on that below). On the other hand, some of these scientists were angry that the Department of Transport seemed to be inviting the misinterpretation of the CIAP finding upon which these attacks were based. They saw this as damaging to their reputations and to the reputation of scientific counsel generally. Among those outraged by the surprising turn of events was Thomas Donahue, the scientist who had led an American Geophysical Union (AGU) review of the CIAP findings; this review had been completed before the findings' release and even before the release of the Executive Summary. Two months after the Department of Transport delivered this supposed summary, Donahue launched into the continuing controversy a seething letter to *Science* explaining the cover-up. He told how the Associated Press news story and the Executive Summary 'had the effect of concealing, and even negating, the fact that CIAP actually supported the predictions' of McDonald and Johnston.

> Thus those who raised the alarm have been effectively discredited and stand accused of providing damaging counsel to this country.[114]

[††] Meanwhile, in 1975 a NAS investigation of the long-term global effects of all-out nuclear war came to the conclusion that, while the local human and environmental devastation would be enormous, the global environmental effects would be small, with one exception: on the strength of the CIAP research, they found that the most outstanding long-term global effect would be NO_x destruction of the ozone layer.[113] This finding was challenged years later, long after the NO_x scare, when a group of scientists launched the nuclear winter scare on Halloween, 1983.

The backlash dragged on through 1975 and into 1976. In fact, the rearguard defence of the CIAP findings prolonged the semblance of scientific support for the SST ozone scare in public forums. These included various inquiries and hearings during the Concorde controversy, where Johnston and Donahue even played the cancer card against its tiny fleet. In April 1975 they testified to the FAA that even if the total of all commercial supersonic jets peaked at the new expected figure of 30 planes, still there would be a price to pay: 1200 to 2100 extra cancers per year in the USA alone.[115]

In July, Congress launched an investigation into whether the FAA had acted properly over its draft environmental effects statement on Concorde. In their testimony, Johnston and Donahue recounted the whole sorry story of scientific opposition to the SST, including the CIAP cover-up, and then they reiterated their views on the skin cancer impact of Concorde.[116] In February 1976, a trial Concorde service to Washington DC and New York City was announced, but the service to New York was challenged at various levels, thus providing further opportunities to caution against ozone destruction. In all of these public forums a small group of vocal scientists were able to maintain the controversy far beyond what was scientifically tenable. *But what of the others? Did any of their fellow scientists try to blow the whistle?* Indeed, while the majority of those involved in CIAP held their tongue, a few experts did step up and try to counter the alarm, and they did so from the very beginning.

4   Scepticism of the SST scares

Angry industry

Of the scientific opposition to the SST scares, the first thing to note is that in the public eye the overwhelming impression was that the protestations were coming from a commercial and political pro-SST lobby. Indeed, surprised and angered as they were when a new front of debate suddenly opened up over the exhaust emissions, this lobby did provide plenty of good copy for the press.

We have already seen Boeing's chief scientist Arnold Goldburg attempting to defend the funding of the Boeing prototypes by discrediting the science and the scientists promoting the ozone scare.* Others would express more general concerns about what was going on, some even claiming they were witnessing the strategy of the 'Big Lie'. The Big Lie is a propaganda technique that was famously ascribed by Hitler and Goebbels to their victims and enemies, but it has since been proposed that the Nazis were only proclaiming their own practice. The idea is that a heavily promoted colossal untruth will succeed where small lies fail. *Why?* Because it would be unbelievable that anyone would have the impudence to distort the truth so infamously. Those accused of using this strategy include not only politicians, but also scientists. Early in the controversy, Senator Barry Goldwater made such an accusation in a despairing opinion piece published by the *New York Times* on 16 December 1970. 'All of a sudden, in the months and weeks prior to the first Senate vote', so Goldwater observes, 'this ten-year-old program…became the legislative monster of all time.' Goldwater claims that sound arguments in support of the SST are being 'drowned in a wave of scare stories, myths, guesses, speculation, half-truths and downright lies'.

> It is not only amazing but downright frightening to see the number of prominent scientists who were willing to lend their name to far-fetched and hypothetical possibilities…[117]

---

* See p. 40 and 44.

In the new year, after Goldburg had reported back to Boeing on the drama at the Boulder meeting, one of the company's SST chief engineers picked up on the 'Big Lie' theme and cut loose in an angry and possibly libellous letter to a leading aerospace trade magazine. John Swihart claims that the simple modelling of the interaction of water vapour, sunlight and ozone is contradicted by the available empirical evidence (i.e. the ozone/humidity trend comparison that Harrison had given Goldburg) and that the scientists involved knew it.

> In other words, the simple theory was in direct contradiction to the measured data, the opponents knew all about this data and they still perpetrated the 'Big Lie' on the American public. The truth is that the forces acting in the atmosphere and stratosphere are extremely complex and no one can predict what those forces are or their reactions. Hence to have publicly condemned the SST by use of a simple model which produced answers in direct conflict with the measured data is, at least, scientifically dishonest if not treasonous.[118]

The allegation of treason was directed squarely at McDonald but other scientists were mentioned, including Harrison, and even London. This letter was followed in the magazine by a 'Letter from Seattle', the home of Boeing, conveying a mixture of despondency and anger about the congressional rejection of the program. 'But the bitterness and concern will never be forgotten, or the names of the men who brought this about.' The magazine editorial named in particular two of these men, McDonald and Proxmire, for 'a deliberate big lie concocted knowingly in the face of measured facts'. For the editor, the playing of the cancer card was a 'cruel and cynical exploitation of the ecological hysteria'.[119]

These fiery invectives against the leading scientists who were promoting the scare served to alienate the industrial expert community from the community of publicly funded scientists who had become reliant on funding consequential to the scare. The polarisation thus established can be conveniently deployed to reduce a narrative of the whole controversy into a battle between (neutral) public science and (biased) industry. This *Goodies-and-Baddies* approach has since been generalised and plays to a receptive environmental movement. *But what of those scientists who do not conform to this narrative?* Indeed, for some academic historians who have taken up this historical Manicheism, these exceptions only confirm the rule. In their heavily marketed book, *Merchants of Doubt: How a handful of scientists obscured the truth on issues from tobacco smoke to global warming,* Naomi Oreskes and Erik Conway use their chapter on the ozone controversies to single out scientists whose expressed doubts cannot be easily written off as the propaganda of industry. These exceptional scientists only serve to exemplify the main objective of their book, which is to reveal a strategy whereby 'a handful of scientists' construct a 'counter-narrative' so as to undermine a clear scientific consensus that is unfavourable to the industries that they (covertly) serve. Yet this moral metanarrative of the scientific debate is strained at best. Moreover, given his previous closer studies of the SST controversy, Conway, at least, should know better.[120] In *Merchants of Doubt*, the failure of Johnston's speculations to hold up under scientific scrutiny passes unmentioned. Coverage of the SST–NO_x scare is left with Johnston and Donahue playing the heroic defenders of science against commercial interests just before their position becomes demonstrably untenable.[121]

It is one thing for these historians of science to deliver the simple message *to beware the peddler of doubt* in direct contradiction to the modern scientific ethos of eternal vigilance against dogma. It is another thing to demonise outspoken sceptics so that, in the new century, it has become almost impossible for them to remain within the publicly funded scientific community.

But in order to reduce the scientific debate to their morality play, these historians must grossly misrepresent the history. Undoubtedly, their cautionary tale would be further confused if there were any more than a hint of a 'counter-narrative' of scepticism before, during and after the SST exhaust scares; moreover, that this counter-narrative was *very much within* the genuine scientific controversy and therefore difficult to reduce to a tale of mere industrial cronyism.

Mainstream scientific scepticism

We will remember that a prevailing scepticism had first to be rejected in order to launch the SST ozone scares. This scepticism first found expression in the SCEP consensus, which dismissed the ozone effect of SST emissions as likely to be 'insignificant'. That view was rejected first by McDonald for water vapor emissions, and then, while that effect was under review, by Johnston in regard to NO_x. After Johnston launched his attack on the scepticism that prevailed among the meteorologists at the Boulder meeting, there still remained a great deal of resistance. Even beyond Boulder, the old view did not just roll over. Soon after hosting that fractious meeting, Kellogg was bold enough to mention in a published article that his discussion with other experts gave him the general impression that Johnston had overstated the effect of SST on ozone 'by a rather wide margin'. This triggered an angry response from Johnston and there ensued an exchange of letters in which Johnston continued to rail against the consensus of 'the Boulder group'.[122] In the summer of 1971 a follow-up to SCEP was organised, and this time Kellogg played an even more prominent role. *The Study of Man's Impact on Climate* (SMIC) involved a smaller but more international selection of scientists, who gathered in Stockholm to specifically address global *atmospheric* effects of human activity. The three scientists who had presented papers on the stratosphere at SCEP and who had all downplayed the effects of SST emissions (Lester Machta, G.D. Robinson and Julius London) were again in attendance, and an extended treatment of Johnston's claims appear in the meeting report. In the end, the SMIC group concluded...

> ...in support of the SCEP judgement, that the radiative effect due to possible changes in ozone would be negligible.[123]

The objection to Johnston was not only that there was insufficient knowledge of the relevant and possible reactions and reaction rates to reach any firm conclusion; it was also that the computer modelling did not consider the moderating effect of stratospheric circulation.

> ...the problem of stratospheric ozone distribution is very complex and must be studied in an atmospheric model in which proper treatment is given to dynamical processes before any definite results can be considered reliable.[124]

As we saw above, from the early days of the controversy, atmospheric circulation was seen by meteorologists as an important factor in the observed distribution and variability of ozone, as well as other trace chemicals. The failure of atmospheric models to incorporate these 'dynamical' or 'transport' processes was therefore seen as a central failing. Indeed, the differences over 'transport' began to define the polarisation of the debate, meteorologists and experimental atmospheric scientists tending to one side and the laboratory chemists (and modellers) tending to the other.[†] But this polarisation was not always so simply defined, and indeed one of the instigators of the SST exhaust scare later explained how he had quietly swapped sides.

Halstead Harrison, the Boeing scientist who had triggered the water vapour scare, was asked by Senator Proxmire to give testimony shortly after the Senate vote had already killed off the Boeing SST program. Writing in hindsight, Harrison claims that his decision not to do so was governed not so much by any

---

[†]This divide was institutionalised as the new field of stratospheric research opened under CIAP. The meteorologists came in under the International Association of Meteorology and Atmospheric Physics, while the International Association of Geomagnetism and Aeronomy was the abode of the laboratory chemists and modellers. When a combined conference on the stratosphere was proposed for 1973, much controversy ensued. In the end, two conferences were held, one for the meteorologists in Melbourne, Australia, and the other for the modellers and chemists in Kyoto, Japan. This latter conference received funding from CIAP and it was there that the Space Shuttle ozone scare began.[125]

loyalty to his colleagues at Boeing but by his 'deep reservations' about his and Johnston's modelling.

> Specifically, our models then assumed a photochemical steady state with a stationary sun over mid-latitudes, and all transport in the vertical, only, parameterized through eddy-diffusivity coefficients. But the chemistry and sun are not stationary, and the dominating currents in the stratosphere are advective, not diffusive, and horizontal, not vertical. The leap from our too-simple models to national decisions seemed to me then, as it does now, more hubristic than wise.[126]

The 'transport' objection is also prominent in another formal review that came in before CIAP. Much less publicised, brief and foreign, it nonetheless provides something of a window into the meteorologists' perspective. Late in 1971, the Australian Academy of Science had been asked to examine the atmospheric effects of supersonic exhaust. A working group chaired by the esteemed meteorologist Bill Priestley reported its findings in February 1972. The Australians agreed with SCEP and SMIC groups that there was great uncertainty surrounding the chemistry of the stratosphere and that the calculations of catastrophe did not adequately allow for 'transport processes in the stratosphere' that would 'ameliorate the impact'.[127]

> It is to be regretted that a number of overseas scientists whose warnings have received much publicity have neglected the dynamic nature of the phenomenon...[128]

This report also makes other moves to moderate the alarm. The cancer threat is put into perspective by drawing attention to how the ozone veil naturally waxes and wanes. Indeed, 'in most parts of the world there has been a general increase in total ozone amounts of up to 10% over the last decade...' Anyway, with the monitoring that the report recommended, the ultimate test of the speculation would come when flights actually commenced, at which time, as soon as any danger was noticed, restrictions could be imposed.[129]

The dangers of 'crying wolf'

Another category of criticism of the SST exhaust scare is worth raising here, although much of it came well after the scare had collapsed. This is where it is remembered in cautionary tales—this time serving to promote scepticism. While other scares about global atmospheric damage were developing, the NO_x story served to exemplify a more general trend of scientists and science embroiled in and embarrassed by public controversies after prematurely promoting fear.

Richard Scorer was the leader of a small atmospheric pollution team in the Department of Mathematics at Imperial College, London, when he entered the fray. Like Johnston, Scorer was an expert on smog and he had even been involved in shaping smog-mitigation legislation. But, already in the mid-1970s he was raising concerns about the various scares over potential global atmospheric disasters. 'The danger of environmental jitters', an article appearing in British science magazine *New Scientist* in 1975, opens with the provocative teaser:

> Scientists who encourage public fears on the basis of incomplete or ill-digested evidence constitute a serious environmental problem.[130]

Scorer warns of the emergence of a new ethic in which the burden of proof is reversed such that 'we should not take a risk if harmful effects are predicted scientifically unless we can disprove the theory in question'. With each of the environmental scares, we face an argument that policy 'should be based on the assumption that the theory is correct until we can confidently prove otherwise'. Scorer compares the situation of the modern scientific age with former times when the Church demanded 'that we should believe a miracle on the say-so of a simpleton on the ground that no one had proved that it never happened, and it was impolitic to ignore a potential message from god'.

On the question of emissions-driven increases in atmospheric carbon dioxide, Scorer notes that when no warming effect is evident (as we shall see below, the global cooling scare was in 1975 already three years old), this did not mean that the theory was abandoned as wrong. Instead, another theory—cooling by aerosols—was invented, which was 'alleged to produce the opposite effect'. What Scorer disputes is not the theory of this or that effect, but the assumption of the simplicity of the causation, as though each mechanism operates in isolation. Moving from the hubris of intentional weather modification to the jitters over inadvertent changes, Scorer says that 'it is not surprising that man should think of himself as a major environmental influence, especially as his ancestors, who were genetically identical, were often inclined to think that sacrifices and abracadabras altered the weather'. He says that 'the Supersonic Transport scare exploits this belief'. At the time that this scare emerged, Scorer says that he gave several independent reasons not to take it seriously, including some to do with atmospheric circulation, but his most important objection was a very general one. It was about 'what sort of place we imagine the atmosphere to be'.

> How could it be alleged seriously that the atmosphere would be upset by introducing a small quantity of the most commonly and easily formed compounds of the two elements which comprise 99 per cent of it? Now,

a few years later, those who proposed the scare are saying that NO_x provides nature's means for controlling the amount of stratospheric ozone. The preposterous presumptuousness of this view is certainly not appreciated by those who put it forward. For them, it is a significant advance on the previous position that NO_x was a danger. It is difficult to bridge the philosophical gulf which separates them from many more mature scientists, particularly meteorologists. To draw attention to the divide by calling them 'laboratory' as opposed to 'outdoor' scientists impedes communication. It seems to create resentment to point out that very clever people can utter foolishnesses...[131]

This new trend in science towards the promotion of scary speculation was also criticised much closer to home. Indeed, cautionary tales about the SST exhaust scare were delivered directly to those embarking on the generously funded Department of Energy 'Carbon Dioxide Program', which had first started to take shape in the late 1970s. By 1980 a plan for a comprehensive decade-long program of research into all aspects of the 'carbon dioxide question' had been produced and a conference of experts was convened in Washington to review it. The nuclear scientist Alvin Weinberg was the founding director of the Institute for Energy Analysis and in that role he was instrumental in getting the program up and running, so it was appropriate that he gave the address at the conference dinner.

Weinberg begins by noting West German Chancellor Schmidt's expressed concerns about carbon dioxide emissions and how these concerns are linked to the Chancellor's interest in promoting nuclear power over fossil fuels.[‡] Weinberg also declares his own interests in that direction. Nevertheless, he calls for caution.

[‡]Schmidt explicitly linked the two issues, including in his opening address to the European Nuclear Conference in Hamburg, May 1979. After making reference to the recent fuss over the Three Mile Island reactor incident (in Harrisburg, Pennsylvania, USA), he says:

> Perhaps even more alarming are the possible catastrophic effects of rising carbon dioxide concentrations in the atmosphere being discussed today by scientists that result from the increased burning of fossil fuels: natural gas, oil, coal, lignite. I am a layman in this area, but I believe it is conceivable that in a few years this problem of carbon dioxide concentrations in the atmosphere and the associated expectations of climate changes—temperature changes especially—as well as the question about exact consequences resulting from such realisations will evoke discussions that are equally as emotional as those about the exact consequences of Harrisburg.[132]

This quotation opens a survey of the global warming debate in Germany by Michael Hatch, which shows how it remained 'intimately linked to the controversy over nuclear power.'[133]

The scientific community really does itself no favours by crying wolf if the wolf cannot be pinned down. The scientific community did cry wolf with respect to ozone and the supersonic transport (SST), despite the fact that the SST turned out to have little effect on the stratospheric ozone layer. The scientific community cannot tolerate many fiascos such as the SST ozone scare. I hope that the CO_2 community will take the SST-ozone affair to heart and refrain from crying wolf too soon.[134]

Another scientist to link the rising interest in carbon dioxide with the SST 'fiasco' was one of the earliest and most vocal sceptics of both: Hugh Ellsaesser, an atmospheric scientist at Lawrence Livermore Laboratory. As early as 1974 he had an opinion piece in the Meteorological Society journal raising concern about the dangers of a 'one-way filter' that permits scientific exploration 'only of those pathways which lead to detrimental effects.' Using as an example early concerns about the potential climatic effects of human industry—warming by carbon dioxide and cooling by aerosols—Ellsaesser argues that the one-way filter is the biggest threat to the future of our civilisation. Thus, research into 'the factors which permit such a ...one-way filter to exist' should be given top priority.[135] Much later, in a 1983 address, he accused William Kellogg of just such filtering.

Back in the early 1970s, one could hardly have accused William Kellogg of applying the one-way filter, not least in his resistance to Johnston's NO_x campaign. But by the late 1970s Kellogg had taken up his own cause: he had come to the view that carbon dioxide warming was the main game. In his 1983 address, Ellsaesser suggests that a recent paper by Kellogg shows that he had learned well from the earlier debates.

Following a script made standard by CIAP, freon aerosol spray cans, increasing anthropogenic emission of atmospheric particles and a host of similar forgotten or as yet unrealized hazards, Will Kellogg has presented the prevailing 'Consensus View' of the CO_2 climate problem and then outlined the cascade of deleterious effects this may produce throughout all aspects of human activity and the global ecology...I would like to have you step back and take a critical look at the process by which the now fashionable 'Interdisciplinary Science through Workshops of a Committee of Experts' operates; how it transforms preliminary guestimates into a consensus supported by a constituency with a vested interest in confirming and perpetuating the problem rather than solving it....I early recognized the adoption of 'The One-Way Filter' approach in which only those pathways of the effects cascade leading to deleterious effects are pursued and described while any possible benefits are studiously ignored. CIAP's evaluation of the SST was comparable to a debate on whether we should continue to have children in which the discussants are allowed to consider only the problem of dealing with bodily excreta....[136]

In the early 1980s, Weinberg and Ellsaesser could have been forgiven for thinking that the rolling ozone scares had run their course. But, as it turned out, we were only at the tail-end of a pause before things really took off. In her book *Ozone Crisis: The 15-year evolution of a sudden global emergency*, Sharon Roan discusses the late 1970s under the heading 'The dark years',[§] while her chapter on the early 1980s is called 'The crisis that wasn't.'[138] By that time, not only had the SST scare collapsed, but the chlorofluorocarbon scare—which had opened with great drama in the mid-1970s—was already foundering. Weinberg and Ellsaesser could never have guessed that the fortunes of this last ozone scare could turn around so dramatically in the mid-1980s. By 1987 it resulted in what has been hailed as the greatest treaty success in the entire history of the United Nations.

The scientist to trigger this ultimate ozone scare was surely the most unlikely. And most unlikely of all was how he did it. He was investigating the stability of natural systems. But this was not to show how their delicate balance is easily upset by humans or otherwise. Quite the reverse. He was developing a hypothesis that would show just how robust and stable natural systems really are. He believed that nothing we could do, not even all-out nuclear war, would come close to threatening life on Earth. Thus, by way of introducing Jim Lovelock and the spray-can scare that he inadvertently sparked, we must first consider an important underlying aspect of the sceptical resistance to these various global atmospheric scares. This comes at a most basic level in our understanding of nature: in our presuppositions about its stability.

The question of stability (What sort of place do we imagine the atmosphere to be?)

How stable is the atmosphere? This question was rarely directly debated during the various atmospheric controversies and yet it underlies the entire postwar discourse over human interventions in atmospheric systems—those purposeful and those inadvertent. Indeed, the question of stability tends to arise, explicit or implicit, with any great advances in the understanding of natural systems.

How stable is the Earth? Isaac Newton's masterful feat was to explain the revolution of the Earth in perfect periodic motion. But the news that the Sun's

---

[§] Roan explains how she appropriated the term from an advocate: 'The period following the [US] aerosol ban was referred to as the "dark years" by Alan Miller of the Natural Resources Defense Council... one of the few who kept the flame burning for additional CFC regulation.'[137]

gravity hurls our globe around at unearthly speeds was most unsettling. *How safe can that be?* Aside from any unscientific concerns that the Sun would eventually drag the Earth into its fiery hub, the stability of the solar system became a serious question for science. When a perfect periodic system is perturbed, it can slip out of alignment. And a small change, even a very small change, could, over a long time, become bigger. This could cause a chain of other effects, and before you know it, Pluto is heading for Earth, or, indeed, Earth is veering into the Sun. There are plenty of candidates for upsetting the solar system, including the constantly changing collective gravitational pull of the other planets (Newton's calculations had treated the orbit of each planet independently). Famously, Newton did not answer the question of stability, leaving it for God to step in from time to time and make the necessary 'reformations'. It was well that he did. While it is easy to determine the orbits of two bodies around each other, when even one more mass enters the equation it becomes just too complicated, and so the problem of stability of the entire solar system reduced to 'the three-body problem'. In 1885, *Nature* announced a prize offered by King Oscar II of Sweden to solve four great mathematical conundrums, the first of which was the three-body problem. The career of a brilliant young French mathematician, Henri Poincaré, was boosted by winning the prize on that first question; and he did conclude, yes, our world is stable. However, while preparing his answer for publication, the editor of *Nature* found an error in the calculations. In his attempt to correct that error, Poincaré discovered that he had overlooked solutions to the problem that lead to non-periodic and entirely unpredictable motion. What he found was chaos.

How stable is the ozone layer? Those early 'aerometers' delivered no neat image of a safety blanket protecting us from the blinding ultraviolet rays. In the first place, the gas itself is in no real sense a layer. The layer is the layer of air where the concentration of this molecule is greatest, at but a few parts per million. When Dobson began measuring the total thickness of ozone between us and the Sun, he calibrated it as though it were concentrated into a layer of pure ozone on the surface of the earth. A membrane a few millimetres thick is not reassuring. But a membrane is hardly the right image for this rare and unstable gas drifting with the weather systems and the seasons. And it is continually being destroyed. Hundreds of millions of tonnes are taken out every day. *How stable is that?*

A simple way to categorise basic classes of stability in equilibrium is often presented through the analogy of a ball on different shaped surfaces, as depicted in Figure 4.1. The idea is to imagine what would happen if the ball was given a small push. In stable equilibrium, the system is like a ball in a curved bowl. Initially it strays, but eventually it returns to rest in the same place. This would be where the system is dominated by negative feedbacks. An uncontroversial example of negative feedback in the stratosphere is where ozone depletion above any layer of air lets more ultraviolet light through and so acts to produce more ozone.

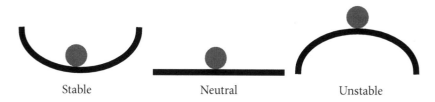

Stable   Neutral   Unstable

Figure 4.1: Classes of stability in equilibrium.

In a system in which *positive* feedback dominates there is no such return: the equilibrium is delicately balanced and a sufficient perturbation results in a 'runaway' effect. This was always the hope of the weather modifiers: somehow their puny interventions would trigger a massive effect. Alas, no such triggers were found.

In global warming modelling, positive feedbacks are also required. This is because the direct greenhouse effect of carbon dioxide emissions is quite small and could easily be damped by systemic negative feedbacks. For there to be any alarming effect, positive feedbacks must dominate. Two standard positive feedbacks were challenged by Ellsaesser in his 1983 critique of Kellogg.[||] The first was the supposition that sea ice melt in the Arctic would mean less reflection of the sunlight and so more warming. . . and so more melting. For Ellsaesser, the systemic response is not so simple and might be the opposite. The other positive feedback that he challenged remains the most controversial. It involves water vapour, which is by far the most dominant greenhouse gas. The idea is that a carbon dioxide increase causes a little warming and this causes more water to evaporate. . . which produces more warming. Instead, Ellsaesser (and many other sceptics) suggested that negative feedbacks might well dominate, including where the greater humidity leads to more shading clouds.[139]

The third class of equilibrium is in between these two extremes of stability and instability, with the feedback neutral or neutralised. In this case the external effect on the system is neither amplified nor damped and so it remains in the system (i.e. the ball shifts to a new position on the flat surface). In neutral stability, an extraordinary perturbation could introduce a 'new normal', or the system could be knocked about constantly by similar effects.

[||] See p. 64.

Presumptions about the general state of a system's stability are inevitable in situations of scant evidence, and they tend to determine positions across the sceptic/alarmist divide. Of course, one could suppose a stable system, in which a relatively minor compensatory adjustment might have an alarming impact on civilisation, like the rapid onset of a few metres of rise in sea level. But it is the use of such phrases as 'disturbing the delicate balance of nature' or 'a threat to life on Earth' that are giveaways to a supposition of instability. The sceptical response goes something like this:

> When a human effect is proposed it would be presumed that this is not the only game in town. Other forces of nature are likely to come into play and dampen such a small external perturbation. This can be expected because any system that has been around for a long time is likely dominated by negative feedback. The atmosphere, and specifically the ozone layer, has remained remarkably stable within certain bounds of human comfort since long before there were any humans around. Therefore, it would be surprising and extraordinary if a small impact could trigger changes outside this normal range. Such an extraordinary proposal would require an extraordinary explanation.

Hence Scorer's incredulity regarding Johnston's leap towards his catastrophic conclusion: 'How could it be alleged seriously that the atmosphere would be upset by introducing a small quantity of the most commonly and easily formed compounds of the two elements which comprise 99% of it?'[140]

The thinking on the other side of the controversy can be very different. Harold Schiff was one of the leading scientists, both in the primary research and on the assessment panels, throughout the various stages of the rolling ozone controversy during the 1970s. He teamed up with a journalist, Lydia Dotto, to present an insider's view of the unfolding controversy under the title *The Ozone War*. Published in 1978, just as the 'dark years' were beginning, their fresh and frank account helps to guide our narrative, although their perspective differs greatly from our own. On the question of stability, their incredulity about Scorer's view is presented in their summary of three myths propagated by the affected industries during the controversy.[141]

Their first myth—that the empirical evidence does not support the depletion theory—we have already discussed. The second myth they call natural 'self-healing'. The final myth is similar to the second, with both coming down to a trust that negative feedbacks will maintain stability.

> Nature has been coping with ozone-destroying processes for billions of years. There are feedback systems, self-righting mechanisms, that enable the atmosphere to continue to cope.[142]

68

This last myth, Schiff and Dotto call 'blind faith'. The use of this term suggests a position of trust based on ignorance, especially a wanton ignorance of the evidence of the senses. To say a scientific position is based on blind faith is surely to denigrate it as fit for summary dismissal. The implicit reference to pre-scientific dogma is made explicit when they quote one of the scientists spreading the alarm saying 'some people have the belief that the Lord designed an atmosphere that can take this kind of abuse. I live too close to Los Angeles to agree with that'. Richard Scorer is singled out as 'one of the most outspoken proponents' of blind faith, with 'his pitch' previously summarised as 'little more than bland assurances about the resiliency of the Earth's atmosphere'.[143],¶

This line of narrative is awkward in the first place due to Scorer's own overt scientistic atheism—his jibes at religious modes of reasoning and scorn for scientists taking on the role of prophets—as well as his leadership of smog research and its legislative mitigation. No other view, not even those of the industry scientists, is treated so carelessly in a chronicle that is otherwise outstanding for its efforts to provide fair treatment to the industrial and political opposition. That Scorer should attract such heightened and misdirected innuendo might best be explained by his becoming something of an apostate in the publicly funded research community after he crossed the Atlantic in the summer of 1975 for an industry-funded speaking tour promoting scepticism. But the terms of this summary dismissal also exemplify what is apparent elsewhere, namely the lack of sympathy among many alarmists for the new ways of understanding systems and the new ways of explaining their dynamics and stability that were being investigated by some of the leading atmospheric scientists at that time. Indeed, during the 1960s and 1970s great advances in the understanding of the generative and regenerative processes of natural systems were emerging within the atmospheric sciences and the impact of these developments would soon reach way beyond understandings of the compensatory effects of atmospheric circulation. New ways to theorise apparent system 'resiliency' and 'self-righting mechanisms' would soon impact profoundly across the sciences, including in entirely new fields such as earth systems science. In what we can now see as something of a paradigm shift, the prevalence of the atmospheric scares has only served to obscure these origins and retard further developments in the atmospheric sciences. However, given the broader and popular applications of the new systems

---

¶ Dotto and Schiff reveal their overriding view of ozone layer stability when closing the final chapter of their book: 'What the ozone controversy has demonstrated is that life on this planet depends for its existence on a very small amount of a very unstable and easily destroyed substance. Ozone is the weakest link in the earth's life-support system. If that seems to you a precarious state of affairs—it is. If we break this chain...it is to be hoped that whatever creature may evolve as a result will be capable of learning from the many grievous mistakes of this human dinosaur'.[144]

theories, it is with the advantage of hindsight that we can more fully recognise their role in the controversies of the 1970s.

When, in 1963, the meteorologist Edward Lorenz planted the seed that destroyed the hopes of deterministic weather prediction,** he had in fact stumbled into the same field of mathematics as Poincaré all those years ago when he was trying to establish the stability of Newton's systems of the world. Summaries of Lorenz's now famous work on non-linear or 'dynamical' systems often give emphasis to 'the sensitivity to initial conditions', or 'the butterfly effect'. However, it was his other butterfly that had implications for our understanding of system stability. Dynamical systems such as his butterfly-shaped 'strange' attractor exhibit a new form of stability (see Figure 4.2). The precise state of the systems at any time might be unpredictable, even indeterminate, but the system itself remains determined and stable. Give it a nudge (or a new initial position) and it will push off in another course through its own phase space without loss of stability. A big enough nudge in the right direction at the right time will undoubtedly send it over the edge, but what is new and important is that Lorenz provides a way to see natural systems absorb perturbations where there is no return to an equilibrium state. The stability is not in equilibrium but in *perpetual disequilibrium*. The implication is this: if natural systems do have at least a degree of non-linearity (as they generally seem to), then this might actually be the seat of their stability.[145] The work by Lorenz and others on dynamical systems thus leads back to the possibility that it might be here, on the edge of the chaos, where the stability of various natural systems can be found, including the systems that determine both Earth's weather and its orbit.

By the time of the SCEP conference of 1970, Lorenz's influence was beginning to impact across the atmospheric sciences, not least on the leading American climatologist J Murray Mitchell. Lorenz's paper for the conference questions the predictive use of mathematical climate models due to 'the extreme nonlinearity of the atmospheric equations'.[146] Mitchell drew on an early version of Lorenz's paper to provide SCEP with background to the problem of climatic change and its causes. Two conceptual approaches to the problem of climatic change are proposed. The first is the 'slave' concept, where 'the average atmospheric state is virtually indistinguishable from an equilibrium state which in turn is uniquely consistent with the earth-environmental conditions at the time'. Thinking in this way, we would expect that when the external conditions change, the system adjusts to its new equilibrium state. In other words, the system changes in response to external bumps. (This conception could be the 'neutral' equilibrium of Figure 4.1.) The other concept he called the 'conspir-

---

**See p. 13.

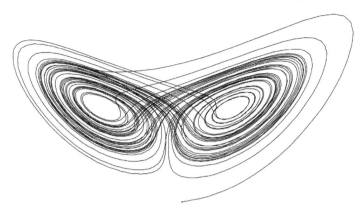

Figure 4.2: The Lorenz 'strange' attractor

The attractor exhibits stable disequilibrium such that the system absorbs perturbations by changing its phase state while yet remaining stable.

ator'. This is where the equilibrium state to which the actual atmospheric state
tends at any given time depends—perhaps to a large extent—upon the history
of the actual state. In other words, the system's response to external bumps is
not straightforward, may not be so great, and might be confused with internally
determined fluctuations.

> In the presence of Lorenz-type transients, the effect of systematic envi-
> ronmental changes on present-day climate (changes, for example, involv-
> ing secular increases of CO_2 or other consequences of human activities)
> might be so badly confounded as to be totally unrecognizable.[147]

If the climate system were a 'conspirator' there would be no easy cause–effect
analysis, irrespective of whether the cause were considered to be variations in
the sun, a volcanic eruption or industrial pollution. Just as short-term changes
(i.e. weather) are treated as mostly changes *internal* to the system (and ultimately indeterminate), perhaps longer-term changes (i.e. in climate) should
also be so treated. If this were 'the sort of place we imagine the atmosphere to
be', then simplistic accounts of inadvertent atmospheric modification would be
viewed with as much scepticism as all those marvellous schemes for weather
and climate modification that preceded them.[††]

Chaos theory was not the end of the systems theory revolution in atmospheric science. In the next few years, another conception of systemic integrity

---

[††] In a review of the SCEP and SMIC publications that appeared in *Nature*, another prominent
climatologist, Hubert Lamb, gave disproportionate prominence to Mitchell's use of Lorenz's work
to support a case against simple proposals for the causation of climatic change.[148]

would emerge. This began with the idea that physical systems in stable states of (chemical) disequilibrium are not only a necessary context *for* biological life, they are also a context created and controlled *by* life.

5 Apocalypse in a spray can

The Gaia hypothesis and origin of the chlorofluorocarbon scare

What is life? In the early 1960s the English medical researcher Jim Lovelock was consulted by NASA during the preparations for an unmanned mission to look for life on Mars. *What should they look for?* As lifeforms might be very different on another planet, the question quickly reduced to a definition of life. Lovelock was surprised to find very little discussion of the topic, but he began with a view that any suggestion of a 'reversal of entropy' might be a sign. Without life, matter would settle into a degraded stable resting state of chemical equilibrium. Thus, from a chemical point of view, a sign of life would be matter that is far from this degraded stability. In our own atmosphere, this is so evidently the case. If all the biological processes that cycle through our atmosphere stopped due to the demise of life on Earth, then its constituent molecular nitrogen and oxygen would no longer be renewed and would quickly react and dissipate, replaced by carbon dioxide. After a series of investigations, Lovelock concluded that the stable state of extreme chemical *disequilibrium* that we find in the Earth's atmosphere would be an indicator that there is life on this planet. His findings had grave implications for the Mars mission. As it was already known that the atmosphere of Mars is mostly carbon dioxide, on Lovelock's argument no mission would be required to answer the question. Mars is a dead planet.[149]

In this way, Lovelock tells the story of how he came to the hypothesis that the biosphere maintains the (non-living) atmosphere according to its own needs, much as an animal maintains its (non-living) fur. Atmospheric oxygen is a favourite example of just how precisely this gaseous coat must be controlled. The concentration of oxygen always remains around 21%. If it were a little less, then respiration would be difficult. If it were as much as four percentage points more, then the whole world would be at risk of conflagration. Life on Earth requires oxygen at just this level and it has been just so throughout eons of evolution. A similar case can be made for salt levels and life in the ocean.[150]

What Lovelock calls 'the Gaia hypothesis' is something of a throwback to

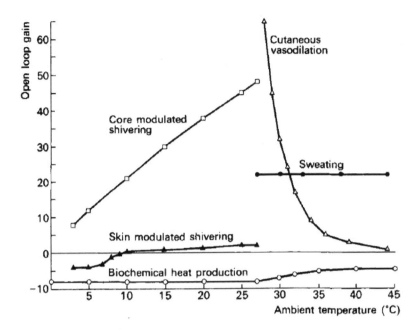

Figure 5.1: Lovelock's temperature regulation analogy.

Lovelock used human temperature regulation as an analogy for the regulation of global temperature by 'Gaia'. Source: *Gaia: A new look at life on Earth.*[151]

the ancient idea that the macrocosm (the world) is the microcosm (the body) writ large, and this is especially evident where he uses the analogy of the human body and its homeostatic control of temperature (see Figure 5.1). The average air temperature at the surface of the Earth has been maintained for eons within the range of a few degrees. Lovelock proposes that the biosphere controls temperature in a range optimal for life.* The bodily analogy extends to the idea that the atmosphere and the oceans act like arteries and veins, transporting vital trace elements from their ingestion sites (in the non-biological physical environment) to where they are needed in the living system. And trace elements might have another important role: their production, control and transportation could be vital to the biosphere's feedback systems. If all this were right, then the necessary controlling negative feedback mechanisms would be there to be found. However, in the early 1970s very little was known about many of these

---

*Contrary to this view is geological evidence suggesting this temperature stability prior to the development of the biosphere and even when the radiative warming of the Sun was significantly weaker.

74

trace elements, how they are transported through air and water, and even if they are present. This is where Lovelock's electron capture detector came in.

Lovelock had invented a device that could detect organic gases in the air and water at extremely low concentrations. In 1972, not long after first formulating the Gaia hypothesis, he took his electron capture detector on a research voyage from the UK to Antarctica so that he could investigate the movement of trace elements through the atmosphere. On this trip, he was especially interested to track the concentration of the spray canister propellants, chlorofluorocarbons (CFCs). This was not because they might have a role in the processes of Gaia, but because they might not. Recently synthesised and unusually stable, they could be used as 'an easily detectable and unequivocal anthropogenic marker' for tracking the circulation and mixing of air around the globe. Returning to the UK in 1973, Lovelock reported his CFC results in *Nature,* including their diminishing concentrations as he travelled south (see Figure 5.2). The concentrations were measured in parts per trillion–equivalent to a few drops in an Olympic-size swimming pool—yet this might represent all the CFCs ever produced. That is, due to their extreme stability and insolubility, most of the CFCs ever sprayed into hair and under arms might still be floating around. Undoubtedly mindful of the earlier fuss over the widespread detection of another persistent synthetic compound—DDT—Lovelock made a point of saying that 'the presence of these compounds constitutes no conceivable hazard'.[152]

Meanwhile, 'Sherry' Rowland at the University of California was looking around for a new interest. Since 1956 he had been mostly researching the chemistry of radioactive isotopes under funding from the Atomic Energy Commission. Hearing of Lovelock's work, he was intrigued by the proposal that nearly all the CFCs ever produced might still be out there. *Were there no environmental conditions anywhere that would degrade these chemicals?* He handed the problem to his post-doctoral research assistant, Mario Molina. Molina eventually concluded that indeed there were no 'sinks' for CFCs anywhere in the ocean, soils or lower atmosphere. Thus we should expect that CFCs would drift around the globe, just as Lovelock had proposed, and that they would do so for decades, even centuries… or *forever? Could mankind have created an organic compound that is so noble that it is almost immortal?*[154,155]

It turned out that photochemistry would be the reaper. Chlorofluorocarbons are not broken down by the weak filtered sunlight that reaches the earth, but they are susceptible to intense ultraviolet light. Any well-mixed gas that is not destroyed or removed by rainfall should eventually drift up high in the stratosphere, where it would be fully exposed to unfiltered sunlight. But this would take some time and it was still early days in the history of these chemicals.

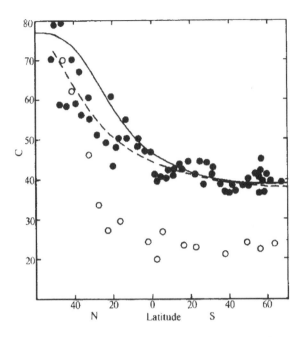

Figure 5.2: CFC-11 concentrations by latitude.

Lovelock's measurements for CFC-11 concentrations by latitude show a declining trend from north to south.[153]

The atmospheric release of appreciable amounts of CFCs had only begun during the Second World War when a new refrigerant, CFC-12, was also found to be an ideal propellant for the newly invented portable spray canisters. These were first widely used by the US military as 'bug bombs'—unscrewing the valve of these torpedo-shaped gas bottles released a mist of insecticide to fill a confined space. Late in the war, CFC-12 was even used to propel DDT in the fight against malaria-carrying mosquitoes. After the war, 'bug bombs' were sold commercially, but the real boom came with the invention of the now-familiar press-valve. This was fitted to lightweight, low-pressure canisters filled with CFC-12, CFC-11 and any household product that could be applied as an aerosol. While CFC refrigerants are only released through leakage and unit disposal, propellants are deliberately released directly into the atmosphere, and so the domestic spray-can boom caused a steady rise in emissions from the 1950s. Nevertheless, by the 1970s, Lovelock's first atmospheric measurements showed that concentrations remained extremely low. Molina and Rowland estimated that decades would pass before significant quantities of CFCs would drift high enough to reach the photochemical zone. But when they did, they would finally break down and give up their chlorine.

We will remember that the halogens, chlorine and bromine, were the 'deozonizers' recommended to Wexler back in 1962 to open the hole in the ozone layer proposed by Chapman. And it was chlorine in the rocket exhaust that might inadvertently produce the same effect.[†] Molina knew nothing of this speculation, indeed he had no background in atmospheric chemistry, but his independent reckoning established that the chlorine released from CFCs would catalyse ozone destruction. The ozone effect that Molina had stumbled upon was different to those previously proposed from rockets and aeroplanes in one important respect: it would be tremendously delayed. Like a hidden cancer, the CFCs would build up quietly and insidiously in the lower atmosphere until their effect on the ozone miles above was eventually detectable, decades later. But when unequivocal evidence finally arrived to support the theory, it would be too late. By then there would be no stopping the destruction of the thin veil protecting us from the Sun's carcinogenic rays. What Molina had stumbled upon had, in double-dose, one sure element of a good environmental scare.

It was in December 1973 that Molina reported back to Rowland his finding of the stratospheric sink and the consequential ozone-depletion effect. Although Rowland also had little experience of atmospheric chemistry, he suspected that Molina had hit on something important. Almost immediately, he decided to seek advice from another laboratory chemist with atmospheric expertise, Harold Johnston.[‡] Johnston was an obvious choice as he was by then at the height of his fame: by the end of 1973, hundreds of scientists around the country and around the world were using CIAP funding to investigate his proposal that NO_x from SSTs would destroy the ozone layer. When Rowland took Molina on a flying visit right after Christmas, Johnston confirmed his suspicion of the huge public policy implications of their find.

The soon-to-be-famous Rowland–Molina paper on this new ozone threat was published in *Nature* the following June. After a few months' delay, this new ozone scare took-off like no other and Rowland stepped up to the media role like a natural. From the beginning, and at every opportunity—and in conflict with other scientists more guarded with their claims—Rowland doggedly campaigned for a complete ban on CFC production. After thousands of hours of press interviews, after fifty appearances at congressional hearings, and after three long decades, his goal would be achieved. Later, Rowland, Molina and another scientist, Paul Crutzen, were rewarded for sounding the alarm with the

---

[†] See p. 36.
[‡] See p. 44.

1995 Nobel Prize for chemistry. The press release announcing the prize presented the ozone layer as the 'Achilles' heel of the biosphere'—a cryptic and perhaps unconscious apology to the hypothesis of system resilience that had inadvertently started it all.[156]

Lovelock strikes back

Back in the mid-1970s, the author of that hypothesis, Jim Lovelock, was annoyed with Rowland's use of his work and with the public excitement already building. He responded quickly, with his first paper appearing in *Nature* as early as November 1974.[157] From that time on, his main strategy remained the same. For the CFC effect to remain alarming, no other major sources of stratospheric chlorine could be found. We will recall that the confirmation in the previous year of significant natural sources of stratospheric NO_x was the first big blow for that scare. In the same way, if natural sources of chlorine were found, then the contribution from CFCs might also prove insignificant. And so, from the beginning and throughout the controversy, Lovelock pointed towards other potential natural sources of this ozone destroyer.

Indeed, for some time prior to the ozone scares of the 1970s, theoretical expectations had already prompted scientists to look for other natural controllers of ozone. This was because calculations suggested that the observed natural levels of stratospheric ozone were too low to be maintained by the direct ultraviolet effect alone. There must be something else up there keeping the ozone layer in check. Long before Harrison, hydrogen from water vapour had been proposed as a possible candidate. Then, shortly before Johnston's intervention in Boulder, natural sources of nitrogen were suggested. Johnston had been unaware of this work when he first triggered the NO_x scare in 1971, but in the previous year Paul Crutzen had published a paper suggesting NO_x derived from microorganisms in soils as the main natural controllers of the ozone layer.[158] The other likely candidates to catalyse ozone destruction were Wexler's halogens: bromine and chlorine.§ Chlorine was especially well-placed, with chlorine ions in the oceans, chlorine compounds throughout the biosphere released to the atmosphere through burning and decay, and with chlorine continually seeping and occasionally exploding from volcanoes.

---

§ Wexler's speculation about bromine and chlorine as deozonizers and about the threat of chlorine from rocket exhaust was never published in scientific journals, but only reported in the press. None of these papers from the early 1970s refer to Wexler's presentations nor to the press reports. However, there is some evidence that the halogen ozone effect was at least informally discussed around the time. For example, G.D. Robinson's SCEP paper quotes the phrase 'punching a hole in the ozone' before it turns to the possible impact of nitrogen oxide.[159]

The main objection to these natural sources was that they carry their chlorine mostly in soluble forms, and so they would be washed out by rainfall before reaching the stratosphere. Nevertheless, Lovelock provided some evidence that methyl chloride, in particular, would deliver large amounts of chlorine to the stratosphere.[||] In that chlorine might arrive in this way, he also argued that the scare over CFCs as an anthropogenic source was entirely disproportionate. If we were to worry about human emissions, then attention should be turned to low-tech sources of chlorine in much greater volumes, including the emissions of methyl chloride from grass and forest fires used in primitive agricultural practices.

To his case against prematurely sounding the alarm over CFCs, Lovelock added another level of attack: he tried to further moderate alarm over anthropogenic ozone destruction by promoting his newly published hypothesis of atmospheric resilience. The first reference to his Gaia hypothesis appears in another missive against the scare, published in *Nature* in July 1975. There he suggests 'an alternative view of the significance of odd nitrogen and odd chlorine'. This arises from the possibility 'that the cycling of gases by the biosphere is not merely passive but represents an active process concerned with homeostasis'. After referencing an early publication of the Gaia hypothesis, he closes with the question: 'Could it be that the biosynthesis [of methyl chloride and nitrous oxide] responds to some function of stratospheric ozone density and acts as a regulator?'[160] The London *Times* interviewed Lovelock for a report on this paper, and the article repeats this line of argument in paraphrase:

> It certainly seems unlikely that the present ozone levels could have been maintained for long enough for man to have evolved if they are in such a delicate balance as the more hysterical opponents of fluorocarbons imply.[161]

Self-stabilising models of the atmosphere were not well known—and certainly not well accepted—outside the circles of atmospheric scientists at the time, and so they could have little polemical effect.[¶] Instead, Lovelock, Scorer and others focused their attack on the oversimplification in the modelling behind the bold and drastic claims about the impact of a little extra chlorine. *New Scientist* picked up on these protestations and became the main vent for scientific

---

[||] Lovelock's view was eventually accepted: the chlorine found in the stratosphere has significant natural sources, and the main one is methyl chloride. In the 1980s, other natural and anthropogenic sources were found and only their relative contribution remained in question. By the 1990s, it was generally agreed that, while natural sources remain significant, the anthropogenic arrivals are dominant.

[¶] See p. 65.

scepticism of the new scare. As early as October 1974—before Lovelock's first intervention and before the controversy had even started—it published a letter from Scorer targeting the demonization of chlorine:

> ...there is much more chemistry going on in the stratosphere than many laboratory scientists have thought of, and to predict disaster from the intervention of a very common and highly reactive element is quite silly.[162]

Soon after, *New Scientist* took up an editorial interest in this scepticism, quoting heavily from Lovelock and Scorer under headlines such as:

> Hysteria ousts science from halocarbon controversy.[163]

Both Lovelock and Scorer made a point of saying that they were not against regulating CFC emissions. What concerned them was that Rowland and other US scientists were leveraging the authority of science on dubious evidence to effect a public panic. Needless to say, the export of British scepticism was not welcomed by the scientific community in the USA. Scorer's 1975 US speaking tour met with great resistance. Depicted as an industry stooge, Rowland refused to debate him, and Scorer went home early.[164] In the same year, Lovelock was called to Washington to testify on the industry side, but no pleas for calm from across the Atlantic were going to hold back this scare. Widespread enthusiasm for policy action would lead the USA inexorably toward a ban on the spray-can propellant within a few years. But the science had to first breakout in the mainstream media, and this did not happen right away. The story of how this breakthrough was achieved takes us back to another minor ozone scare, this time over NASA's proposal to continue space exploration using the Space Shuttle.

Space Shuttle chlorine

On their late-December visit in 1973, Johnston introduced Rowland and Molina to the intense discussions already swirling around CIAP and NASA circles on ozone destruction by chlorine pollution. This had begun with the environmental impact assessment of the Space Shuttle program. Assessing and reporting on the potential impacts of such hi-tech programs was legally required under the same legislation that instituted the EPA. Early research had cleared the Space Shuttle exhaust of any major pollution concerns; chlorine emissions had been considered, but not for their impact on the stratosphere.

Then two physicists took up on this work for NASA. Ralph Cicerone and Richard Stolarski had previously been investigating an extreme outer layer of the atmosphere, the ionosphere, but they were now keen to get in on the stratospheric research field that was opening up under CIAP. Their Space Shuttle investigations concluded that the main deleterious impact would be where chlorine in the exhaust attacks stratospheric ozone. The impact would be ever so slight, although there was still concern at NASA to contain any discussion of this finding (which is entirely understandable given the public sensitivity to ozone threats recently demonstrated during the demise of the SST program). However, the scientific significance of the finding was in the relevant reactions and their relative rates.

The two scientists were keen to present this aspect of their work to their peers. Fortunately it was not necessary to mention the Space Shuttle as the source of the chlorine. This is how Stolarski came to present their findings to a CIAP stratospheric chemistry conference in Kyoto, in September 1973, on the supposition that the source was natural—volcanic—not manmade. And this is why Stolarski's paper came under attack from Mike McElroy, an atmospheric chemist from Harvard, for the improbability of this volcanic source. And this is how a tussle between two research teams began.

After the Kyoto conference, Stolarski and Cicerone submitted their paper to *Science*, but it was rejected following a critical peer review by McElroy's collaborator, Steven Wofsy. For some time McElroy's team had also been studying the chlorine effect, but the paper McElroy delivered at the CIAP conference was, not surprisingly, mostly about NO_x. Early in the winter, around the time that Rowland and Molina were visiting Johnston, the Space-Shuttle–chlorine link was starting to come out. When McElroy sent the final version of his paper for publication in the conference proceedings, it was all about the chlorine effect. Sensing that McElroy's team was about to steal priority, Stolarski then made a late push to have a new version of his rejected *Science* paper appear in the conference proceedings too. NASA went public in February 1974, just before both papers appeared in the proceedings with explicit discussions of the Space Shuttle.

Luckily for NASA, interest was shifting to another synthetic source of chlorine, namely CFCs, and the tussle between the two groups continued on this new front.[165] Rowland and Molina's now famous publication of June 1974 received only minor press attention that summer. The *New York Times'* Walter Sullivan, who had been an influential promoter of the NO_x scare three years earlier, was holding back this time around. Meanwhile, Cicerone and Stolarski picked up on this new and much greater source of stratospheric chlorine and submitted another paper to *Science*. This time it was accepted and scheduled for publication in September. But, the day before it was due out, a story appeared on the front page of the *New York Times* that led with new findings by their Harvard rivals, McElroy and Wofsy. According to Walter Sullivan, they had calculated that spray-can CFCs...

...have already accumulated sufficiently in the upper air to begin depleting the ozone that protects the earth from lethal ultraviolet radiation.[166]

On current emission trends, 30% of the ozone layer would be destroyed as early as 1994. This was no longer a story about saving the sky for our grandchildren. These scientists had found an effect, already in train, with 'lethal' consequences for all living things during the lifetime of most of the *New York Times'* massive and influential readership. Sullivan brought in the work of Cicerone and Stolarski in support of these alarming findings, and he also used Rowland and Molina as background, but the team from Harvard had won the scoop.

Sullivan's story was picked up by a much grander media icon, Walter Cronkite, on the CBS network television news, and this new scare soon came alive in the media across the country. Perhaps it helped that the Watergate scandal was out of the way, with the resignation of Nixon the previous month. More likely, the esteem in which Sullivan and the *New York Times* were held was the catalyst that enabled the scare to break out across the mainstream media. It was also that McElroy had provided good copy: here we had Harvard scientists suggesting that hairspray destruction of the ozone layer *had already begun*. Verification of the science behind this claim could not have played any part in the breaking of the scare, for there was nothing to show. It turned out that McElroy and Wofsy had not shown their work to anyone, anywhere. Indeed, the calculations they reported to Sullivan were only submitted for publication a few days after the story ran in the *New York Times*. By that time already, the science did not matter; when McElroy and Wofsy's calculations finally appeared in print in February 1975, the response to the scare was in full swing, with spray-can boycotts, with 'ban the can' campaigns, and with bills to that effect on the table in Congress. When their paper was published, the theatre of Congressional hearings had already run a full winter season on the Hill.[167]

The breaking of the CFC story in the autumn of 1974 sent the rolling ozone scare into a peak of excitement that winter, sustaining momentum through 1975 and into 1976. At the beginning of this period there were many layers to the scare and some confusion. By the end, the other threats had receded in the public consciousness, as the powerful movement against the spray can came to dominate. Remember that, coming into the winter of 1975, the scientists were awaiting the release of the CIAP report. When it came out in January, all the recriminations and retributions that ensued might have been dreadfully important for those involved, but for everyone else they were little more than the background grumblings over a fast-fading, might-have-been hi-tech fantasy. With

Tests Show Aerosol Gases May Pose Threat to Earth

By WALTER SULLIVAN

Two scientists have calcu-lated that gases released by aerosol cans have already ac-cumulated sufficiently in the upper air to begin depleting the ozone that protects the earth from lethal ultraviolet radiation.

The calculations, by scientists at Harvard University, follow the recent discovery that these gases, used as aerosol propel-lants for hair sprays, insecti-cides and the like, while inert chemically, are high efficient in ·r·

the development of the ozone layer late in the earth's his-tory. The lethal wavelengths cannot penetrate water.

. The most prevalent concern, however, is not for total loss of the ozone. which is broken down and restored in a com-plex sequence of day and night chemical reactions. Rath-er, it is a fear of sufficient de-pletion to cause widespread skin cancer and other effects.

Furthermore, because ultra-absorption by zone

Figure 5.3: McElroy's unpublished findings are aired in the press.

This 26 September 1974 front page story in the *New York Times* is based on un-published work by McElroy and the Harvard group. The manuscript of the paper was not even received by *Science* until three days later. Four months later it was published.

the new scare coming on apace, defenders of Concorde (and also the Space Shut-tle) only encouraged it along, deflecting claims against their respective machines by pointing to the much greater damage effected by spray cans.

That winter, the spray can came under attack from all directions. It was not only environmental groups calling for a full ban; politicians at various levels of government and of all shades of opinion joined the campaign. By the summer, following well-publicised hearings, a ban had already been achieved in the state of Oregon. The momentum of the public and political response was nothing short of astounding.

The US spray-can ban

Calls for legislators to take immediate action met with an overwhelming response. Within a year, 11 states had legislative proposals for CFC regulation and 16 bills affecting aerosol production had been introduced to the US Congress, including one to grant NASA funding for a program of research along the lines of CIAP.[168] Alas, there would be no grand research mission legislated this time around. In part, this was due to the greater level of urgency. A compromise was soon reached with those calling for immediate action; one year, at least, could be spared to confirm the science. In fact, a plan to establish the scientific grounds for action by way of a NAS panel was already on the table when the first Congressional hearings began in early December 1974.

These hearings, before the Subcommittee on Public Health and Environment, were set to consider action proposed in two bills that were already before the House: one to amend the Clean Air Act for regulation of CFCs so as 'to prevent any increased cancer risk' and the other to prohibit CFC manufacture and importation. However, neither of these bills pronounced on the science. Instead, they deferred to the NAS and a study (already in the planning) that they were now formally instructed to undertake. If the Academy found that CFCs 'present no substantial danger to human life, agriculture, or the national environment' then no ban would be imposed.[169]

These early bills never got through, but the idea stuck that the Academy would have effective judicial responsibility over the decision to ban CFCs. This principle was reinforced in various quarters, including, after the NAS study had already begun, a federal interagency taskforce on the Inadvertent Modification of the Stratosphere (IMOS). IMOS conducted extensive hearings from early 1975, but when its report came out in the summer, it only deferred policy advice to an 'in-depth' study by NAS.[170]

Never before had such direct policy responsibility been imposed on NAS, which came under increasing pressure to produce a quick response. Its Climatic Impact Committee had established a panel, which was now given one year to answer the scientific questions about CFCs and advise on a ban. It was on track to deliver its findings by April 1976 when it was hit with the shocking discovery of a new chlorine 'sink'. On receiving this news, it descended into confusion and conflict and this made impossible the timely delivery of its much-anticipated report.

The new 'sink' was chlorine nitrate. When chlorine reacts to form chlorine nitrate its attack on ozone is neutralised. It was not that chlorine nitrate had previously been ignored, but that it was previously considered very unstable. However, late in 1975 Rowland concluded it was actually quite *stable* in the mid-stratosphere, and therefore the two most feared ozone eaters—NO_x and CFCs—would neutralise each other: not only could natural NO_x moderate the CFC effect, but hairsprays and deodorants could serve to neutralise any damage Concorde might cause. Rowland notified the committee of his findings and this prompted a frenzied review of all the related reaction rates and then frantic re-calibrations of the models. The early results suggested massive reductions in the CFC effect, with some even suggesting that CFC transport of chlorine into the stratosphere might serve to *increase* ozone levels. *After all the public excitement and the race to legislative action, had the scientists now found they had raised a false alarm?*

In the confusion that followed, the NAS committee struggled for more time and more funding. In April, when the much-anticipated report was due, a cryptic press release cited the consideration of 'new evidence' as the reason for delays. By May, details of this new evidence emerged and there was widespread speculation in the press that the scare might be over (see Figure 5.4).

McElroy described it as 'the most embarrassing thing this field has seen in a long time' and it caused further conflict among the scientists involved, not least between McElroy and Rowland.[171] They had already clashed the previous year, after McElroy expressed reservations and doubts in the IMOS hearings. That was in early 1975, after Rowland had stolen back the running on the scare on

Ozone Could Be Safe From Spray Cans

New research indicates that the danger of depletion of the earth's ozone layer by the fluorocarbons used in many spray cans and refrigerating systems may not be so grave as had been thought.

The ozone layer, 10 to 30 miles above the earth, screens out much ultraviolet radiation from the sun. Since 1974, scientists have feared that depletion of the layer by fluorocarbons might lead to an increase of skin cancers and harmful effects on climate and crops.

near future and population will soon peak, making the world's resources sufficient to meet the food, energy and raw material needs of the expected population of 15 billion by the year 2176. What is required, if disasters are to be averted, is better management and distribution of the resources, the group says.

The Hudson researchers read the evidence on supply and demand differently than do the doomsday theorists, and stress the long view rather than the short. They reject the idea that the world's resources are fixed; rather, they see an expanding potential in better use and recycling of existing resources and development of new ones as the world progresses economically, educationally, technologically.

Figure 5.4: The chlorine nitrate story in the press.

The NAS assessment of the CFC threat was already one month overdue when the reason for the delay was announced in the press (*New York Times*, 9 May 1976).

the back of his publishing priority and by providing the media with the simple message that the science was already solid enough to proceed towards a full CFC ban. As for McElroy, only months after his alarming calculations of imminent destruction featured on the front page of the *New York Times*, and just as the scare was spiralling towards mass hysteria, he was pulling back. Emphasising the uncertainties in the calculations, he argued against rushing into a ban: '...if we stop using chlorine compounds within five years', McElroy said, 'we will not have done irreparable harm'.[172] It was not only that Rowland and other advocates of a ban were trying to present CFC science as a settled consensus. In 1975 the messaging of scientists in the ozone-fear game was crowded and confused, and other work that McElroy was undertaking at the time was only making it more so. On top of the planned full-fleet SST, the Concorde landings, the Space Shuttle proposal, refrigerant leaks and spray cans emissions, McElroy introduced yet another threat to the ozone layer: crop fertilisers.

The steady rise in US agricultural production rates since the Second World War was due to a number of factors, including the application of new pesticides and the propagation of new plant varieties. But the most direct effect on field production rates came through loading the soils with powerful new fertilisers. One class of these new fertilisers massively supplements soil nitrogen, and it was these that became the target of the new scare. Picking up Crutzen's original work on the ozone layer effect of the NO_x emissions of soil microbes, McElroy (and others) showed how the ever-increasing application of these nitrogen fertilisers would massively augment natural soil emission rates of this ozone destroyer. McElroy calculated that the current trend in nitrogen fertiliser usage would result in an effect, noticeable by 2050, that would eventually reduce stratospheric ozone by up to 30%. The fertiliser–ozone scare did garner some interest, but perhaps less than had the alarm sounded a few years earlier, when little was known about the stratospheric chemistry moderating the drastic NO_x effect envisaged by Johnston. Now the new research into the chlorine effect reduced even further the expected depletion by way of this and other nitrogen emissions. In fact, McElroy had not even formally published his fertiliser calculations when the chlorine nitrate crisis disrupted the NAS assessment of the spray-can threat. When the new chlorine nitrate finding arrived, it impacted on scares about emissions of both nitrogen and chlorine—both McElroy on fertilisers and Rowland on CFCs—but Rowland was not for backing down on the need for an immediate CFC ban and this only further angered McElroy.

When, in May 1976, reporters started dialling up their scientific contacts to find out what was up with the NAS panel, some, like McElroy and Ralph Cicerone, agreed with sceptics like Lovelock that the speculation behind the scare had received a major blow. Rowland and Molina did not exactly disagree, but instead of pulling back on CFC alarm, they surged ahead, only switching to a new tack. Far from exonerating CFCs, Molina and Rowland would explain to reporters that the involvement of chlorine nitrate only worsened their impact on climate. They were referring to one implication of the chlorine nitrate effect widely discussed during the crisis: this was that chlorine nitrate only protects the lower stratosphere, while chlorine-driven ozone destruction would continue above. This would allow more ultraviolet radiation to penetrate to the lower stratosphere, which would further augment the ozone levels there. Thus the overall effect would be less on the total column of ozone and much more on its vertical profile. With more of the photochemical production (and destruction) of ozone in the lower stratosphere, its warming effect would shift downwards. Rowland and Molina argued that this would distort the climate of the stratosphere and also, potentially, the climate below.

What exactly would be the overall climatic effect? This was never fully elaborated or explored, neither by these chemists nor by meteorologists. Indeed, just such an anthropogenic effect on the vertical profile of ozone had been considered by meteorologists at the SCEP conference back in 1970. Back then, when SST exhaust was the concern, they dismissed such a change as 'of little significance'.[173] Back then, Harrison, McDonald and Johnston would soon shift SST concerns from climatic change to cancer. Now, at the height of the spray-can scare, there was a shift back to climate. This was reinforced when others began to point to the greenhouse effect of CFCs. An amazing projection, which would appear prominently in the NAS report, was that CFCs alone would increase global mean temperature by 1°C by the end of the century—and that was only at current rates of emissions![174] In all this, McElroy was critical of Rowland (and others) for attempting to maintain the momentum of the scare by switching to climatic change as soon as doubts about the cancer scare emerged. It looked like the scientists were searching for a new scientific justification of the same policy outcome.[175]

One enduring effect of the chlorine nitrate crisis was that it served to link the public discourse of spray-can ozone depletion with climatic warming due to greenhouse gases, a linkage that was not unhelpful in the transition from the ozone treaty to the climate treaty during the late 1980s. Back in the 1970s, the transition was incomplete. While the scare was now about climate *and* cancer, cancer-causing ozone depletion remained the dominant policy concern. Indeed, when the NAS panel belatedly reported its findings in September 1976, it found that the re-evaluation of chlorine nitrate formation was not as threatening to the perpetuation of the scare as the initial re-calculations had suggested. Nevertheless, their previously drafted estimates were nearly halved and were much more uncertain. At 1973 emission levels, ozone depletion would peak in the following century somewhere between 2% and 20%. Such a vague finding only reflected the report's uncertainties about the net effect. It should be no surprise then that the NAS Climate Impact Committee's recommendation was against proceeding immediately to a CFC propellants ban. Given their panel's own experience during the assessment, they could hardly say that the science was settled, and so they recommended two more years of research before making that call.[176]

The chemical industry made the most of the Academy's recommendation against regulation 'at this time', quoting it directly in large newspaper advertisements. Yet, at this time the industry was besieged by an upsurge of opposition. The press mostly interpreted the NAS findings in support of the scare, and the regulators were not hiding their competitive eagerness to proceed. Within a few weeks, despite its previous undertaking to follow NAS, the interagency IMOS taskforce unanimously recommended for regulation. Then one of the regulators, the Food and Drug Administration, got the jump on the EPA and announced plans for a ban on the 'non-essential' usage of CFC spray cans under its jurisdiction. Alexander Schmidt, the FDA administrator, won lots of press attention with his plain assessment of the situation:

> The known fact is that fluorocarbon propellants, primarily used to dispense cosmetics, are breaking down the ozone layer. Without remedy, the result could be profound adverse impact on our weather and on the incidence of skin cancer in people. It's a simple case of negligible benefit measured against possible catastrophic risk, both for individual citizens and for society. Our course of action seems clear beyond doubt.[177]

Of course, not everyone involved thought the case so 'simple' and the need for immediate action 'beyond doubt'. Certainly, it is hard to find any connection between this view and the conclusions of the NAS panel or the recommendation of the NAS committee. One who chose to point this out was Herb Gutowsky, a member of the committee and also the chairman of NAS's permanent panel on atmospheric chemistry. He was quick to launch a public protest that included a formal letter of objection to the FDA. Gutowsky would only follow the committee's recommendation in saying that proceeding to regulation at this stage would be an overreaction, that time was needed for further investigations, and (if necessary) to find safer substitutes.[178] But this protest was hardly heard above the clamour to *ban the can*, and so, despite the legislated authority of its findings, the NAS recommendation to delay action was ignored by the regulators. Instead, the FDA administrator presented a very different view of the science when announcing the ban. According to Schmidt, the science was already settled and the decision was simply to act against a bunch of frivolous cosmetics to save the people from the scourge of cancer. With the destruction already underway, the situation is sufficiently urgent that the commencement of the phase-out cannot await further research. This was not only Schmidt's view but also that of the other regulators. Just days later, the EPA announced moves towards a ban on the use of CFCs propellants in its domain of regulation. Negotiations and planning continued into the next year, and then, finally, on 11 May 1977, all the relevant regulators—the FDA, the EPA and the Consumer Product Safety Commission—jointly announced a timetable for the complete phasing-out of spray-can CFCs. In August, an amendment to the Clean Air Act passed Congress, this time giving extraordinary judicial powers, not to NAS, but to one of the regulators. To ensure that the stratosphere could be saved for the health of the people, the EPA administrator was to regulate 'any substance... which in his judgment may reasonably be anticipated to affect the stratosphere... if such an effect may reasonably be anticipated to endanger public health and welfare'.** The experiment to delegate judicial power to an Academy of Sciences committee had been a political failure, and so, by this amendment, the power to judge any newly discovered ozone threat, and to act on this judgement, was returned to where it had been with the politically successful decision on the banning of DDT.

Declining enthusiasm and the dark years

The ban on the non-essential uses of spray-can CFCs that came into force in January 1978 marked a peak in the rolling ozone scares of the 1970s. Efforts to sustain the momentum and extend regulation to 'essential' spray cans, to refrigeration, and on to a complete ban, all failed. The tail-end of the SST-ozone scare had also petered out after the Franco-British consortium finally won the right to land their Concorde in New York State in 1977. And generally in the late 1970s, the environmental regulation movement was losing traction, with President Carter's repeated proclamations of an environmental crisis becoming increasingly shrill (more on that below). Eventually, in 1981, Ronald Reagan's arrival at the White House gave licence and drive to a backlash against environmental regulation that had been building throughout the 1970s. Long before Reagan's arrival, it was made clear in various forums that further regulatory action on CFCs could only be premised on two things: international cooperation and empirical evidence.[179]

** Clean Air Act as amended August 1997, Section 157(b) 1977. In 1990 this section was amendment to become CAAA #615.

If we consider firstly the demand for international cooperation, it must be said that the domestic US movement for the spray-can ban was remarkable for how little attention was paid to the need for global action. An international industrial pollution problem ultimately requires an international solution, whereas unilateral regulation risks disadvantaging domestic industry. If we compare the unilateral decision on the spray-can with the decision on DDT, it is true that there the USA also acted unilaterally, but then the supposed environmental and health effects of DDT were mostly local. As for the SST, the supposed environmental effects were global and this did influence public opinion at the eleventh hour, but it never had a chance to prevail in a decision predominantly based on objections to the program's costs blowout (see Chapter 3). Thus, the US action on spray-can CFCs is outstanding as a unilateral policy response that would only be effective if it were universal. The decision was not strictly unilateral, as Canada acted in concert to some extents. Sweden and Norway soon followed with similar bans. But still, the US regulators seemed content to lead from the front, and the result was that elsewhere in the world production and emissions continued unabated. Efforts to persuade the rest of the world towards urgent policy action came almost as afterthoughts, as though they were attempts only to appease local industry and to advance the stalled domestic campaign.[††] When their diplomatic efforts began properly in 1977, US negotiators were faced with the other premise for further action increasingly demanded on the home front, namely the need for direct evidence.

On the international stage, the most important bloc that needed to be persuaded came under the newly empowered diplomatic unit, the European Economic Community. Two of its most powerful members, France and Britain, were strongly resistant to taking 'precipitative' action. Their position was most demonstrably in support of their industry. While the aggressive US diplomatic evangelism was overtly driven by the regulators, mostly the EPA, equally prominent on the French and especially the British side was the industrial lobby. Indeed, with their governments holding out against matching the US spray-can ban, European CFC manufacturers were starting to win increased market share in the global trade.[181] But there were other reasons for European resistance.

[††]In the US diplomatic campaign, equalising the trade situation for US industry subject to the spray can ban would remain important right through to the Montreal meeting in 1987. This is indicated in a comment proffered by the head of the British delegation in response to an account of that meeting by the head of the US delegation Richard Benedick. Qualifying Benedick's account, and against the general view that the USA was pushing for the treaty against enormous European resistance, Fiona McConnel claims that the EEC 'was the true architect of the Montreal Protocol' and that the USA 'nearly wrecked agreement' because of its 'obsession' with restricting the use of CFCs in spray cans over all other uses.[180]

The fight over Concorde landing rights had tested sympathies towards US excitability over environmental threats. In Britain, the anxious press monitoring of the US debate gave the ozone-cancer allegations a prominence they had never achieved in their own domestic supersonic debate. Across Europe, the overwhelming concern had remained the direct effects of the sonic boom over land. Once overland flights were ruled out, the press and the public on both sides of the channel became decidedly uninterested in the environmental case against Concorde. When next the US came calling—for global regulation to mitigate the threat of CFCs—neither the press, the public, nor the scientists in France and Britain gave their governments any reason to accept the US overtures.[‡‡] As with the SST, so with the spray cans, the grassroots campaigns conducted by US-based environmental groups failed to gain anything like the traction they achieved back home.[*] Amid all this apathy, indifference and outright opposition to US enthusiasm, scepticism on the science did play a significant role, if only by providing an authority to support government resistance for whatever political end. And this scientific scepticism only reinforced what the industrial lobby in the USA had been saying all along, namely that the scientific case needed to be stronger before any action could be justified.

To some extent, the demand for better science had always been resisted. From the beginning, advocates conceded that direct and unequivocal evidence of CFC-caused depletion might be impossible to gain before it is too late.[†] But concerns over whether the science was adequate went deeper. The predictions were based on simple models of a part of our world that was still remote and largely unknown. As the investigations progressed, and knowledge of strato-

---

[‡‡] While the British were closely watching the Concorde hearings in Washington, *New Scientist* ran an editorial entitled 'Environmental imperialism'. Given the prospect that the refusal of landing rights could 'kill off Concorde', according to the editor Jon Tinker, the US hearings 'spotlighted the growing phenomenon of environmental neo-colonialism'. For Tinker, the environmental case against Concorde presented at these hearings was exemplar of 'the growing habit of the US to export its environmental conscience'.[182]

[*] According to Richard Benedick, 'not until early 1987 did the efforts of some US environmentalists in the United Kingdom begin to pay off in the form of television interviews, press articles, and parliamentary questions...Indeed, these American private citizens were so successful that Her Majesty's Government in April 1987 asked the US Department of State to restrain their activities.' To support this latter claim, Benedick cites a cable in his personal collection.[183]

[†] Even after the discovery of the ozone hole, some advocates of regulation rejected the requirement for empirical evidence of depletion. While debate continued over the cause of the hole, Lee Thomas, the EPA administrator, made clear that empirical evidence was not required before a decision to regulate. In March 1986, he explained to a workshop that 'EPA does not accept as a precondition for decision, empirical verification that ozone depletion is occurring'.[184] As already noted on p. 89, the 1977 ozone layer protection amendment to the Clean Air Act only required that he have a reasonable anticipation of danger.

spheric dynamics and chemistry improved, every attempt to improve the simulations only delivered further complications and further doubts. The descent into doubt continued through to 1984–5, when it seemed to many that the scare was all over. As Roan plays it in her narrative of ultimate heroic triumph, the 'dark years' of the late 1970s was followed by the descent into 'a crisis that wasn't'.[185] And yet, if the public excitement of the mid 1970s is put to one side, we find that the message of 'no apparent crisis' was pretty much the official scientific advice *on both sides* of the Atlantic *all along*.

The elusive science

Before the NAS report belatedly appeared in September 1976, the British Department of Environment had published its own assessment of the effect of CFCs on ozone. This arrived, with very little fanfare, in April, right when the NAS report was due, and just weeks before the chlorine nitrate crisis went public. Although the re-evaluation of chlorine nitrate is not covered, the British report is otherwise informed by much the same science as the NAS report. What is different is that it achieves some critical distance from the excitement in the USA, starting out on a sceptical footing and continuing in this vein throughout its discussion of the science at every level.[186]

The discussion of its findings opens thus:

> It is easy to produce hypotheses which forecast dire consequences; it is much more difficult to obtain data which support the acceptance or rejection of such hypotheses...
> [I]nadequately informed opinion is easily influenced by highly coloured and emotional views and such material is a poor basis on which to take decisions.[188]

The SST-ozone scare is then used to exemplify an extensive investigation that has shown how 'original fears were exaggerated'. Likewise, with CFCs, further investigations are required. And they are required to extend beyond evidence of chlorine-driven ozone destruction, which is not sufficient cause for alarm. The report accepts Lovelock's view that natural chlorine might always have been a substantial destroyer of ozone before any chlorine arrived from the breakdown of CFCs.[189]

The modelling discussed in this British report predicts a similar ozone impact as would later be quoted in the NAS report, with current rates of CFC release effecting a maximum depletion of around 8% after about 100 years. But the British are more sceptical of the models because of their oversimplifications, especially because they do not incorporate 'atmospheric transport mechanisms'.

92

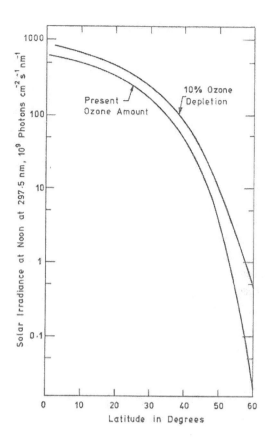

Figure 5.5: Effect of ozone depletion on solar irradiance.

The first British CFC assessment report of 1976 emphasised the minimal impact on the intensity of ultraviolet radiation of a projected ozone depletion of 10%.[187]

Then, even if the modelling were right, the report next addresses anxieties over ozone depletion by putting the possibility of ultraviolet penetration increases into some perspective: if the 8% depletion were to increase ultraviolet levels by 16%, this would be the equivalent of moving from the north to the south of England, 'an increased risk that most people would accept without further thought'.[190] This comparison is repeated several times in the report and also by the British negotiators, to the consternation of some opponents at home and across the Atlantic.[191]

When next the discussion turns to the cancer connection, the British report is dominated by concerns over the confusing and contradictory state of the science. 'Tenuous assumptions' and great uncertainties mean that medical evidence 'does not allow a prediction to be made with any confidence of the increased number of skin cancers for a given increase in UV radiation'. Epidemiological studies conducted in Australia present significant discrepancies when compared to those conducted in the USA. Of the US findings, the British report notes that IMOS had reported predicted impacts of a 1% increase in radiation that varied by a factor of seven. Even then, IMOS only considered the increase in non-malignant cancers. As for deadly malignant melanomas, the British quote a study which concludes that the relationship with ultraviolet radiation is 'complicated and in some respects obscure'. All considered, the report finds that there is sufficient time to investigate further and reduce some of the uncertainties. It concludes that 'there appears to be no need for precipitate action' and recommends 2–3 years' more research to resolve some of the outstanding issues.[192]

Undoubtedly, this is a more sceptical assessment than that released by the NAS panel a few months later, and yet the overall findings of both reports are much the same. Both present the science as anything but settled, and the situation anything but urgent. And both reports resulted in similar policy recommendations: a few more years' research before considering any action—which is precisely the recommendation ignored by US regulators.

What did happen with the science in the next few years is interesting, and the course of the scientific developments is tracked well by subsequent NAS assessments. Two years later the NAS panel produced an update, with subsequent updates issued approximately every two years thereafter. These biennial reviews of the science were undertaken at the request of the EPA, which itself was required to update Congress on the status of the research into ozone-destroying pollution generally. The NAS panel's first update in 1979 raised the estimate of depletion (at the standard 1973 emission rates) to 16.5%. This would be the highest estimate of any official panel. Its main policy recommendation was not for further domestic restrictions, but for the USA to lead international cooperative action. This was already underway at the UN, where the EPA was already on a collision course with the British, who were by then leading the European resistance to regulation. Tit-for-tat, there soon appeared a second British scientific assessment. This raises doubts about the modelling results behind such high depletion findings as those given in the NAS update. But, anyway, CFC usage is now declining, so it notes, and this trend is expected to continue. With emission rates on the decline and 'in the light of the many uncertainties still prevailing', the second British assessment finds no need for intervention. Pending further research, it concludes that 'strict regulation is not warranted at present.'[193,194]

As the science then progressed through the early 1980s, many of the uncertainties over stratospheric reactions and their rates began to resolve in the wrong direction for US negotiators. A second NAS update in 1982 included significant revisions that resulted in a reduction of the depletion estimate back to around 7%.[195] When a third update came out two years later, it reduced this estimate even further, coming in at a record low of any official report. Five years after the record high of 16.5%, NAS was now estimating a peak of only about 3% depletion. It also found that there had been 'no discernible' change in ozone levels throughout the 1970s, and it even predicted that due to the influence of natural causes the total ozone readings might rise by 1% over the following decade.[196]

Given the vast range of natural variability, a predicted peak of anthropogenic impact at 3% late in the following century was hardly going to drive governments to immediate action. As if to emphasise the trend away from alarm, this NAS panel included a graph giving the trend in the US projections. Starting with Molina and Rowland's first calculations back in 1974, the graph shows a peak for their 1979 report, followed by a steady decline down to its own latest minimal result (see Figure 5.6).

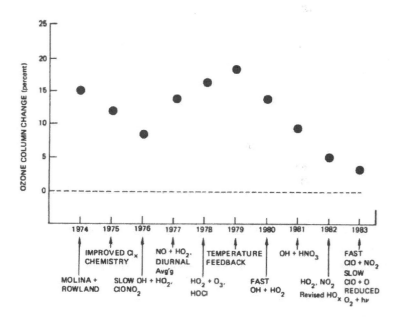

Figure 5.6: Assessments of the effect of CFCs on the ozone layer.

The NAS 1983 assessment update shows how the projected effect had declined since the 1970s.[197]

With the release of this third update in February 1984, there were again headlines announcing that the threat might be over. One of the panel members told the *Los Angeles Times* that they had found the threat in the foreseeable future to be so very small that if asked 'Should we put lids on the use of fluorocarbons?' the answer would be 'a resounding no'.[198] But by this time, interest had waned to such an extent that such a call was hardly news. A front-page editorial in the *Wall Street Journal* saw fit to first remind its readers of the ozone scare before then sticking the boot into the EPA for rushing into regulations based on the theorising of a few scientists:

> Well, here we are a few years later and new evidence shows that the ozone layer isn't vanishing after all; it may even be increasing. What's vanishing instead is the credibility of those who urged dubious policy decisions based on premature scientific evidence.[199]

'Very Small Threat to Public Health

New Study Downplays Ozone Depletion Peril

By LEE DEMBART, *Times Science Writer*

The threat to public health from depletion of the Earth's ozone layer is substantially less than previously thought, according to a new report by a panel of the National Research Council.

In its third biennial study for the Environmental Protection Agency, the research council, an arm of the National Academy of Sciences, predicted that continued release of chlorofluorocarbons into the atmosphere at current rates would reduce the ozone in the upper atmosphere by 2% to 4% over the next century, a substantial decrease from previous estimates.

"The threat in the foreseeable future is very small," Frederick Kaufman, a chemist at the University of Pittsburgh and a member of the

estimate, released Wednesday, further reduces that prediction.

While fluorocarbons are no longer used in spray cans in the United States, they continue to be used as refrigerants, in dry cleaning and in the manufacture of plastic foam. The State Department is attempting to persuade other countries to follow the U.S. lead in the use of aerosols.

The revised forecast of ozone depletion is based on new mathematical models of the complex chemistry of the upper and lower atmosphere, according to Barbara Jorgensen of the National Academy of Sciences. Previous models dealt only with fluorocarbons, she said, while the new models include trace gases such as methane, oxides of nitrogen and oxide.

Figure 5.7: The end of the ozone scare announced.

With the release of the third NAS assessment update it was starting to look like the end was nigh for the CFC scare (*Los Angeles Times*, 25 February 1984).

Later that year, Lovelock attempted an impartial review of the whole controversy after first declaring his interest as 'a veteran of the ozone war'. For Lovelock, the third NAS update represents a return to measured scientific sobriety, and something of a reconciliation with the past. Likening it to the memoirs of generals, it displays 'detachment from the blood and tears of war' where 'the heat and ultra-violet light of what a previous chairman of the same NAS panel [Schiff] called "the ozone war" have faded in the mist of objectivity'. Accounting the costs of the war, Lovelock finds some damage to industry (although much less than many anticipated), while the clear winners are science and scientists. Vast sums have been dispersed for atmospheric research that would never have been available but for the ozone scare.

In this observation, Lovelock's analysis presents an implicit moral dilemma. 'Had we known in 1975 as much as we know now about atmospheric chemistry', he reflects, 'it is doubtful if politicians could have been persuaded to legislate against the emission of CFCs'. Doubtful too is that they would have invested so much money in the atmospheric research that brought them to this assessment. Having criticised the means, but now recognising its end, Lovelock's review becomes an offer, if somewhat backhanded, to accept the report as conciliation. He closes thus:

> In the early days of this affair, I was repelled by the unbridled ambition of those who broke every rule of scientific conduct in their mad scramble for fame and funds. The cool excellence of this report suggests that the war was worthwhile, even if it was a messy and gaudy way to gain public support and money for scientific research.[200,‡]

What is most remarkable about these dark years of the ozone scares is that international negotiations begun in 1977 were sustained throughout. Indeed, the year following Lovelock's sobering assessment, in January 1985, a series of diplomatic meetings began as a new initiative to prepare a convention to protect the ozone layer from the ravages of CFCs. The prospect of persuading the nations of the world towards unified action at that time could hardly have been worse. The leader of the US delegation, Richard Benedick, prefaces his memoir of ozone diplomacy with a reflection on this initiative in order to emphasise how subsequent events make it easy to forget just how difficult was the diplomatic task in 1985. And much of this difficulty was because of the state of the science. As he explains, the objective of the gathered diplomats was 'to craft an international accord based on an unproven scientific theory'.[203] The situation with

---

‡ After the Antarctic ozone hole was discovered, Lovelock came to the view that it is at least partially caused by the impact of anthropogenic chemicals. However, he never backed down from his criticism of the conduct of the scientists. In *The Ages of Gaia* he says 'the excuse that "Ozone Warriors" had right on their side does not excuse their discourteous and unscientific behaviour'.[201] For more on his views, see *The Ages of Gaia*.[202] Lovelock also came out in support of the idea that CO_2 emissions were causing catastrophic global warming, although he retreated from this position very late in life.

the science had not improved when in March the leader of the UN Environment Programme (UNEP) opened the meeting in Vienna where the Framework Convention would be finalised. There, Mustafa Tolba pitched their work as an important precedent for 'an anticipatory response to many environmental issues', in that it deals with 'the threat of a problem before we have to deal with the problem itself.' Indeed, they would be acting for future generations who 'could be threatened'.[204] Perhaps not surprisingly, the gathered delegations could not be moved to settle their differences, put aside the usual horse-trading and agree to some real action. At least the framework was in place when, in the summer of 1986, they attempted again to construct an action protocol. Yet again, Benedick reminds us just how difficult the diplomatic task was, and how the difficulty remained with the flimsy scientific ground upon which the framework for policy action had been built:

> The science was still speculative, resting on projections from evolving computer models of imperfectly understood stratospheric processes— models that yielded varying, sometimes contradictory, predictions of potential future ozone losses each time they were further refined. Moreover, existing measurements of the ozone layer showed no depletion, nor was there any evidence of the postulated harmful effects.[205]

The year ended with only five countries having ratified the Vienna Convention. But by then the political situation was rapidly transforming. Benedick begs to disagree, but Tolba, Rowland, McElroy and most other actors and observers will say that there was already an important change in the air, and that this would quickly drive the politics towards an international accord that had previously seemed unachievable.[§] This change came with the discovery in Antarctica of a 'hole' in the ozone layer. Real and direct evidence had arrived at last.

[§] Richard Benedick challenges Rowland's claim that the news from Antarctica was the 'driving force' behind the negotiations in Montreal. Benedick says the ozone hole was 'never discussed at the negotiations, which were based solely on the global models'.[206] However, the absence of any explicit discussion does not mean there was no political impetus. Tolba notes that early news of the Antarctic discovery had already arrived before the Vienna meeting in 1985 and that it 'electrified the delegates'. Yet this had little impact on the outcome (which included the Framework Convention). For Tolba, it was the 'mobilisation of public opinion' from December 1986 that made the difference and drove the success in Montreal and beyond.[207] Benedick's dissent from this view might be partly explained by a conscious US strategy during the negotiations that is also explained in his memoir. This was not only that the American diplomats were careful to distance themselves from exaggerated claims of doomsayers, but that 'American diplomats and scientists also deliberately did not stress the Antarctic situation, because of the major scientific uncertainties... They worried that linking the US position with the ozone hole would risk its being undermined if that phenomenon turned out to be unrelated to chlorine.'[208]

6 The Antarctic ozone hole

> Industry always said that we'd have plenty of advance warning of any ozone problems, but now we've got a hole in our atmosphere that you could see from Mars.
>
> Sherry Rowland, quoted in the *Wall Street Journal*, 13 August 1986

The story of the Antarctic ozone hole picks up from Dobson's success in establishing a record of ozone above the southern continent during preparations for the International Geophysical Year.* By the early 1960s he had discovered and explained the Antarctic anomaly, where the poleward flow of stratospheric ozone is blocked by the polar vortex throughout the dark southern winter (see Figure 6.1). Since then the British Antarctic Survey had continued taking ozone readings at Halley Bay and these continued to show this southern anomaly. Other records, including eventually early US satellite data, filled in the picture of ozone building up outside the vortex before flooding in during its spring breakup. But then, in the early 1980s, the small stratospheric research group of the British Antarctic Survey noticed a secondary anomaly that had emerged in the late 1970s. Each October, just as the Sun began appearing low on the horizon, they noticed a further dip in ozone. Ozone levels would return to their normal range with the breakup of the vortex in November. But in the weeks beforehand, the already depleted Antarctic ozone layer thinned still further. During the early 1980s the monthly mean ozone levels were lower than any in the Antarctic record, and even below the equatorial norm.

Joe Farman had been involved in British Antarctic research since the International Geophysical Year and now led the stratospheric research group. When, in the austral spring of 1984, the group's new Dobson spectrometer delivered the

---

* See p. 34.

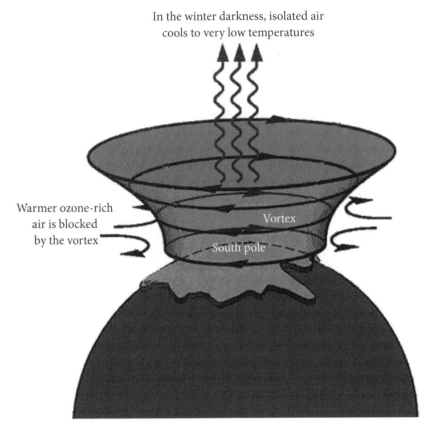

In the winter darkness, isolated air
cools to very low temperatures

Warmer ozone-rich
air is blocked
by the vortex

Vortex

South pole

Figure 6.1: The Antarctic vortex.

The diagram depicts the vortex in the stratosphere over Antarctica during winter
and early spring.

lowest readings yet, Farman decided they had sufficient evidence to confirm the trend. But in the article he submitted to *Nature*, he did more than that.[210] Farman not only presented the anomaly but proposed its cause. Using data that his group and others had collected on other trace chemicals above the southern continent, he argued that 'additional chlorine might enhance [ozone] destruction' in the unique and extreme conditions of the Antarctic spring. The likely source of this 'additional chlorine' was manmade CFCs.

Farman published on the new anomaly and its cause in May 1985. This was only weeks after the meeting in Vienna that produced the Framework Convention. With the benefit of hindsight, this convention could be seen as one giant

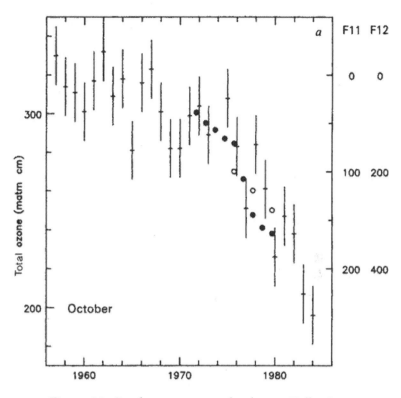

Figure 6.2: October mean ozone levels over Halley Bay.

Farman's chart (from the first reading in the International Geophysical Year until 1984) shows an inverse correlation with the likely source of the increasing chlorine, CFC-11 (F11) and CFC-12 (F12). The ozone is measured in Dobson units while the southern hemisphere measurements of CFCs are given in parts per trillion.[209]

leap towards Montreal, and an international agreement on emissions reduction targets. In fact, it was no more than a frustrating failure, with all hope for binding emissions reduction targets grinding to a halt. But between Vienna '85 and Montreal '87, there was a transformation in the diplomatic climate, which led to the agreement at Montreal, its subsequent ratification, and the later steps to strengthen the treaty through to the 1990s. The bad news from Antarctica, and other shocking scientific evidence, produced a steady stream of headlines announcing an ever-worsening and threatening situation. From 1986, environmental groups reengaged with the issue, and this mobilised public opinion and political will. In the summer of 1988, Margaret Thatcher underwent her dramatic conversion to the cause of global environmentalism, while in the US presidential election campaign, George Bush used that same cause to set himself apart from Reagan. In what Rupert Darwall has dubbed the second wave of global environmentalism (the first wave peaked in the early 1970s), there was broad agreement that finishing the job of protecting the ozone layer would be a forerunner to action on climate.[211] Yet, behind the scenes, the application of the ozone science to the policy debate was never straightforward.

The first problem with Farman's proposal was that it had not been predicted nor even vaguely implied by the modelling upon which the CFC ozone scare was originally based. Indeed, the models could not come close to explaining why there would be such a dramatic effect at that time and in that place, *but nowhere else*. Whether or not Farman's argument on 'additional chlorine' had any merit, the attribution to CFC-derived chlorine was not going to be easy.

In many ways, CFCs were unlikely culprits. Firstly, the Antarctic had the lowest concentrations of CFCs anywhere in the world. Then, when the Antarctic stratosphere was investigated further, the depletion was found to have happened almost exclusively in its lower parts, whereas (especially after the chlorine nitrate crisis) the models expected it to be almost exclusively high up in the photochemical zone. Meanwhile, natural explanations beckoned. Hints of natural causes were found in unusually high levels of volcanic particles, unusual circulation, unusual solar activity and unusual cold. There was broad agreement that the cold anomaly was important in one way or another, as ozone minimums are strongly associated with cooling of the lower stratosphere. While cooling could be an *effect* of less ozone, in turn, cooling can *cause* other changes at the poles. Sometimes, in the dark of winter, the air gets so very cold that it can no longer hold even the very little water vapour it contains. Clouds of ice crystals appear and these also contain other trace compounds frozen out of the thin air. These strange clouds are found nowhere else in the stratosphere and at no other time. They are more common in the colder South Pole, and, indeed, extensive coverage of such clouds was noted over Antarctica during the early 1980s. Since Dobson's time, ozone depletion has been associated with the presence of suspended solids, including volcanic dust, but also ice in clouds. Cloud-catalysed ozone destruction is one of the known reasons little ozone survives descent into the troposphere. So with clouds, volcanoes and other known influences in play, there were lots of reasons to suspect that the recent spring downturn in ozone levels might be a fluctuation due to natural processes.

The first reactions to Farman's paper were muted. Press coverage in the UK consisted of reports in *New Scientist*, the *Guardian* and then later one in the *Economist*.[212,213,214,215] The story was entirely neglected in the US press until November 1985, when Walter Sullivan wrote a feature article for the *New York*

Times.[216] Sullivan's angle on the story was news that a NASA scientist had taken 'a quick look' at the new satellite data for the previous month and found the 'hole' appearing once again that year. NASA had not previously picked up the anomaly in their analysis of the satellite record. For the scientists at NASA, Farman's findings were a complete surprise. Indeed, many US scientists had never heard of Joe Farman, nor the British Antarctic Survey, when they first read of the ozone hole in *Nature*. The circulation of that article prompted a re-examination of the satellite data, which, sure enough, not only corroborated the British ground-based readings, but also confirmed that the phenomenon extended across the interior of the vortex. When the October 1985 data came though, they found it again, and hence Sullivan's story.

So once again we find Walter Sullivan belatedly breaking what looks every bit like yet another environmental scary story. However, this time his treatment is noticeably restrained and sceptical. 'Even under normal conditions', he explains, 'the ozone layer is subject to wide variations, and whether the recent depletion is part of a long-term trend is difficult to establish'. Sullivan notes that the NASA expert who confirmed the hole in the satellite data, Donald Heath, was also sceptical of Farman's anthropogenic attribution. Sullivan then cites an alternative explanation involving 'sulphur compounds and other particles ejected into the stratosphere by the 1982 eruption of El Chichón in Mexico'. This effect had already been proposed by a Swiss scientist to explain a record minimum in one of the longest continuous ozone records. In 1983 the ozone layer above the alpine village of Arosa was on average thinner than it had ever been since the first spectrometer readings back in 1925. A further challenge to the anthropogenic attribution in Sullivan's article comes where he recounts the weakening of the scientific support for the CFC scare ever since the US spray-can ban:

> In 1977 a ban was imposed on fluorocarbons as spray-can propellants, but it became evident that the ozone varies in response to a variety of interacting natural and human influences. By 1984 an Academy report estimated ozone reduction, due to fluorocarbons, at only 2 percent to 4 percent.[217]

Perhaps it was due to Sullivan's reticence this time around that the story was not picked up; it remained dormant in the US media throughout that winter.

Meanwhile, interest *was* stirring among US scientists. By the spring of 1986 this interest percolated through to the press, and by the summer environmental groups were again involved. Three important developments came that June. An issue of *Nature* contained papers by two groups—one involving Rowland, the other McElroy and Wofsy—giving new and varied processes by which additional chlorine might work with the stratospheric clouds as they dissolve in the spring sunshine so as to enhance ozone destruction.[218,219] Just before their proposals were published, some of these scientists gave congressional testimonies on what was happening down south, and on what it might portend. Rowland was among the first to come out and publicly blame the ozone hole on CFCs. For him, it only confirmed the urgency of moving immediately to a full global ban.[220] Other scientists were much more reticent, pointing to many gaps in the data and many uncertainties. They all called for further investigations. Two weeks later a US expedition to Antarctica was announced.

Driving the hasty organisation of the expedition was British-born chemist Robert Watson. At the time, Watson was the leader of NASA's stratospheric research program. He also led the Coordination Committee on the Ozone Layer, a group formed in 1977 to give periodic scientific updates to the international negotiators.[†] Watson had built a career around the ozone scares after completing his PhD in London in 1973 and then moving to California to work on CIAP under Harold Johnston. The young leader of the Antarctic expeditioners on the ground also had experience in Johnston's laboratory; the only woman on the expedition, Susan Solomon had completed her PhD under Johnston in 1981 before taking a position at the National Oceanic and Atmospheric Administration (NOAA) in Boulder. The rapid advance in the careers of both these chemists would continue into the global warming scare, where both would transition into global leadership roles.

For this first US ozone hole expedition, twelve scientists were flown down to the base on McMurdo Sound in the depths of the southern winter. There they spent two months under the forming hole. As the Sun's rays began again to penetrate the dark Antarctic air, they investigated the various proposed causes, both natural and manmade. By the middle of October 1986, the expeditioners were persuaded by their preliminary analysis that the phenomenon was not caused by unusual atmospheric dynamics but by a 'chemical process'. Solomon expressed this view via a scratchy satellite link to a press conference in Washington, and it received extensive press coverage.

The trouble was that by this time the debate over the hole's cause was already deeply polarised. Atmospheric scientists persuaded by natural explanations found Solomon's announcement incautious and premature.[221,222] A special supplement dedicated entirely to discussion of the hole and its possible causation was then in preparation for the November 1986 issue of the prominent journal, *Geophysical Research Letters*. The editors of this supplement were them-

---

[†]The Coordination Committee on the Ozone Layer was made up of representatives of national, intergovernmental and nongovernmental organizations, and scientific institutions. It met annually from 1977 to 1986, reporting to the Governing Council of the UNEP, and producing periodic assessments of ozone-layer depletion and its impacts.

selves inclined towards some kind of natural explanation, and a symptom of the polarisation of the debate at this time was that the 46 papers in the supplement were almost entirely given over to various arguments for natural causes.[223] But then, as the new evidence from Antarctica was compiled and analysed, some of the dynamical proposals published in the supplement came crashing down. Still, many of those on the other side of the debate, including Watson, agreed that the science remained inconclusive. More evidence was required, and another expedition was soon in the planning for the following austral spring. This would be a more expensive affair, and for some time funding was not forthcoming. In the end the expedition went ahead after a substantial contribution from the chemical industry. ‡

This time a specially designed and fitted high-altitude aeroplane would fly repeatedly through the forming hole. It turned out that these flights would provide what has since been hailed as the 'smoking gun' of the CFC effect. The aeroplane was fitted with several sensors, including one newly developed to detect chlorine monoxide at its extremely low stratospheric concentrations. Chlorine monoxide is the tell-tale product of chlorine reacting with ozone. During August and September 1987, samples were taken on flights penetrating the vortex into regions of the lower stratosphere affected by stratospheric clouds. These showed a strengthening correlation between the forming ozone hole and the presence of chlorine monoxide (see Figure 6.3).

At the very time when these critical samples were being collected high over Antarctica, diplomats in Montreal were debating the terms of the now-famous protocol. However, it was only after the text of the Montreal Protocol had been agreed that the scientific results were officially announced. Following their release, some scientists on the dynamical side did hold out; others went quiet, while those who had suspected a chemical process all along declared the science settled: *chlorine derived from CFCs and other manmade sources is primarily responsible for the hole.*

Yet, the Antarctic hole still sat awkwardly with the original scientific messaging behind the scare. The depletion every Antarctic spring was certainly no catastrophe. With the Sun barely rising above the horizon in October, there was little risk of excessive ultraviolet causing any damage. No penguin would be blinded or burned by the weak light cutting a tangent through the frigid southern air. If ozone were to be sacrificed in order to remove excess chlorine, then there could hardly be a better place or time to do it. Most awkward for the scientific messaging was that the Antarctic effect could never be predicted by

---

‡As Benedick points out, throughout the scare the chemical industry provided significant financial support for the research into the environmental effects of CFCs.[224]

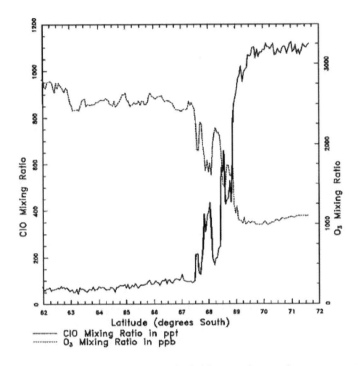

Figure 6.3: The 'smoking gun' of chlorine-driven destruction.

This chart shows ozone and chlorine monoxide concentrations measured as the aeroplane passes into the interior of the stratospheric vortex. This striking inverse correlation was detected during a flight on 16 September 1987.[225]

the global modelling. Nor could an event occurring in the extraordinary circumstances of the Antarctic spring go very far toward supporting the idea of a generalised depletion. There was no evidence elsewhere of ozone depletion beyond natural fluctuations, and nor could any evidence be found for the very basis of the scare, namely an increasing incidence of deadly ultraviolet rays in populated areas. Indeed, there was evidence to the contrary. Some of this came out in February 1988 when the US National Cancer Institute published a report on ultraviolet radiation records from eight meteorological stations across the United States between 1974 and 1985. They showed a general downward trend.[226]

Thus, in order to sustain the scare, the annual brief thinning of the ozone veil over Antarctic would not be enough. More evidence of scary consequences would need to be provided than the ozone hole could ever deliver. And so it was: over the next two years, successive reports of frightening new evidence

106

were delivered into the gathering political excitement that became global environmentalism's second wave.

Ozone depletion has begun over the populated north

During 1988, the second wave of global environmentalism would reach its peak in the USA, with CFC pollution its first flagship cause. Mid-March saw the US Congress voting *unanimously* to ratify the Montreal Protocol. It was only the second country to do so, while resistance remained strong in Europe. The following day, NASA announced the results of a huge two-year study of global ozone trends. Within a week, Du Pont, the leading producer of CFCs, made an historic announcement. Earlier in the month it had pointedly rejected calls from US senators to voluntarily cease production. Three weeks later the company reversed this decision and agreed to a production phase out. How much the announcement of the NASA study results influenced Du Pont's decision is open to speculation. What we do know is that the ozone trend results were unavoidable in the public debate. So shocking were the findings that they made headlines around the world and on front pages of major newspapers across the USA. We also know that these results caused the UNEP leadership to speed up the process of implementing the emissions protocol. UNEP Executive Director Mostafa Tolba called for the four assessment groups specified under the Montreal agreement to convene early, before they could be authorised by the assembled parties to that agreement. This legal breach would be endorsed *post-facto* by the parties, who, as Tolba later explained, 'agreed that action was urgent in the face of the new scientific evidence'.[227]

The new scientific evidence came from a re-analysis of the ozone record. This found that the protective layer over high-population areas in the mid-latitudes of the northern hemisphere had been depleted by between 1.7% and 3% from 1969 to 1986. These trends had been calculated after removing the effect of 'natural geophysical variables' so as to better approximate the anthropogenic influence.[228] As such, these losses across just 15 years were at much faster rates than expected by the previous modelling of the CFC effect. Only three years back, in 1985, a NASA–UNEP report predicted that CFC emissions at current rates would deliver a peak depletion of 4.9–9.4% sometime late the following century. This prediction was up from that given in the NAS report of the previous year, which had given a figure as low as 3%.[§]

Now came this new study of the actual ozone record. The 'International Ozone Trends Panel' had been organised through the collaboration of three US

[§] See p. 87.

Ozone Depletion Far Worse Than Expected

Long-Awaited Study of Loss in Protective Layer Raises Key Health and Environmental Concerns

By THOMAS H. MAUGH II, *Times Science Writer*

Man-made chemicals have depleted the Earth's protective ozone layer by about 2.3% over most of the United States since 1969, and by significantly larger amounts near the north and south poles, scientists disclosed Tuesday in a long-awaited report.

In some places in the Northern Hemisphere, the ozone layer drops by as much as 6.2% below 1969 levels during the _, _ _ dine t·

20% increase in skin cancer in those areas over the next two decades. In addition, scientists said, the depletion may already be adversely affecting agricultural and fishery production in temperate regions of the world.

Researchers said there is little doubt that the ozone destruction is caused b the continued release of synthr chemi'; called chloro- r 'R' s, v' ich are

Figure 6.4: Alarm sounded on ozone depletion.

The *Los Angeles Times'* front page on 16 March 1988 sounds the alarm on the discovery of the ozone depletion trends over the heavily populated areas of the northern hemisphere.

government agencies and two bodies of the UN. It involved more than 100 scientists from 12 countries and was specifically mandated to inform the treaty process. It found that much of the depletion expected in the far distant future *had already occurred.* In other words, the empirical data showed that the depletion rate in the real world had been drastically underestimated by the models. The leader of the study, Robert Watson, was quoted explaining the implications of the findings:

> At the moment, it would not be unreasonable to say our models are not doing a good job to predict ozone changes. Therefore one would have to question whether we're underestimating the rate of change of ozone in the future.[229]

'Ozone depletion far worse than expected' read the headline on the front page of the *Los Angeles Times*.[230] In the *Boston Globe*, Watson is quoted as saying:

> This is a very, very important result...This is the first time scientists are saying there are measurable meaningful changes that are not attributable to natural causes...[231]

Another scientist involved in the study, John Gille of the National Center for Atmospheric Research, said that while earlier data had given scientists a 'smoking gun', this report had given them 'the corpse' because 'we're actually seeing a decrease in a very credible and verifiable way'. The *New York Times* also supported the claim that the human influence had been detected in these trends, quoting Watson saying the study shows that ratification of the Montreal Protocol was only 'an essential first step' and that the UN should now look at more 'draconian' measures to stabilise the protective ozone shield.[232]

The statements of the scientists (at least as quoted) made it clear to the press that this panel of experts had interpreted the empirical evidence as showing that a generalised CFC-driven depletion *had already begun*, and at a much faster rate than expected from the modelling used to inform the Montreal Protocol. The executive summary released at the press conference is, however, rather more ambivalent. In its 'Key findings' section, far from attributing the observed changes since 1969 to the CFC effect, it says that the model calculations are 'broadly consistent' with these observations. Where the observed changes are the same or greater than modelling expectations, the summary says that these changes '*may* be due wholly, *or in part*, to the increased atmospheric abundance of trace gases, *primarily* CFCs'.[||,233,234] Weasel words and grammar had been employed to rhetorically suggest anthropogenic attribution, but yet avoid any direct claim. Today we can take a quick look at the body of the report and see why (which we will shortly). But at that time the full report had not been released. Four months later, in a review of the executive summary, an atmospheric scientist, Kevin Trenberth, reported that 'the full report of the Panel is not yet complete and may not be available for several months'. Until it is released, Trenberth complained, it is impossible to assess the complex work undertaken by the panel.[235] This delay also made difficult any timely challenge to the headlines, although some did try.

Fred Singer had built a respectable career in atmospheric science, and especially in the development of the satellite monitoring of ozone. He had been recruited to the government side during the SST controversy, where he made some attempts to moderate ozone alarm. Since then he had held various high-level government offices. At the time of the International Ozone Trends Panel press conference, Singer's scepticism was consolidating, helped in no small way by the rejection of a number of his critical letters to prominent scientific journals. Eventually, in June 1989, he was able to air his views on the science behind the CFC–ozone scare via an essay in a political magazine.[236]

As previously with Lovelock and Scorer, Singer makes it clear that he is also

---

|| Emphasis added.

not against CFC emissions control. His main concern is with 'the poor state of the scientific evidence' that was being used to legitimise the rush to policy action. When he comes to the International Ozone Trends Panel findings (now more than a year after they were announced), he repeats Trenberth's complaint about the ongoing delay in the release of the full report. Until the report is published, he says, there can be no 'independent check of their analysis'. Singer also echoes some of Trenberth's reasoning about why such an analysis was especially urgent: the extreme variability, patchiness and occasional unreliability of the primary data, the many natural influences on ozone fluctuations across various timescales and their unknown impacts, and the fact that other reports give significantly different trend rates to those of the panel. Of particular concern to Singer was that any measured trend appears strongly dependent on the choice of time period.[237]

Today we know that each one of these concerns was justified, if only by reading the report itself. It was quietly released in January 1991, nearly three years after its executive summary.[238] What it shows, firstly, is how the original scope of the assessment had been greatly reduced by the lack of adequate records. A major setback was that the satellite data available from 1979 was found to be corrupted by instrumental degradation. This meant that only the unevenly distributed Dobson station records could be used to provide generalised trends. Even then, there were only sufficiently reliable time-series datasets to provide trend-lines for the northern mid-latitudes and the Antarctic. Secondly, the report confirms that a number of natural influences on ozone variability are identifiable, but only two of them—the solar cycle and the 'quasi-biennial oscillation'—were removed to give the depletion trends quoted in the executive summary and in the press. The known impacts of major volcanic eruptions—including the huge eruption of El Chichón just before the absolute low point (1983) of the best and longest records—remained in the data. Thirdly, the computer models had not only underestimated the rate of depletion since 1969, there were also major discrepancies with the observed pattern of depletion. These discrepancies brought into question the models' predictive power. Most alarming of all, the observed vertical profile of the depletion was almost the reverse of what had been predicted. As with Antarctica, observations showed the depletion to be mostly in the drifting clouds of ozone in the lower stratosphere,¶ while the depletion effect of CFCs was expected mostly higher up. Thus there was good reason for caution towards any anthropogenic attribution. Such caution is found in the analysis within the full report, and it is still traceable in the guarded language of the executive summary. Yet, it is all but abandoned by

---

¶ See p. 84.

Watson when he ventures to pronounce on the policy implications of the assessment at the press conference—a conference held long before the report had been finalised, long before it was published and thus long before it was available for scrutiny. Nevertheless, the dubious grounds of any human attribution is not the most striking aspect of the widely quoted global depletion trends announced by Watson.

What is most surprising is the date-range of records chosen for discussion in the executive summary and for delivery to the press. The 'Key findings' section of the executive summary only discusses the depletion trend after 1969, but the full report shows how the trends for northern latitudinal bands had been carefully statistically constructed from those Dobson records that were sufficiently reliable, stretching back beyond 1969 and right back to Dobson's own efforts to establish standard monitoring across the globe in the International Geophysical Year. A chart of results starting in 1957 clearly shows in each of four latitudinal bands an upward trend through the 1960s towards an absolute peak in 1969 (see Figure 6.5). This corroborates Julius London's much earlier work to establish a general trend, work that was discussed at the SMIC conference in 1970 and eventually published in 1974. Based on a much smaller dataset, London had found a general thickening of the northern hemisphere ozone cover from 1957 to 1970.[239,240] By comparison with the new study, London's 7.5% thickening comes in rather high, but a general thickening trend through the 1960s remains undisputed. Nevertheless, in the executive summary of the new study, this positive trend up to 1969 is not even mentioned. Reporting results only from 1969 of course means that there is no need to explain what caused the earlier positive trend, no need to explain what caused the turnaround in 1969, nor any need to account for what appears to be a general sine wave pattern. There is also no need to report on the difficulties experienced in attempting to simulate this observed trend in computer models. All this is discussed in the full report, but any debate over the interpretation and analysis of the complete trend from 1957—a debate that both Trenberth and Singer were keen to take up—would surely have confused the messaging at the press conference. Instead, a one-way trend of depletion after 1969 could be described as attributable to 'the increased atmospheric abundance of trace gases, primarily CFCs' and thereby as representing the beginning of expected continuing CFC-driven destruction, if only *at a faster rate*.

When the full *Report of the International Ozone Trends Panel* was eventually released, the policy debate was long-since won. In the meantime, the findings announced to the press in March 1988 remained a standard reference for the treaty process.[242] They provided empirical confirmation that the depletion caused by CFCs was not limited to the icy southern wilderness: a more general

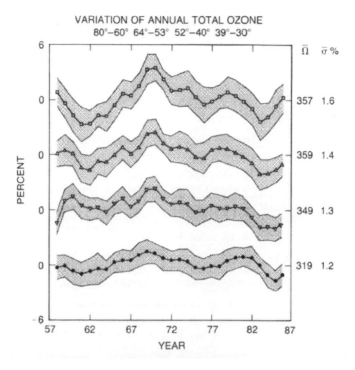

Figure 6.5: Total ozone trend along four northern latitudinal bands.

The graph shows the total ozone trend along four northern latitudinal bands from 1957 to 1986, each showing a peak in 1969. This graph is from p. 282 of Volume 1 of the *Report of the International Ozone Trends Panel*. However, the full report was not released until nearly three years after the press conference delivered its executive summary. In this summary, the rise to 1969 went unmentioned, while in the press conference the decline since 1969 was presented as conclusive evidence of faster than expected CFC-driven depletion. The depletion story was announced in front-page headlines across the USA.[241]

depletion was already well underway in the skies above the densely populated regions of the north hemisphere.

The ozone hole is spreading across the southern hemisphere

Meanwhile, in the southern hemisphere, there soon began speculation that the Antarctic hole might expand towards, or drift over, populated areas. When the size of the hole was found to be larger in 1987, fear spread that it would advance across the skies over Australia and South America. Of course, this fear was entirely unfounded, as the hole is only ever as large as the inside of the polar vortex, and this never comes close to the populated mid-latitudes. Nevertheless, a subtle link was found between the hole that year and unusually low ozone recordings over southern Australia and New Zealand a few months after the vortex breakup, and this is what triggered a scare.

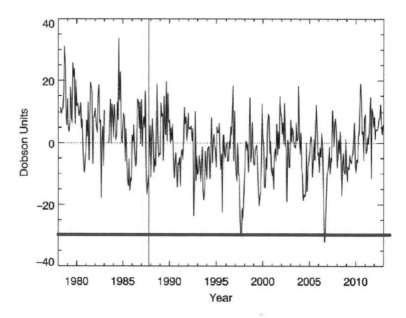

Figure 6.6: Total ozone measured from a Dobson station in Melbourne.[243]

A vertical line indicates December 1987, which was the time of the event that triggered the Australian and New Zealand ozone hole mythology. Also marked is 10% below the monthly mean (horizontal line).

During the last two weeks of December 1987, most of the daily ozone readings over Melbourne were the lowest on the satellite record for that month. Analysis of the data later showed how this thinning was associated with the breakup of the polar vortex, where a finger of ozone-depleted air drifted north across southern Australia and New Zealand.[244] While the link with the vortex breakup is scientifically interesting, there is no strong case that the event is extraordinary or alarming. After all, the satellite record was then less than a decade old and daily readings as low as these were not uncommon over Melbourne later in the summer, nor were they particularly low when compared to other populated areas around the globe. Figure 6.6 puts the December 1987 anomaly in the context of the longer Dobson record.

During the following year, reports from Australian and New Zealand scientists of the discovery of this link were generally moderate in tone. The most alarming analysis from a scientist came from Ralph Cicerone. At a presentation he gave to the AGU in December 1988, Cicerone is quoted saying that 'physically, this is serious, it is important, and the numbers are big'. He said 'we feel very much shaken by this'. One newspaper reported his comments under the headline 'Dramatic move of ozone hole to Australia 'shakes' experts'.[245] However, such alarming coverage cannot be found in the Australian press. Indeed, this may be a case where scientists did little to encourage a popular distortion. And what a distortion it was! A colourful and evocative mythology quickly emerged of a deadly hole in the natural ozone shield drifting across the Antipodes.

This linking by scientists of the breakup of the southern vortex with low ozone readings in southern Australia during December 1987 morphed into the idea that the ozone hole itself had moved over southern Australia. All sorts of further exaggerations and extrapolations ensued, including the idea of the hole's continuing year-round presence. An indication of the strength of this mythology is provided by a small survey in 1999 of first-year students in an atmospheric science course at a university in Melbourne. This found that 80% of them believed the ozone hole to be over Australia, 97% believed it to be present during the summer and nearly 80% blamed ozone depletion for Australia's high rate of skin cancer.[246]

The Arctic ozone layer is primed for destruction

With the ozone hole appearing in the polar vortex of the southern hemisphere, it is understandable that questions would arise about the northern vortex. *Could the hole appear there also?* Such an event would likely cause greater alarm given the dense populations of Europe and North America close by. Indeed, back in the early 1960s Dobson had noted a second anomaly, localised to the high latitudes of northern Canada. But the northern vortex lacks the intensity of the strong westerly winds blowing uninhibited around the Antarctic continent and it never holds together long enough to allow anything so grand as the depletion inside the persistent southern vortex. Dobson noted the Canadian anomaly to be weaker and more erratic in its timing.[247] As for any winter depletion descending further into a springtime 'hole', Farman had anticipated this question in his original article when he noted that 'comparable effects should not be expected in the Northern Hemisphere' because its vortex 'is less cold and less stable'.[248] Later, Watson found the importance of the stability of the southern vortex confirmed by the flight data of the second Antarctic expedition, which sug-

114

gested that the vortex acts as 'a kind of "restrainment vessel", in which the perturbed chemistry could proceed without being influenced by mixing in more normal stratospheric air from outside or below'.[249] The vortex continuing to trap stale air over the Antarctic for some weeks after the sunlight returns is key to the formation of the spring hole. This is according to the prevailing explanation, where sunlight is critical to the chemical process that causes the hole. Just as critical is that the south-flowing ozone-rich air does not enter, and so continues to build up on the other side of the vortex. Nothing like this happens in the north. The northern vortex is a leaky and unstable vessel that is generally only able to sustain itself for extended periods in the coldest period of the winter darkness. Even this is not enough to hold back the ozone. In fact, while the Antarctic ozone hole was making headlines around the world, the ozone blanket over the Arctic was on average as thick as almost anywhere in the world.

Despite these observations, and despite the uncontroversial explanation of the difference between the two poles, some scientists still chose to emphasise one component of the prevailing account of the hole-causing chemical process that was not entirely unique to Antarctica, namely the presence of polar stratospheric cloud. Some atmospheric chemists began to express concern that such cloud formation over the Arctic during colder winters might be sufficient *in itself* to allow CFC chlorine to gouge out a northern hole. Acting on these concerns, millions of dollars would be invested in northern expeditions essentially aimed at testing the views of Dobson, Farman and Watson on the critical importance of the strong and persistent southern vortex.

The first declarations of an Arctic ozone hole were made by Wayne Evans of Environment Canada. He announced to the press (and later published) that the ground-based data collected from far north-western Canada confirmed a hole also evident in the satellite record.[250] When he first reported it in early 1987, attention remained focused on the efforts to establish causation down south. But then, following the success of that second US Antarctic expedition, more scientific attention shifted to the north. During the winter of 1988, groups of scientists from countries including France, Norway, Sweden, West Germany, Britain and the USA were sent to bases within the Arctic Circle in Greenland and Sweden. These expeditions found evidence of the same chemical processes that were causing the hole in the south, as well as a relative depletion in the lower strata of the usual thick winter ozone cover, but they found nothing like the southern hole.[251,252,**]

** The Soviets were also collecting evidence at this time. This work was raised in the ozone treaty talks of 1989 by the Soviet negotiators, who claimed that no scientists had provided evidence to attribute Arctic depletion to non-natural causes. With their insistence on observations before they believed, *Le Monde* likened their scepticism to the doubting of St Thomas.[253]

THE GUARDIAN
Saturday February 18 1989

Scientists find significant levels of chlorine monoxide

'Smoking gun' clue to Arctic ozone loss

Tim Radford
Science Correspondent

A TEAM of scientists from four nations who have have been studying the Arctic stratosphere since January 1 confirmed yesterday that the manmade chemicals which destroy the ozone layer are 50 times greater over the North Polar regions than ̇d 'ted.

The Antarctic destruction is a consequence of a series of chemical reactions in an atmospheric "vortex". In the intense polar cold, over a period of about six weeks, the ozone layer is destroyed by a complex chain of reactions involving ice crystals, chlorine, ozone and the first sunlight of the spring.

The scientists, who surveyed the Arctic with a converted U2 spyplane with sensing instruments and a ̇ packed with

spheric chemist who took part in the study. "We do have the same machinery working in both hemispheres, but the machine in the northern hemisphere gets switched off much earlier."

Some 200 British, American, Norwegian and West German scientists gathered in Stavanger. The two aircraft flew 14 missions, spanning arcs from Greenland to the Pole. The scientists emphasise their

Figure 6.7: Rapid ozone destruction over the North Pole.

The *Guardian* of 18 February 1989 reported to its readership that the NASA Arctic expedition results showed that the ozone destruction over the North Pole was 50 times faster than that expected in the modelling that was used as the basis for the Montreal Protocol for CFC emissions reduction.

Nevertheless, the speculation about a northern hole continued. The following winter a large and expensive airborne expedition was funded and organised by NASA. Operating out of Stavanger in Norway, one of the expedition leaders was again Robert Watson and, from the beginning, he downplayed the idea that they would find a northern hole. As work got underway in January, he was quoted saying that they expected 'no quick changes over the North Pole' where there was 'not yet a hole in the ozone layer'.[254] This was a direct challenge to Evans and the Canadians, who were conducting their own research that winter on Ellesmere Island at the northern extreme of the vast Canadian territories inside the Arctic Circle. A month later, on 15 February 1989, a Reuters news report announced that the Canadians had 'discovered dense ice clouds similar to those that have helped tear a huge hole in the ozone layer over Antarctica'. Many weeks before the return of the Sun, Evans declared that 'the stage is set for ozone depletion in the Arctic'. He said that 'all of the conditions that are required for ozone depletion in the Antarctic are now present in the Arctic'. It looked as though the Arctic-hugging northerners had got the jump on their

116

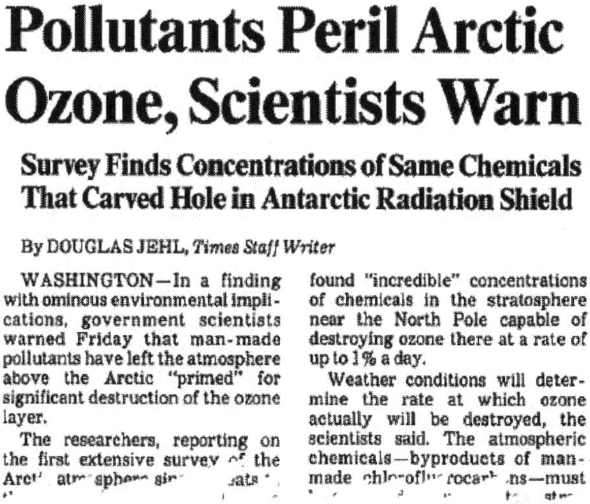

Pollutants Peril Arctic Ozone, Scientists Warn

Survey Finds Concentrations of Same Chemicals That Carved Hole in Antarctic Radiation Shield

By DOUGLAS JEHL, *Times Staff Writer*

WASHINGTON—In a finding with ominous environmental implications, government scientists warned Friday that man-made pollutants have left the atmosphere above the Arctic "primed" for significant destruction of the ozone layer.

The researchers, reporting on the first extensive survey of the Arctic atmosphere since ... .ats ... .

found "incredible" concentrations of chemicals in the stratosphere near the North Pole capable of destroying ozone there at a rate of up to 1% a day.

Weather conditions will determine the rate at which ozone actually will be destroyed, the scientists said. The atmospheric chemicals—byproducts of man-made chlorofluorocarbons—must ...

Figure 6.8: The press responds to the NASA press conference.

The coverage of the threat of an Arctic ozone hole included this story appearing on the front page of the *Los Angeles Times*, on 18 February 1989.

richer southern neighbours.

The following day, the leader of the NASA expedition, Michael Prather, responded by disputing the Canadians' claim for a hole, saying that what they had found was only ozone 'pushed to one side by a severe storm'. Prather said that the NASA team had returned home and would be announcing their results the next day.[255] The Canadian story was published widely in Canada but otherwise received very little attention. The press conference organised by NASA in Washington on 17 February was an altogether different affair: with Watson once again at the fore, it received some coverage in Canada, significant coverage in Europe, and was front page news across the USA.

At the Washington press conference, Watson announced they had found the stratospheric chemistry over the Arctic to be 'highly perturbed'. There were 'incredible' concentrations of chemicals built up in the Arctic stratosphere such that it was 'primed for a large destruction of ozone' over the coming days.[256] Losses of up to 1% per day over the next several weeks are possible as sunlight returns to the Arctic, the *Washington Post* reported Dan Albritton of NOAA saying: 'We've always had ozone being created and destroyed in the atmosphere, like a bathtub with the water running and the drain open, it comes in and goes out and you have a balance,' but now 'what we've done with these man-made chemicals is put another drain in the bathtub.[257] Watson did provide one caveat to his prediction of an imminent ozone apocalypse: it would only occur if the Arctic vortex sustains itself through the springtime warming as the sunlight returns.[258] Nowhere in the reports of the news conference is there any suggestion that Watson or the other scientists indicated just how unlikely that would be.

Remarkable as the coverage of this story was, it was all the more remarkable that nothing much was heard again of this impending catastrophe. The Canadians (and others) had indeed found pockets of depletion in the lower stratosphere over northern Europe and over Canada.[259,260] But the NASA team was right that these were hardly 'holes'. In something of a compromise, the Canadians came to refer to them as 'notches' or 'craters'. As for the NASA expedition, it turned out that they had discovered, not the precursors of destruction, but its product. They had found chlorine monoxide in high concentrations around regions of stratospheric cloud formation, just as the (much cheaper) ground-based expeditions had found the previous winter. But the NASA expedition found no significant depletion. In other words, and to use their metaphor, the gun had been primed and it had gone off. They had found the gunsmoke but they had failed to find a corpse. *A weapon, a culprit but no crime.* Chlorine was at work but it wasn't creating any hole. It seems that the airborne sampling ceased in early February, when any hope of finding Antarctic-like chlorine-driven depletion disappeared with the evaporation of the stratospheric clouds even before the spring sunrise. The evidence they had gathered only suggests that Dobson and Farman were right all along: the critical meteorological element for a steady-state depletion is Watson's so-called 'restrainment vessel', and such a vessel is absent in the North.[261,262] The NASA expedition was reported to have cost $10 million. Following a script rehearsed by the Canadians, Watson and the NASA expedition leadership had managed to conjure an attention-grabbing story from some very unremarkable results.

Politically, the headline news that the Arctic was 'primed for ozone destruction' was ideally timed, coming as it did just two weeks before the 'Save the Ozone Layer Conference' was set to commence in London. A piece of political theatre organised by the Thatcher government in cooperation with UNEP, this conference attracted representatives from no less than 124 governments.[263] On the eve of the conference, news spread among the delegations of a meeting in Brussels at which the European Economic Community had just committed to speeding up CFC reduction and entirely phasing out production by the end of the century. How much the extensive press coverage of NASA's warning of an imminent Arctic catastrophe influenced this decision we cannot say, but we do know that a sudden collapse of resistance propelled the members of the Western European bloc out of their long-held bastion of scepticism and into policy leadership over the crusading Americans. They arrived in London to witness their host, the once-recalcitrant Margaret Thatcher, hold up a British torch, *a Tory torch*, of ozone policy leadership.

The ozone layer is saved

After the London 'Save the Ozone Layer Conference', the campaign to save the ozone layer was all but won. It is true that a push for funding to assist poor-country compliance did gain some momentum at this conference, and it was thought that this might stymie agreement, but promises of aid were soon extracted, and these opened the way for agreement on a complete global phase-out of CFC production.

By the end of the 1980s, those raising the alarm on the ozone emergency were delivered a spectacular triumph. Leading this movement were a string of chemists who would not have succeeded if they had not pushed their way into the domain of the meteorologists. Johnston had paved the way for Rowland's success in the USA during the late 1970s. Then, in the late 1980s, Watson had succeeded in directing global attention towards the ozone crisis. For their efforts, Rowland and Watson would be well rewarded. Rowland won a Nobel Prize in 1995. As for Watson, after a successful transition into leading the assessments of global warming, he won a knighthood. Mustafa Tolba also benefitted. He and his UNEP had guided the policy development to the success of a global treaty. The level of agreement between nations achieved in this treaty was extraordinary, and in many respects unprecedented. Consequently, his previously obscure UN agency soared to prominence.[††] The heroic success of these leaders came despite reluctance, reticence and even outright opposition from within the meteorological establishment, and *at every step of the way*.

Perhaps it was right then that no such rewards came to the WMO and its equivocating membership. Throughout the process, it was forever dragging its heels and then left to play second fiddle to UNEP. Indeed, even before the ozone treaty was in the bag, there were some in the WMO Executive Council who recognised this, who saw it as their failure and who wished to ensure that it did not happen again. Against an overriding mood of cautiousness, they wanted to make sure that the meteorological organisation led from the front with the next big meteorological scare.

By the mid-1980s this new scare was already emerging and it was again entirely within the scope of the WMO. By the summer of 1988 everybody knew

[††] Tolba's parallel and subsequent leadership on the climate change treaty was not so well rewarded, as we shall see.

about it. That year, while Watson and others were promoting the scary new evidence of global ozone destruction, and while Tolba was using this evidence to speed up the ozone-treaty process, worldwide attention shifted dramatically towards the much greater task of saving the climate. Tolba had long envisaged a new treaty, this time aimed at protecting humanity from the ravages of self-inflicted climatic change. By 1989 the process of its development and negotiation was already mapped out. International negotiations were to commence in Washington at the invitation of an enthusiastic new US president, George Bush. But the treaty negotiations could not begin until January 1991. Despite the eagerness of world leaders to get things moving, the policy process was being held up, awaiting the completion of a full scientific assessment. An assessment panel had first been proposed by the WMO, but it was implemented under the joint direction of the two rival UN bodies. In this assessment, under the banner of the IPCC, the meteorologists would once again struggle to come up with evidence that would verify the model predictions of a new catastrophe. Indeed, they had been struggling to do so for a long time, with few signs of progress across their many previous assessments. We now take up this science-policy story from its beginnings in the oil crisis of 1973.

Part II

Climate

'Climate change' means a
change...attributed directly or
indirectly to human activity...which
is in addition to natural climate
variability...

Definitions, UN FCCC, 1992

Climate change...occurs...either for
natural reasons or because of human
activities. It is generally not possible
clearly to make attribution between
these causes.

Glossary, IPCC Working Group 1
Second Assessment, 1996

7 The carbon dioxide question

Why did the greenhouse effect become important? Why did carbon dioxide take centre stage? Why this issue at this time? In the USA during the late 1970s, scientific interest in the potential catastrophic climatic consequences of carbon dioxide emissions came to surpass other climatic concerns. Most importantly, it came to surpass the competing scientific and popular anxiety over global cooling and its exacerbation by aerosol emissions. However, it was only during the late 1980s that the 'carbon dioxide question' broke out into the public discourse and transformed into the campaign to mitigate greenhouse warming. For more than a decade before the emergence of this widespread public concern, scientists were working on the question under generous government funding. Indeed, while there are many apparent similarities to the ozone scares, where the global warming mitigation movement differs is in the broader and enduring scientific interest long before the scare took off, and this is also what distinguishes it from the various other great environmental campaigns that it would eventually overwhelm or consume.

Our initial concern is therefore this early development of interest within the scientific community during the late 1970s, long before the greenhouse scare began. Just how developed was this interest already by the end of that decade is well illustrated in the proceedings of a conference that took place in Washington early in 1980. We have already touched on this 'Carbon Dioxide and Climate Research Program Conference' with a quote from the conference dinner address by Alvin Weinberg, in which he recalls the SST-ozone scare as a warning against prematurely 'crying wolf'. His hope and plea is that 'the carbon dioxide community will take the SST-ozone affair to heart and refrain from crying wolf too soon'. * Incidental to Weinberg's warning, and what interests us here, is his reference to 'the carbon dioxide community'. This is no rhetorical flourish. Already by 1980 there was an identifiable network of scientists with investment in this issue. The formation of this community had been facilitated by a coordinated

---

*See p. 63.

A NATIONAL PROGRAM ON CARBON DIOXIDE, ENVIRONMENT AND SOCIETY (Cont'd)

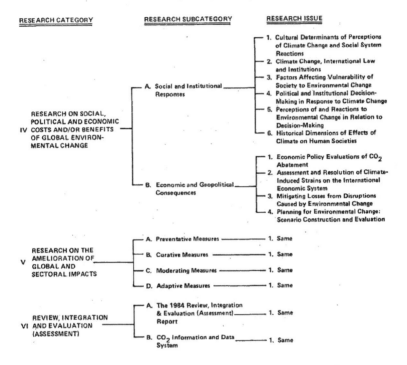

Figure 7.1: Part of a scheme for the National Carbon Dioxide Program.

This is the second half of the scheme, as it appears at the front of the 1979 annual report of the DoE 'Carbon Dioxide and Climate Research Program'.[264] It includes the component on social and economic cost–benefit analysis of global environmental changes, about which the report says the DoE is 'currently working with literally hundreds of environmental and social scientists to define the question in these new and difficult areas'. This component of the program would be curtailed in 1981 by the incoming Reagan administration, but other components would continue undiminished.

government funded investigation. The conference in Washington was part of that initiative, which, as the conference convenor did there boast, *was already five years old*.[265] And even if many of the scientists whom Weinberg addressed were not yet entirely committed to the cause, they were given every reason to expect that this new research community would continue, prosper and expand.

One hundred experts from universities, laboratories and research centres around the country had been called to the national capital to discuss a proposed comprehensive national research program. Called the 'National Program on Carbon Dioxide, Environment and Society', it was otherwise known as the 'Carbon Dioxide Effects Research and Assessment Program', but mostly it was simply referred to as the 'Carbon Dioxide Program'. Already in the previous few years a study group and various workshops had been organised, with 33 research projects already commissioned and underway.[266] Now, at this Washington conference, leaders in the many relevant research fields were to critically analyse a newly drafted plan for a major expansion. The various versions of this plan show just how ambitious and comprehensive it was. Ongoing scientific investigations of carbon dioxide's global warming effect was only one part of it. Also included were the environmental, social and economic impacts of the warming, as well as investigations into the various possible policy responses (see Figure 7.1). It was a grand vision to involve hundreds more scientists and social scientists across the country for a decade and more. It would include periodic assessments every five years; the first two were already scheduled for submission in 1984 and 1989.[267] And it was much more than some pipe dream of science bureaucrats, for the ambition for the program had grown upon considerable financial encouragement. Already for 1978, $1.5 million had been assigned. By the end of 1980 (the year of the Washington conference), a total of $12 million would have been distributed to 52 individual (sub-)programs at various science institutions, including 17 at universities, 12 in the national laboratories and others at federal agencies and private research institutes. The massive expansion beyond 1980, as outlined in the new national program plan, would be financed by an annual budget as high as $30 million.[268,269,270]

That was the plan. As it turned out, this program was not implemented exactly to plan, although annual funding rose to $10 million for 1981, and continued at that level through to 1985 when it grew to a whole new level again.[271] This was a government funding commitment that quickly surpassed the Senate-sanctioned commitment to stratospheric research in 1971 and its circumstances were also completely different.

We will remember that the $21 million assigned to the CIAP stratospheric research program came in the context of a mature campaign against supersonic transportation. Moreover, it only came after alarming press stories injected news of Johnston's frightening discovery into this campaign. Indeed, the publicity surrounding the various startling discoveries by Harrison, McDonald, Johnston, Molina, Rowland, Farman and Watson triggered and maintained the excitement around the ozone issue, with the effects felt beyond CIAP in the funding of the various other scientific assessments, investigations and polar expeditions.

By contrast, in the late 1970s, the science relating to carbon dioxide was old news. There had been no dramatic discoveries in the field since a rising trend in the background atmospheric concentration of carbon dioxide had been confirmed in 1960.[272] That finding did not attract much interest, let alone funding.[†] It was 15 years before funding did flow to this obscure field, and it expanded with the expansion of the Carbon Dioxide Program in 1980 while public interest in the issue remained subdued. Yet while CIAP was set to wind-up after three years, this carbon dioxide funding stream was sufficiently secure for scientific institutions to reorganise their research portfolio so as to attract it, and for scientists to redirect their careers accordingly. This reliable funding continued for a full decade before global warming came alive in the political sphere. The importance of the program to the later development of the warming scare can also be seen through its similarities to the international assessment program that took shape a decade later under the IPCC. There were other important developments during the late 1970s, as we shall see, but if ever there were a beginning to the global warming scare, then this is it. So this part of our story begins with the Carbon Dioxide Program and its arrival as an unexpected child of the US 'energy crisis'.

[†] This will be covered later in this book. See p. 138.

8 Energy and climate

The brief recession of 1970 ended ominously. As US production rates recovered, unemployment only slowly improved, while inflation shot up. Then the recovery began to falter. Just before Christmas 1972 the stock market jittered and crashed, remaining bearish through to January 1973 when production started to fall away again. But what really marked the beginning of the great stagflation of the mid-1970s was a brief conflict that broke out in the Middle East that October. This time, Saudi Arabia successfully deployed its 'oil weapon', placing an embargo on sales to the USA (and some of its allies) that held firm among the other Arab supplier states. The shock waves that shot through the US economy impacted on the unprecedented lifestyle expectations cultivated during the long postwar boom. The Nixon administration responded to the emergency by taking control of private oil stocks so that they could implement restrictions on oil consumption, including prioritisation and rationing. The embargo was lifted in March 1974 but in the meantime production quotas agreed among Arab suppliers brought about a quadrupling of the market price of oil. This high price was sustained by the OPEC cartel until 1979 when the Iranian Revolution triggered another price hike, as high again as the first one.

President Nixon announced his drastic response to the oil supply emergency in November 1973, only weeks after the embargo had been announced and while he was already deeply mired in the Watergate scandal. But after the embargo was lifted, and after he resigned the following August, emergency regulation of the US oil market remained in place. It continued under President Ford, and then Carter, and right through to 1981, when Ronald Reagan honoured an election pledge to deregulate. Many commentators argue that the government controls that Reagan abolished had only exacerbated and prolonged a crisis that was not sustained elsewhere.[273] But crisis it was, and it triggered a vigorous public debate over US energy policy for the best part of a decade.

The oil crisis was not the only motivation for this reappraisal of energy policy. It had come on top of anxiety that the population 'explosion' and resource depletion would soon lead to a critical global scarcity of non-renewable resources; in particular, fossil fuels.* In this view, any solution to the oil cri-sis that involved returning the USA to the boom-time carefree consumption of cheap non-renewable energy would be a disaster. That would only bring for-ward a bigger crisis already forewarned and now looming even closer.

When Jimmy Carter came to power in the snowy winter of 1977, this broader 'energy crisis' was launched as a dominant theme of his presidency; one that would be sustained until he left office. Indeed, the energy crisis began to take on the character of a personal crusade, as he searched for ways to persuade Congress, industry and the public of its gravity. His efforts included a series of television addresses to the nation. In one of the early broadcasts, he called the energy crisis a 'national catastrophe' and the 'moral equivalent of war'. Victory could only come through the nation pulling together with the urgency of all-out warfare. The American people must drastically reduce their everyday energy consumption, while energy production must shift to alternative sources, espe-cially renewables.[276] Only in this way could the nation achieve what is urgently required, namely, viable and enduring energy independence. Carter's pleading was reinvigorated by the second oil shock of 1979, but in the twilight of his one-term presidency it looked like desperation. Certainly this is how it appeared when Reagan came to power (on a falling oil price) and proceeded to free up the market.

Nevertheless Carter was only continuing a campaign on energy security be-gun even before the first oil shock hit. In Nixon's November 1973 televised re-sponse to the oil embargo, he invoked the national war effort as a model for his comprehensive response. But this announcement came after he first explained how the oil supply shock only served to bring attention to the 'urgent energy problem'. Over the previous two years he had tried to persuade Congress that the energy supply situation was pressing, but to no avail. Now it seemed this new crisis would provide sufficient impetus for actions that they had previously been unwilling to take. The aim of Nixon's so-called 'Project Independence' was not only to free the USA from dependence on foreign oil, but to achieve com-plete energy self-sufficiency by 1980. Just as with the Manhattan Project and the Apollo program, American science, technology and industry would pull to-gether and rise to the challenge. And there would be no shortage of funds: in this first response to the oil crisis, Nixon promised the American people that cash to support the rush to energy self-sufficiency would be 'far in excess of the funds that were expended on the Manhattan Project'.[277]

Nixon's timeframe for self-sufficiency must have seemed no less ambitious

---

*See for example Paul Ehrlich's *The Population Bomb*[274] and the *Limits to Growth* report of the Club of Rome.[275]

than Kennedy's timeframe for the lunar landing. And it would surely require a commitment no less than that which won the race for the atomic bomb, especially given the lack of development in the renewable energy sector. But another option beckoned. While at the time nuclear power contributed only a few percent to the total national supply, this was still early days; after all, prototype reactors had only come online in the late 1950s. Since then, enormous government investment had been rewarded with great advances through the 1960s and into the 1970s. Indeed, during 1973, orders for new nuclear power plants had been coming in apace. They would end up totalling 41 for the year; a new record. And yet, even this level of expansion was dwarfed by the ambition of the Atomic Energy Commission (AEC). It was projecting that nuclear energy production would be around 80 times greater by the end of the century.[278] That would see as many as 1000 plants in operation around the country. To help realise this exponential growth, Nixon called on the AEC to fast-track licensing and construction. In all, it seemed that the response to the oil crisis could only invigorate an already booming nuclear industry.[279] Alas, these aspirations were about to come crashing down.

The 1960s nuclear expansion had been accompanied by growing public concern over safety standards and inadequate environmental protection. Then, opposition to nuclear power *per se* became one of the major anti-technology themes of 1970s environmentalism. (We will recall that this was the other big foundational campaign for Friends of the Earth.) By 1974, the much-criticised AEC was disbanded. But the troubles in the industry only multiplied, and many of those plants ordered in 1973 soon ran into construction delays and cost overruns. Projects were cancelled as power utilities opted for what they saw as the cheaper and safer option of coal or gas. The situation was not helped when the anti-nuclear movement was re-invigorated by publicity around an accident at the Three Mile Island reactor in early 1979. As it turned out, after 1973, no reactor that was ordered would come into operation.

The Three Mile Island accident, coming just as the Ayatollah was replacing the Shah in Iran, only accentuated what was already apparent a year earlier; that is, that the energy production reappraisal throughout the oil crisis years was as tough for nuclear as for oil. It was not that nuclear electricity production did not continue to grow. It was only that this growth was much slower than had been anticipated. The power source that came out looking good through all the fuss over oil and nuclear was an old stalwart: coal. After the War, coal consumption went into decline due to the decline of coal fired heating and steam rail locomotion, but a boom in the electricity market drove a recovery from the 1960s. Cheap and with plentiful domestic supplies, it had always been a major threat to nuclear expansion.

But in the struggle between nuclear and coal, the proponents of the nuclear alternative had one significant advantage, which emerged as a result of the repositioning of the vast network of government-funded R&D laboratories within the bureaucratic machine. It would be in these 'National Laboratories' at this time that the Carbon Dioxide Program was born.

The National Laboratories

When the AEC was closed, its regulatory functions would be transferred to a new specialist commission, but there were still the R&D responsibilities to consider. These would move to a new body called the 'Energy Research and Development Administration' (ERDA), which would thereby inherit the national laboratory system that had grown up under the urgent and secretive nuclear program of the Second World War.

After the Manhattan Project delivered a spectacular ending for that war, this network of laboratories had continued and expanded across the length and breadth of the country under the civilian authority of the AEC. Separate from the university system, it was believed that the laboratories could rapidly realise development goals through the concentration of the required expertise in mission-oriented programs. Nuclear power development had become the peacetime priority, although the prodigious public spend was also justified by widely celebrated advances in basic (atomic) physics and spin-off applications such as nuclear medicine. Now, however, ERDA had an expanded mandate covering all forms of energy, and suddenly potential new fields of research concerning other means of energy production opened up to the thousands of scientists stationed across this R&D archipelago.

One laboratory prominent in the nuclear bomb and nuclear power story was built on green fields near Knoxville, Tennessee. At Oak Ridge Laboratory, Alvin Weinberg led nuclear power research for 18 years until 1973. He was then sacked by the Nixon administration following a disagreement in a (still ongoing) controversy over reactor design. But then, after Nixon was ousted, Weinberg was recruited to Washington. For a brief period in 1974 he directed the new Office of Energy Research and Development, where he facilitated the establishment of ERDA. After that, he returned to 'Atomic City', as his old R&D hometown was often called, to establish the Institute for Energy Analysis. As founding director of this policy think tank, he had the freedom to get more involved in the debate over the relative merits of fossil and nuclear fuels.

In this debate there were already some well publicised negative environmental impacts on both sides. For the nuclear option, these all related to radiation pollution, concern for which was growing. As for fossil fuels, ever since the early

130

days of coal-fire heating, they had carried a ghastly reputation for localised air pollution, especially black carbon soot and sulphate aerosols. Then, postwar, with the intensified use of petrol-powered vehicles came the lingering clouds of photochemical smog. By the early 1970s, industrial sulphate emissions were being blamed for 'acid rain', sometimes so remote from the offending smoke stacks that it came down hundreds of miles away, in another country entirely. In those pollution-conscious 1970s, fossil fuels had a bad rap.

Yet just as broader public concern arose, so too did technological fixes. This meant that 'clean energy' appeared something of a realisable goal, achievable through legislative controls, some of which had already been successfully implemented. Thus where anti-pollution campaigns targeted coal they were more about strengthening 'clean air' regulations and much less about the banning of new plants or shutting down existing ones.[†] *Not so for the nuclear option.* Demands for the immediate closure of all nuclear plants may have remained at the political margins, but as the anti-nuclear campaign gained momentum, the more moderate demand for a moratorium on new plants went mainstream, with petitioning in the US Senate. In the face of this growing opposition, the promoters of nuclear energy argued that it could be safe and also clean; cleaner than coal.

This was certainly Weinberg's view. While still the director at Oak Ridge, just before the oil crisis, he famously promoted the nuclear option as a solution to the anticipated energy crisis. Nuclear power development does require that society make a significant and long-term commitment, he said; this was his much-quoted and sometimes mocked 'Faustian bargain'. But once this commitment has been made, reactors can provide 'an inexhaustible source of energy' cheaper than fossil-fuel burners, and 'when properly handled... almost non-polluting'. While fossil-fuels 'emit oxides of carbon, nitrogen, and sulfur dioxide', he could find 'no intrinsic reason why nuclear systems must emit any pollutant except heat and traces of radioactivity'.[280] Naturally, this 'clean' portrayal of nuclear energy only outraged its opponents, and Weinberg retreated from this line of argument in the oil-crisis years. Instead, he gave more time to expanding on the possible failings of the competition. This was especially evident when the debate over relative polluting effects took off on a completely different tack.

Around 1974, there was a shift towards consideration of the impact of energy production on climate. Whether or not Weinberg initiated this discussion,

---

[†] Note that, even in the climate change mitigation movement, campaigns for a moratorium on new coal power plants (and for the shutting down of existing ones) have come late, and the idea has really only taken off in recent years.

he certainly helped it along. In October 1974, he published a brief opinion piece in *Science* on 'The global effects of man's production of energy'.[281] These 'global effects' are all about climate. One of them is direct heat, a consequence of all energy production, including nuclear. This, Weinberg believed, might eventually have a global effect if the postwar trend in production were to continue through the next century and beyond. His article also discusses the possible climatic effects of aerosols and carbon dioxide produced by the burning of fossil fuels. Sometime in the next two centuries, these various climatic effects might impose limits on global energy production, and perhaps in the next 30–50 years the world's energy policy might require adjustment to take them into account.

> Unfortunately, the science of climatology is unable to predict the ultimate consequences for the earth's climate of man's production of energy. At what rate of energy production would the ice caps melt? Will the carbon dioxide or dust thrown into the atmosphere by the burning of fossil fuel threaten the stability of the weather system? How does the geography of man's energy production affect weather in various parts of the world?[282]

Weinberg acknowledged that research in this field is already underway, but he proposed that a much greater effort is required because 'answers to these questions may eventually dominate long-term energy policy'. We need to improve our understanding of these effects and improve our general understanding of global climate so that, 'say, 20 years from now, we can base energy policy on a much sounder understanding of the [climatological limits on energy production]' To this effect, he proposed...

> ...an institute (or even institutes) of climatology be set up with a long-term commitment to establishing the global effect of man's production of energy. Such an institute should be assured long-term stability, since the question is a long-range one that simply will not go away. The institute would naturally serve to focus the efforts of smaller groups of climatologists, working on more general, basic aspects of climatology; but the institute itself would also contribute to our general understanding of the dynamics of the world's climate.[283]

The Institute for Energy Analysis that Weinberg was establishing at this time was not this proposed 'institute of climatology', but it did move quickly to emphasise one climatic impact exclusive to fossil fuels. And it did this explicitly in defence of the nuclear option. In 1976 it published a major report on the economic and environmental implications of the proposed moratorium on US nuclear power plant construction.[‡] Not surprisingly, the report finds that one of the most likely implications of a ban on new nuclear power plants would be an ever-increasing reliance on coal to fuel the expected expansion of the energy market. In its consideration of the environmental implications of energy production generally, the report does discuss the potential climatic effects due to direct heat and aerosol emissions. But aerosol emissions are controllable and direct heating remains insignificant at the global scale. Thus attention should turn to the effects of carbon dioxide emissions from fossil fuels, and specifically from coal, which has the highest rate of emissions.[285] Even though the climatic effects of carbon dioxide are not likely to be significant before the end of the century, the IEA sees this as no cause for complacency, as that is not so far away in the timeframes required for energy planning. The report advises that consideration of the carbon dioxide question should begin without delay because 'decisions made now on the nuclear/non-nuclear issue will have an impact reaching many years into the future'.[286]

This Institute for Energy Analysis report and Weinberg's earlier opinion piece show that there was some effort within the nuclear energy lobby to promote research into the detrimental climatic effects of fossil fuel burning, especially those due to carbon dioxide. Further investigation may establish just how much more the nuclear lobby petitioned (formally or informally) for a coordinated research effort on the carbon dioxide question. Whatever the case, we know that by the time Weinberg's think tank report was released, there were already moves afoot within the national laboratories towards this end. In fact, by 1976, preliminary approval to proceed had already been given.

The US Carbon Dioxide Program established

The 'Biomedical and Environmental Research' division of the former AEC had mostly been concerned with the potential environmental and health effects of radiation emitted with the release of atomic energy. Now under ERDA, with its expanded 'energy' responsibilities, it was natural for the division to also consider the environmental and health effects of the main sources of energy: fossil fuels.

Its annual late-winter meeting in March 1975 was focused on sulphate and acid pollution, and would mark the beginning of a research program to investigate the impacts of these emissions from the burning of coal, oil and gas. But

---

[‡] According to this report,[284] one of the authors, Ralph Rotty, had also just produced two IEA reports developing the environmental impacts aspects of the main report: 'Environmental Implications of a US Nuclear Moratorium' and 'Energy and Climate'. Rotty, an expert on emissions from energy technologies, also attended the Villach 1980 conference (see p. 192.) as an IEA 'consultant'.

at the meeting, an atmospheric modeller from Lawrence Livermore Laboratory named Mike MacCracken suggested to the convenor that there was another problem they needed to add to their portfolio of investigations, namely the carbon dioxide effect on climate. MacCracken was invited to write a letter introducing the issue to Rudy Engelmann, the division's Deputy Manager of Environmental Programs. To MacCracken's two-page letter, Engelmann attached a covering letter, circulating both documents to eight leading experts for their comment. They replied with varied views as to the importance of the topic, but mostly with support for an ERDA research initiative.[287] A workshop was proposed and this was approved by the ERDA administration, as was a study group. Weinberg was elected to the chair of the study group and the Institute for Energy Analysis became heavily involved in it, as well as in the preparations for the workshop, which was scheduled for Miami, Florida, early in 1977.[288]

This five-day workshop on 'the global effect of carbon dioxide from fossil fuels' had a strong contingent from Oak Ridge and other laboratories. There were also experts from NASA, from the National Center for Atmospheric Research (NCAR) and from NOAA—including the workshop chairman, Lester Machta. Some international participants also arrived, from Australia, Canada, Sweden and West Germany, but the vast majority of the 70 scientists assembled for the first time to discuss specifically the global effects of carbon dioxide emissions came from the US university sector.

Of the four formal panel deliberations, only the final one was set to cover 'the climatic effects'. Others discussed carbon biological and oceanic effects. However, this division is misleading because the overarching consideration was how these effects would impact climate, for example where the impact on climate is moderated through biotic and oceanic uptake of carbon dioxide. In their presentations and discussions, it was reiterated that much more research was required for a proper consideration of the threat, and this overriding recommendation of the workshop lent additional support to the proposal for a full research program.[289]

The Carbon Dioxide Program began with the first assignment of funding—$1.5 million for the 1978 fiscal year— approved soon after the Miami workshop. This was still in the early days of the Carter presidency and just when ERDA was expanding to become the new—and even more generously supported—Department of Energy (DoE). In 1979, plans for the comprehensive national program were circulated and then reviewed at the 1980 Washington conference.

Thus, over the course of four years, what had begun as Weinberg's study group came to look like a much more substantial affair. Weinberg and his Oak Ridge Institute for Energy Analysis remained important, preparing the program's reports and acting as a hub for information and research, while managers

Chapter 8. Energy and climate

March 27, 1975

Dr. Rudolf J. Engelmann
Division of Biomedical and
  Environmental Research (DBER)
U. S. Energy Research and
  Development Administration
Washington, D.C. 20545

Dear Rudy:

   With the increasing pressures for greater domestic energy production, ERDA is proposing and many utility executives are urging that this country's coal reserves be rapidly developed to serve as the backbone of United States' energy resources for at least the next hundred years. In the atmospheric sciences, these plans have prompted formation of a research program to investigate the potential sulfur/sulfate concentrations which may result. But at this crucial juncture in energy planning, there seems to have been no critical assessment similar to the CIAP weighing the relative merits of fossil fuel usage and, for example, nuclear or solar energy generation in which the implications and consequences of continuing reliance on combustion of fossil fuels and the resulting increase in atmospheric carbon dioxide concentrations have been carefully considered. Rather, an implicit assumption has apparently been made that increasing atmospheric carbon dioxide is of no significance; a decision made without assessing the current indications that small changes can cause large effects (e.g., Grobecker, Coroniti, and Cannon, 1974).

   Carbon dioxide is the one atmospheric constituent for which there is a very good evidence that it has been increasing from about 290 ppm in the mid 1800's to more then 325 ppm at present (see, for example, Callendar, 1958; Revelle, 1965; SCEP, 1970; SMIC, 1971). And based on extensive studies of the carbon cycle, it seems quite clear that the approximately

Figure 8.1: MacCracken's proposal for a carbon dioxide research program.

and scientists from across the various national laboratories coordinated activities. Organised in this way, the program engaged all the major groups of relevant specialists outside the national laboratory system, with nearly three quarters of the funding going to US universities.[§]

This surge of new funding meant that research into one *specific* human influence on climate would become a major branch of climatic research *generally*. It was a targeted involvement of a government department not directly interested in atmospheric research and it began to influence the research portfolio

---

[§]In his introductory speech to the Washington conference in 1980, the program director, David Slade, made a point of saying that, of the $12 million already allocated in the first three years, almost three quarters had gone to universities.[290]

135

and priorities of the scientific establishments involved in atmospheric research and other areas of geophysical science relevant to the climatic change question. As with stratospheric research and the Department of Transport, so now with climate research and the DoE, the opening of this line of funding was broadly welcomed, but it came with particular controls and constraints. *Perhaps this was the scientists' own Faustian bargain.* Not all scientists and science administrators were entirely comfortable with the encroaching tutelage of the DoE and its national laboratories. The plan for a comprehensive national program on the climatic influence of carbon dioxide, undertaken in the style of a military mission, was sometimes roundly criticised in the program meetings and elsewhere. As Spencer Weart explains:

> Other agencies disliked the DoE plan, however, and their complaints went beyond the normal bureaucratic defence of turf. Prestigious scientists on a Climate Research Board, newly created by the Academy, criticized the plan as poorly designed and over ambitious. Meanwhile an Interdepartmental Committee on Atmospheric Science, as well as NASA, NOAA, and the Department of Defence, were each developing their own research plans. The meteorological community and its friends in the bureaucracy were determined to push for a better-designed consolidation of climate research.[||,294]

Weart's portrayal of this criticism is most apposite where he says that others were pushing 'for a better-designed *consolidation of climate research*'. They might have been so pushing, *but why would the Carbon Dioxide Program plan even challenge that?* Weart's tacit assumption is that the DoE plan was something of a push to takeover climatic research generally. Today we might pass this over for the simple reason that the 'carbon dioxide question' has long since come to dominate the entire field of climatic research—with the very meaning of the term 'climate change' contracted accordingly. But this had not yet happened when DoE got involved. However, maybe that is the point. If not *in intent*, then *in effect*, this was the impact of the DoE's comprehensive plan; it brought the carbon dioxide question to dominance in climatic research. Maybe, even for some of the actors back then, a take-over was their very *intent*, and those

---

[||]Spencer Weart, who was awarded a doctorate in physics in 1968, researched the history of global warming science while resident historian at the American Institute of Physics. His accounts are largely based on discussions with US scientists investigating the physics of global warming, including those involved during the years when the DoE funding dominated. The views of one, Wallace Broecker, feature prominently in Weart's influential history, *The Discovery of Global Warming.*[291] Impatient with the pace of research, Broecker was a vocal critic of the DoE program while he tried in various ways to drum-up a greater sense of urgency.[292,293]

complaining knew it. That this domination actually came to pass makes the *effect* of the DoE program, and any hint of any expansionist *intent*, all the more important for us to consider.¶

Already in the first three years of DoE funding, the millions of dollars distributed throughout the various research institutions was shaping the economy of this emerging research field. An illustration of this effect can be found at the Climatic Research Unit (CRU) at the University of East Anglia, UK, where it was accentuated by a leadership change. Founded in 1972 as only the second research institution in the world specifically dedicated to climate, CRU's funding base was never secure and its first director, Hubert Lamb, was forever struggling to retain a handful of untenured researchers. This financial insecurity continued through to his retirement in 1978. The following year, the new director, Tom Wigley, won a substantial grant from the DoE Carbon Dioxide Program. Recurrent funding continued and increased over the following years so that by 1983 the Carbon Dioxide Program had become the major baseline sponsor of the unit, providing it with financial stability for the first time.**

While Lamb had always been vocally sceptical about the increasing attention paid to the possibility of greenhouse warming, Wigley rapidly reshaped the unit's research profile towards this new funding source. This meant that it was well placed to take a leading role in the late 1980s, when the new field really took off, and when Wigley was invited for a weekend at the Prime Minister's retreat at Chequers to help Margaret Thatcher plan her dramatic policy conversion. More global warming funding then came on line and the unit had quadrupled its staff by the time Wigley departed in 1993. Undoubtedly, the 'climate change' research leadership that the unit had established by the beginning of the 1980s— a leadership that continues to this day—would not have been possible without the early and continuing involvement of DoE.[298]

¶In the course of the research for this book, it has become evident that participant accounts and commentaries often confuse the development of interest in the following:

- climatic change and its human impacts
- manmade climatic change
- global warming due to emissions of carbon dioxide and other greenhouse gases.

Just as in Weart's history, their narratives sometimes present as though each leads naturally to the next. This is clearly not the case. This confusion and conflation is only of interest to us in that it suggests a typical victor's view of history. For our purposes, the way of transition from one to the other is of great interest, and so we are careful to distinguish each of them and examine the transition between them.

**According to the CRU annual reports, the first grant in 1979 was $37,000.[295] By 1982, the DoE Carbon Dioxide Program was its major sponsor, with $205,000 in grants to date.[296] By the end of 1984, CRU had received nearly half a million dollars from DoE.[297]

Wigley was one of those invited over to Washington in 1980 to discuss the expansion of the research program to which he was already committed. The vision laid before him and the other scientists at that meeting would have it stretching out at least to the end of the decade. This seems to fit well with Weinberg's vision, back in 1974, of a research initiative with 'long-term stability'; one that would not only address the specific problem of energy production, but also contribute to 'our general understanding of the dynamics of the world's climate'. Thus, whether the atmospheric scientists outside the national laboratories liked it or not, there is no denying that something close to the vision of this atomic scientist had triumphed. And even if, in the process, the science was transformed in ways not always welcomed, nevertheless the funding pathway opened up by the laboratories was finally delivering substantial and stable funding across the atmospheric and oceanic sciences. This funding success is all the more significant given the background of so many other attempts to secure programmatic climatic research funding having ended in disappointment.

The US National Climate Program

Since the International Geophysical Year, there had been widespread recognition of the need for an expanded and coordinated program of climatic research, but successive attempts to win substantial and reliable funding for such a program would all fail. One indicator of the precarious state of the funding during this time is found in the efforts to support the monitoring of the background concentration of atmospheric carbon dioxide. Roger Revelle of the Scripps Institution of Oceanography was a master at winning grants, and through International Geophysical Year funding provided via the Weather Bureau he was able to support the establishment of a monitoring station high on a mountain in Hawaii.[††] When the International Geophysical Year ended, the Scripps CO_2 program had continuing support from the Weather Bureau, so that in 1960 the scientist driving the work, Charles David Keeling, could publish the first indications of the rising trend.[301] However, in 1963 funding was cut to any Weather Bureau research not directly related to weather forecasting and so the program

---

[††] At this time, in an article with Hans Suess on the exchange of carbon dioxide between atmosphere and ocean, Revelle coined the famous phrase where industrial emissions are referred to as 'a large scale geophysical experiment'. This was later adapted to a message of alarm, including in the opening of the 1988 'Toronto Statement' (see p. 221). During the rush towards a climate treaty that followed, Revelle expressed concern about prematurely taking drastic policy action. However, since his death in 1991, his ambivalence towards global warming alarm has often been misinterpreted as unambiguous support, while his original reference to a 'geophysical experiment' has often been misinterpreted as an early sounding of the alarm, when it clearly was not.[299,300]

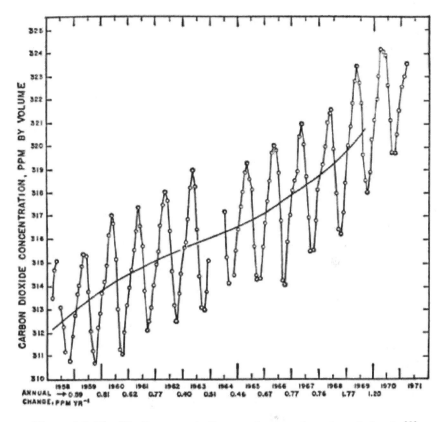

Figure 8.2: The 'Keeling curve' of atmospheric carbon dioxide levels.[303]

This version of the graph appears in the SMIC report of 1971.

had to be abandoned until finance was found elsewhere.[302] This is why the now-famous 'Keeling curve' appeared in the 1971 SMIC report with a gap around that time (see Figure 8.2).[‡‡]

Later in the 1960s, an expectation of expanded long-term climatic research funding was raised when the International Council of Scientific Unions (ICSU)[*]

[‡‡]The gap was later closed using data from other locations. Soon after this funding crisis, Revelle would have been writing an assessment of the carbon dioxide threat that appeared in 1965 and contained a most alarming projection on sea-level rise (see p. 152).

[*]The ICSU is an international non-government organisation with a membership mostly consisting of national academies of science; it also includes international associations of various scientific specialities. It became actively involved in global environmental issues during the preparation for the Stockholm 'Human Environment' Conference in 1972. It is now called the International Council for Science.

and the WMO responded to UN General Assembly resolutions calling for international cooperation in atmospheric research. The Global Atmospheric Research Program (GARP), as it was called, was launched in 1967 in the hope of taking full advantage of satellite technology. This was a great advance in globally coordinated atmospheric research. However, funding was drawn more towards short-range weather predictions than towards GARP's so-called 'second objective', which was to study the physical basis of climate.

The peak of climatic interest under GARP was an international climate conference held in Stockholm in 1974 and a publication by the 'US Committee for GARP' the following year. But both of these initiatives went no further than the formulation of plans for a comprehensive climatic research program. The US GARP report was called 'Understanding climate change: a program for action', where the 'climate change' refers to natural climatic change, and the 'action' is an ambitious program of research.[304]

It was around this time that US federal government interest in developing a coordinated and comprehensive 'national' climate program was driven by alarm about global cooling (more about that below).[†] This is when the DoE began to promote its Carbon Dioxide Program as a 'national program'. The idea seemed to be that the specific DoE program might come under the auspices of the general climate program coordinated by NOAA.[305] Not only did this never happen, but the great ambitions for the National Climate Program were disappointed. This is why the DoE initiative on the carbon dioxide question was seen more as a competitor than a willing supplicant, just as Weart had portrayed it. Weart also explains that the new Climate Program Office at NOAA...

> ...had only a feeble mandate and a budget of only a few million dollars. Scientists did not get as well co-ordinated a research program as they had called for. Without the backing of some unified community or organisation, their movement had been impeded by the very fragmentation it sought to remedy.[306]

We have already seen how, in those grey days of high unemployment and budgetary constraint, the supersonic jet controversy attracted funding to stratospheric research, and how CIAP became a haven for atmospheric researchers. At the same time as the ozone threat demanded this urgent investigation, there was also the great energy crisis and a succession of governments more than willing to lavishly fund efforts to resolve it. An approach to the crisis involving

---

[†] This idea was not realised until late in 1978, when the passing of a National Climate Act finally led to the establishment of a Climate Research Board at the NAS, a National Climate Program Office housed at NOAA, and a Climate Program Policy Board for interagency policy planning.

geophysical science surely held the promise of substantial funding. It would be surprising, then, if there were no attempts outside the laboratories to attract programmatic climatic research funding using the link between energy production and climate. Indeed, some such attempts were made.

In December 1974, only a few months after Weinberg's proposal in *Science* of the new institute for energy and climate, the AGU held a symposium on the possible climatic constraints on energy usage. The National Research Council had already established a Geophysics Research Board, under which it then convened a 'Panel on Energy and Climate'. These moves represented a significant new trend in consideration of climatic issues within major US science-policy and funding institutions. Previously, climatic change had been considered predominantly in terms of natural variability, global cooling and purposeful weather/climate modification.[‡] With the new interest in 'energy and climate', the warming effect of carbon dioxide would come to dominate the climatic discussion in these forums. *But not yet.* Not much was done about it until 1977. This was the same year that Jimmy Carter arrived in the presidency, hyping the energy crisis as 'the moral equivalent of war'. He came with drastic solutions in mind, for example the mass production of oil from coal. 1977 was also the year of the ERDA-funded conference on carbon dioxide in Miami. And it was not until that year that the papers from the 1974 AGU symposium were published. They formed the basis of a report of the Panel on Energy and Climate published by the NAS.[§]

The symposium papers that make up the body of NAS *Energy and Climate*

[‡] During the mid-1970s, anxieties in the USA about deliberate weather and climate modification peaked and then receded. The peak included discussion of the unintended consequences of such projects as the diversion of north-flowing Arctic rivers and the damming of the Bering Straits. Anxiety about weather control as a weapon of war was fuelled by the revelation in the Pentagon Papers that the USA had used cloud seeding during the Vietnam War, and this was admitted in the US Senate in May 1974. Two months later, Nixon and Brezhnev signed the Joint Statement Concerning Future Discussion on the Dangers of Environmental Warfare, which expressed the desire to limit the military use of environmental modification where the effects would be 'widespread, long-lasting and severe'. Although the latter qualification would not likely impact cloud seeding and perhaps not even defoliation with Agent Orange, the agreement did seem to quieten alarm, as did the similarly qualified UN Convention on the Prohibition of Military or Any Other Hostile Use of Environmental Modification Techniques (ENMOD), which was open for signature in May 1977, but which did not come into force for the USA until 1980.[307]

[§] There were other explorations of the energy–climate link in other countries in the mid-1970s. For example, in his memoirs, Bert Bolin mentions an 'Energy and climate' report he was commissioned to write for the Swedish government in 1976. Bolin explains that this was followed by a government bill 'concerning future Swedish energy policy', which concluded that 'it is likely that climatic concerns will limit the burning of fossil fuels rather than the size of the natural resource'.[308] The following year, the privately funded Beijer Institute was established in Stockholm to investigate these concerns (see p. 215).

report also discussed the problems of direct heat and aerosol emissions, but, according to the foreword…

> The principal conclusion of this study is that the primary limiting factor on energy production from fossil fuels over the next few centuries may turn out to be the climatic effects of the release of carbon dioxide.[309]

The foreword explains, 'with the end of the oil age in sight, we must make long-term decisions as to the future energy policy'. The intention of the report is, so says the foreword,

> …as a preliminary step in a process, which will require a number of years to complete, aimed at placing in the hands of policymakers credible information on the most likely climatic consequences of major dependence on fossil fuels…[310]

It goes on to explain that 'even at this early and somewhat uncertain stage, the implications warrant prompt attention'. The report was released at a press conference, where the chair of the panel, Roger Revelle, placed the emphasis even more on carbon dioxide emissions, and specifically those from the burning of coal. This resulted in a front-page headline in the *New York Times*:

> Scientists fear heavy use of coal may bring adverse shift in climate

Under this headline, the *New York Times'* science reporter, Walter Sullivan, explained by way of context that the recent warnings about the 'disastrous effects on climate' of 'increasing, long-term dependence on fossil fuels, notably coal', had been 'seized on by advocates of nuclear energy'.[311] That might have been overstating it, but the headline recommendation of the report itself would have given comfort to such advocates. It says that the possible climatic effects of fossil fuels 'should be given serious prompt consideration'. But this would not be through a carbon dioxide research program. Rather, the report recommends that this issue be addressed through a comprehensive climatic research program and through new institutional arrangements, including a peak 'Climatic Council'. Thus the report explicitly uses the carbon dioxide issue to push the case for the coordination of climatic research generally.[312]

Indeed, this is what was eventually achieved with the National Climate Program. But it was a hollow victory, bereft of funds, as we have just seen. Moreover, it is hard to see how these concerns about the impact of energy production went very far towards achieving that outcome (as we shall see below, anxiety about natural climate cooling was the main driver). What their 'movement' lacked was the urgent warnings of catastrophe promoted by the likes of Harold Johnston.

According to Johnston, soon after the SST fleet took to the skies, all animals not wearing goggles would go blind. In Sullivan's popular interpretation, these aircraft would spell the 'end of ozone'. Compare this alarm-raising to the NAS *Energy and Climate* report's talk of an effect that 'may turn out' to be important 'over the next few centuries'.[313] Such moderate and doubting language would not cut it. Sullivan could only embellish so much a vague warning that was so transparently aimed at winning more general funding.

The proven trigger for the release of funding was to forewarn of catastrophe, to generate public fear and so motivate administrators and politicians to fund investigations targeting the specific issue. The dilemma for the climatic research leadership was that calls for more research *to assess the level of danger* would fail unless declarations of danger *were already spreading fear*.

Nobody in the field—not at NAS, NOAA, NCAR or NASA, nor any grouping of meteorologists—was declaring an urgent need to investigate the possibility of a catastrophe due to carbon dioxide. And if any rogue insider *had* raised the alarm, the resistance would surely have been as strong as that met by the outsider Johnston. This is because most of the leaders of atmospheric research during the 1970s – in US institutions and around the globe – simply could not attach any level of urgency to the carbon dioxide issue. Invariably their early circumspection would resolve into scepticism as the scare took off.[314] One such sceptic was Robert White. The former head of the US Weather Bureau was, in 1970, appointed the founding director of NOAA and in 1979 he was elected the inaugural chair of NAS's Climate Research Board. From these positions, he dominated the US discussion around the development of climatic research during the 1970s, promoting investigations into global human influences so that the level of danger could be properly assessed. However, he became less and less convinced of any urgency as the research progressed or, rather, as it did not. In April 1989, with the call for a global climate treaty broadcast from the highest levels of global politics, White was among those leaders who raised concerns. In an address to the Annual Meeting of NAS he warned of the small and precarious scientific base; the climate change mitigation proposals were balanced, he said, on 'an inverted pyramid of knowledge'.[315, ‖]

Back in the 1970s, it was not that White and the rest of the research leadership were generally against investigating the human effect. It was never that they would reject such funding when it came their way. It is rather that they were not

---

‖ That same year, he made a more extended case for scepticism. This article was published alongside sceptical essays by an outstanding selection of research leaders in the physical sciences, including some who feature in our story: William Nierenberg, the director of Scripps, who had chaired the large NAS assessment of 1983, and Frederick Seitz, who would trigger the 'Chapter 8 controversy' in 1996.[316]

actively whipping up the scare, and indeed some of them actively downplayed it. The point is only that there should be no surprise that they failed to attract significant funding on the energy–climate link, whether specific or generalised.

Yet, the national laboratory scientists did win funding. To release millions of dollars of funding under their Carbon Dioxide Program, no doomsayer scientists were required. There was no public scare and, for them, there was no need for one. With little effort, Weinberg and MacCracken were able to introduce a huge research program specifically addressing carbon dioxide. The result was that, from the late 1970s, this 'comprehensive' Carbon Dioxide Program would flood the climatic research community with millions of dollars of new recurrent funding and, by the early 1980s, the carbon dioxide question came to figure large in the expansion of climatic research. After that, when more coordination came, the carbon dioxide question was already dominant. After that the public scare took off. After that, other triggers perpetuated it. But, *before* the 1980s, it is hard to see any other way that momentum could have been achieved sufficient to launch the public scare. Whether this could have come later by some other means, we will never know.

Anyway, this is how it happened. The scare that would eventually triumph over all preceding global environmental scares, and the scare that would come to dominate climatic research funding, began with a coordinated, well-funded program of research into potentially catastrophic effects. It did so before there was any particular concern within the meteorological community about these effects, and before there was any significant public or political anxiety to drive it. It began in the midst of a debate over the relative merits of coal and nuclear energy production. It was coordinated by scientists and managers with interests on the nuclear side of this debate, but it was remote from the bulk of the scientists with interests in fossil fuels, most of whom worked for the large global corporations that dominate the industry. If political or public alarm was driving the funding of this program, then this was only indirectly, where funding due to energy security anxieties was channelled towards investigation of a potential problem with coal in order to win back support for the nuclear option.

When the Institute for Energy Analysis folded soon after Weinberg retired in 1984, the DoE-directed funding remained dominant. It is true that the program's wings had been clipped with the arrival of the Reagan administration; the parts of the program covering the environmental, sociological and economic impacts were removed, as were the investigations of policy responses.¶ How-

¶Before this trimming of the program's scope, a major interdisciplinary workshop on the environmental and societal consequences of possible CO_2-induced climatic change was held in Annapolis, Maryland in April 1979.[317]

ever, the DoE program continued to coordinate much of the modelling and empirical research on the carbon dioxide question. A series of peer-reviewed 'state-of-the-art' reports produced in 1985 bear witness to the fact that Weinberg's 'CO_2 community' was at that time thriving.[318,319,320,321]

It was also in 1985 that some of the key scientists in this community were invited to a conference in an Austrian alpine village that would be instrumental in affecting the conversion of this scientific interest into a global policy campaign within the UN. By 1985, the groundwork had been done. From there, the scare would build into a monumental public-policy campaign in the USA, elsewhere, and, most importantly, at the UN. For this to happen, the carbon dioxide question had to come to dominance in the UN's ongoing discussions of climatic change. As we shall see, it was the global cooling scare that brought attention to the climate issue, both in the USA and in the UN. The response and reaction then paved the way for the warming scare to storm in and take the global stage.

9 Meteorological interest in climatic change

The conventional meteorological practice of climatology obliterates any evidence of climatic change. This is because the standard statistical technique for determining the climate of a locality from its weather measurements removes all variation over time. If climate is average weather, then that is what it must do. If weather is also considered to vary randomly about its climatic norm, then the statistical analysis of random behaviour also gives standard variability and the frequency of extreme events.

A major application of this sort of climatology is to establish the likelihood of particular weather outcomes. This might be the chance of sunshine on parade day, the chance of a crop-destroying drought in a particular farming region or the chance of exceeding a flood level during the lifetime of a bridge. The climates of different places can also be compared. The only trouble is that weather data has been recorded for different places over different periods. As regional and global climate maps were being established from the records of standardised weather stations, the statistical results were known to be influenced by the date range used in the analysis. Thus a standard period was also required. This is why the Commission for Climatology of the International Meteorological Organization (the predecessor to the WMO) was pushed to establish a 'climatic normal period'. In 1935 the first normal period was declared as the first three decades of the 20th century.[322] Such a three-decade reference period remained in use throughout the 20th century, only shifting forward as the decades proceeded. Even though demand for this convention can be seen as an implicit recognition of climatic variability, its institution only served to reinforce the effect of the methodology in obscuring it; and climatology came to be disparaged as the 'book-keeping' department of meteorology.

It had not always been so. The institutional triumph of this methodology extinguished an early long-running and lively scientific debate in the New World over the influence of forest clearing and settlement on climate trends found in local records.[323,324] However, just as a new dogma of unchanging climate came to dominate the emerging national and international meteorological institutions, it was met with a challenge. This came from a scientific discipline that pays no heed to the changing conditions of the air above, concerned itself only with the secrets hiding in the rocks below.

In many ways, geology comes to the problem of change from the opposite direction, and this is most strikingly so in terms of scale. While statistical meteorologists were building their climatic norms from local daily records, those exploring the history of the Earth in its rocks and soils were finding hard evidence of dramatic global changes across tens of thousands and millions of years. Some of these large-scale changes were primarily in climate. Indeed, at least for the most recent geological past, climatic change would come to define the actual timeframes of the geological, biological and archaeological records. Thus the current geological epoch, the Holocene, is defined as beginning at the end of the last glaciation, when a lower sea level created land bridges for plants, animals and humans to cross. The Holocene itself has archaeological significance in that it is the stable warm period in which agriculture-based civilisation developed.

The idea of successive tides of glaciation gained broad acceptance in the late 19th century. Soon after that, evidence was found of smaller-scale climatic cycles and fluctuations across the Holocene. Such changes became evident not only through very recent geological indicators (for example, changing sea and lake levels or the thickness and composition of sedimentary layers), but also in palaeontology (the rise or fall of climate-sensitive species) and archaeology (for example flourishing civilisations in places where there are now deserts). As interest in recent climatic changes increased, other indirect ('proxy') climatic indicators were developed besides those hidden in the ground, including those botanical (such as variations in the width of tree rings) and those documentary (such as variations in harvest and sea ice records). This work was mostly initiated by non-meteorologists, but it soon extended the 'paleoclimatology' record of climatic change into the period of the meteorologists' weather records. In this way, the geologists challenged the dogma and method of conventional climatology by introducing the idea that the climate has been changing constantly on all timescales and therefore that we should expect it to change now and in the future. Rather than statistically obliterate or ignore evidence of climatic shifts in the instrumental record, we should look for them. The implication for forecasting is that prediction could be improved through a better understanding of these changes and their causes.[325,326,327]

It was not until the 1950s that interest in general climatic trends began to take hold among meteorologists. In particular, a generalised warming from the

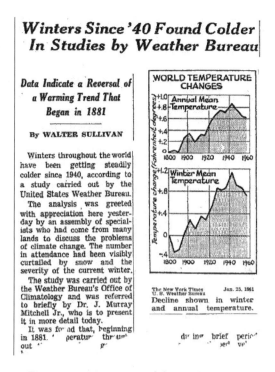

Figure 9.1: Coverage of the cooling trend.

1961 was a year when interest in recent climatic changes developed both outside and within the meteorological establishment. The climate conference reported in this *New York Times* article (25 January 1961) took place in New York during a particularly cold week of a particularly cold winter.

late 19th century was found in the instrumental record. Then, just as this was becoming accepted, the record for the previous decade suggested that this trend might have stopped or even reversed. By the early 1960s there was broad agreement that a reversal had taken place around 1940.

Meanwhile, pressure to address the problem of climatic fluctuations was coming from another direction: from those concerned with the impacts of climatic changes on human welfare. Drought is one short-term climatic change that everyone agrees about. During recent times and back through the ancient past, infamous droughts extending across multiple seasons in vast semi-arid regions have caused mass starvation, social breakdown, armed conflict and waves of migration.

Questions have often been asked about the frequency of droughts, whether they are random or predictable, whether they might be increasing, and, if so, why. In the 1950s, the United Nations Educational Scientific and Cultural Organization (UNESCO) was concerned to address general problems of welfare in regions of marginal human habitation. In 1957 it established an Advisory Committee on Arid Zone Research, through which it soon collaborated with the WMO to organise a symposium on 'the changes of climate with special reference to the arid zones'. The UNESCO–WMO 'Changes of Climate' symposium was held in Rome in 1961, a year that marked something of a beginning to governmental and intergovernmental interest in climate variability, prediction and control. Attracting 115 scientists from 36 countries, the Rome conference effectively brought together all the disparate research already underway in Western Europe, in the Soviet Union and the USA. Of the 45 papers presented, at least one quarter address the topic from the geological perspective, while others are evidently informed by it. Most papers show meteorologists open to the idea of recent climatic changes extending into the period of the instrumental record.[328]

This new interest in the impact of climate fluctuations was difficult for the meteorological establishment to ignore, and some inroads were achieved. During the planning for Changes of Climate symposium, the Commission for Climatology had agreed to establish a working group to consider ways to identify and investigate 'climatic fluctuations'. The seven meteorologists eventually elected to this group, all leaders in this new field, had presented papers at the UNESCO–WMO symposium and at other climate conferences during the early 1960s. When they submitted their report, the approval was such that the Commission decided to publish it in full. Appearing in 1966, it established standard terms and definitions for this new field of endeavour. Most importantly, it directly challenged the conventional statistical approach to climatology, with the bulk of the report surveying various statistical methods for identifying *non-random* climatic fluctuations in the weather record.[329]

And so, while resistance to the idea of recent (natural) climatic change remained strong in the meteorological climatology field right through the (man-made) global warming scare, by the mid-1960s the study of these changes had gained some legitimacy. Already at this time, attention was being drawn to possible mechanisms of change and their causation, including global changes influenced by human activities.

The idea that humans could cause *local* climatic changes had never been controversial. The influence on local climate of the draining of marshes, the removal of forests or the planting of windbreaks might always have been disputed, but the argument was only ever about how much or in what way. The local warming effect of cities has never been controversial. But then in the 20th century there was speculation that eventually some local effects might become so large that they would have a general impact, such as where dust released by agriculture might cool the atmosphere over vast regions and thereby impact climate globally. Waste heat from energy production might one day reach such intensity that it would warm the entire globe. Speculation on this latter effect among Soviet scientists during the early 1960s led to the proposal that it might be mitigated by geo-engineering, with aeroplanes spreading a veil of sulphate aerosols in the stratosphere.[330] By the mid-1960s, Revelle was already speculating about the distribution of reflective particles over the oceans and the seeding of clouds to mitigate possible future carbon dioxide warming.[331] Otherwise, and as we have already seen, reviews of proposals to purposefully control weather and climate—including discussions of their possible unintended consequences—led to further discussions of climatic interventions that were entirely inadvertent. Most of the inadvertent effects discussed in this way were local and regional, such as the effect of artificial lakes, of (subsonic) contrails and the desertification effect of overgrazing that had been much discussed at the 1961 UNESCO–WMO meeting.

Possible *global* human influences first came under consideration, even before the 1960s, through research spinoffs from the race to solve the mystery of the ice ages. This was the case for two effects of volcanic emissions. One was the cooling effect of volcanic 'fog', or aerosols. Interest was soon drawn to sulphate aerosols, especially where these were propelled above the clouds by massive explosions so they could not be rained out. The human effect—a more constant injection into the troposphere of agricultural dust and industrial sulphates—came to be referred to as the 'human volcano'. The other effect of volcanic emissions to feature in the great ice age debate was warming by carbon dioxide. It was first proposed late in the 19th century that a *decline* in atmospheric concentrations during periods of low volcanic activity could moderate the (natural) greenhouse warming and thereby trigger the slide into an ice age.[332] The human effect is, of course, opposite, with an *increase* causing warming. These two human effects, one warming and the other cooling, would soon dominate the discussion of the global human influence, where they would become associated with two opposing prospects for the global climatic future.

By the end of the 1960s, and in these various ways, possible human influences were integrated into a rather academic discussion of the causes and mechanisms of recent climatic change. External to this discussion—among those interested in the impacts of change, whether the UN, governments or the public and so the press—a bottom-line question started to demand an answer: *Should we expect further cooling or warming?* Given the complexity of influences and the overwhelming ignorance of their relative importance, few scientists would make that call. A few did wager on moderate warming or cooling in the near future, while even fewer raised the possibility of catastrophes, but usually in some far distant future. With the appetite for apocalypse of those times, any fore-warning of catastrophe tended to amplification in the press. The trouble was that there were two directly opposing views. An indicator of this predicament is found in a 1968 *Time* magazine feature that waxes philosophical on pollution as a modern malaise. A shift to consider 'systemic' impacts opens thus:

> It seems undeniable that some disaster may be lurking in all this, but lay-men hardly know which scientist to believe. As a result of fossil-fuel burn-ing, for example, carbon dioxide in the atmosphere has risen about 14% since 1860…scientists fret that rising carbon dioxide will prevent heat from escaping into space. They foresee a hotter earth that could melt the polar ice caps, raise oceans as much as 400 feet, and drown many cities. Still other scientists forecast a colder earth (the recent trend) because man is blocking sunlight with ever more dust, smog and jet contrails. The cold promises more rain and hail, even a possible cut in world food.[333]

The idea that sea level might rise 400 feet probably comes from an obscure short assessment by Roger Revelle that was buried in a submission to the new president (Johnson) in early 1965, and published that November. This subsec-tion of an appendix to a general report on pollution contains perhaps the most alarming claims ever by a scientist on the impact of carbon dioxide warming. Revelle first conceded that the world has cooled since 1940, during which time 40% of all carbon dioxide emissions have occurred. Still, 'we must conclude', ac-cording to Revelle, that climatic noise from other processes has at least partially masked the warming effect. Addressing an (unreferenced) suggestion that this warming would melt the entire Antarctic ice cap, he calculates (rather symmet-rically!) that this would raise sea levels 4 feet per decade until the melt finished after 1000 years with a rise of 400 feet.[334,*]

Revelle's reckoning of a modern diluvian catastrophe was missed (or per-haps avoided) in the press coverage of this pollution report, but some diligent folk must have read it. An aide to President Nixon sounded the alarm in 1969 using figures that seem to be derived from Revelle's calculations:

---

*The report informed a 'Special Message to the Congress on Conservation and Restoration of Natural Beauty' by Lyndon Johnson, delivered three weeks after his inauguration on 8 February 1965. In this wide-ranging speech, carbon dioxide is mentioned as a global-scale atmospheric pollutant:

> Large-scale pollution of air and waterways is no respecter of political boundaries, and its effects extend far beyond those who cause it. Air pollution is no longer confined to isolated places. This generation has altered the composition of the atmosphere on a global scale through radioactive materials and a steady increase in carbon dioxide from the burning of fossil fuels.

> It is now pretty clear that the CO_2 content will rise 25% by 2000. This could increase the average temperature near the earth's surface by 7°F. This in turn could raise the level of the sea by 10 feet. Goodbye New York. Goodbye Washington, for that matter.[335]

Hubert Heffner, deputy director of the Presidential Office of Science and Technology, replied in January 1970:

> The more I get into this, the more I find two classes of doomsayers, with, of course, the silent majority in between. One group says we will turn into snow-tripping mastodons because of the atmospheric dust and the other says we will have to grow gills to survive the increased ocean level due to the temperature rise.[336]

Heffner was right that most experts, when questioned, would respond with varying degrees of ambivalence, while the few 'doomsayers' were divided at both extremes. But then, 'doomsayer' is perhaps too strong a term because the speculation mostly consisted of vague warnings of possible negative (as well as positive) consequences in the distant future .

That this situation held at least up to 1971 is supported by a reflection on the SMIC conference of that year by one of its convenors, William Kellogg. We will recall that this 'Study of Man's Influence on Climate' was a scientific conference scheduled as a follow-up to SCEP. It was more specific to atmospheric issues, more international in its participation, and convened in Stockholm in the hope of informing the United Nations 'Human Environment' conference scheduled for that city the following year. Kellogg later reflects how, as the three-week study was drawing to a close, the organisers tried to bring the group to some consensus on the overriding climatic threat:

> ...we decided to call an evening meeting to thrash out a consensus, and to decide (if we could) whether we would predict a net cooling or a warming in the decades ahead due to man's activities. It would clearly be useful if we could make such a prediction with some degree of conviction. However, the impasse prevailed, much to my disappointment.[337]

Within five years Kellogg himself would come out pushing warming alarm, but not yet. He explained that the 'impasse' at SMIC accounted for the weakness of the consensus statement, but the tone of that document suggests outright resistance to the push to declare a global impact of any kind for mankind (emphasis added):

> While it is conceivable that man *may have* had *a small part* in the most recent climate changes we have just described, it is clear that natural causes

must be sought. In fact, *as has been frequently pointed out*, it will be *difficult to identify* any man-made effect because, first, with our present state of knowledge, *we do not know* how to relate cause and effect in *such a complex system* and, second, man-made effects will be *obscured by* the natural changes that we know *must be occurring*.[338]

The consequence of this resistance was that neither this nor any other group of scientists delivered a message of climate alarm to the policymakers' meeting the following year. Nor was there much policy interest at this time in the impacts of natural changes. But this was about to change. A series of events that very year would propel global cooling alarm into such prominence that a concerted effort was required to rein it in.

10 The global cooling scare 1972–76

Little Ice Age return

By the late 1960s, historical climatology had developed from its geological roots into a distinct field of research that had picked up on two century-scale fluctuations in the well-documented climate of Europe over the last millennium. The first was a relatively stable warm period during the high Middle Ages, which extended growing seasons and allowed cropping to shift northward. This 'Medieval Warm Period' was followed by a generally cooler period that included times when repeated severe winters extended across vast regions. At times during this 'Little Ice Age', glaciers were seen to extend across mountain meadows, growing seasons were shortened and episodes of 'bad' weather brought famine.

The general 'recovery' of the climate around the close of the 19th century was seen to mark the end of the Little Ice Age. But then, just as this warm–cool–warm trend was coming into acceptance, any presumption that Europe might be at the beginning of another extended warm period started to look decidedly premature when a series of extreme winters in the early 1960s was followed by a general cooling trend. Recognition of this change led to speculation that the brief modern warming might only have been a short respite in the cooling of the preceding centuries; in other words *that the Little Ice Age has not yet run its course.*

The concern was that the great population growth and economic expansion since the late 19th century had been supported by increasingly favourable climate, but that a return to a cooler climatic regime might bring a grave reckoning. Hubert Lamb, the most eminent historical climatologist at the time, would often express this view while raising the prospect of more severe winters in the coming years. However, as the 1970s progressed, he changed his forecast. This was due to his early investigations of a climatic oscillation in the northern North Atlantic. Using documentary evidence to supplement instrumental records, he had tracked an oscillation with a period of roughly one century. The minimum in this oscillation often brings severe winters to Britain, and what he noticed

155

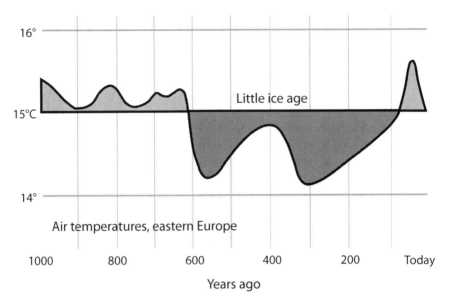

16°

Little ice age

15°C

14°

Air temperatures, eastern Europe

1000   800   600   400   200   Today

Years ago

Figure 10.1: Temperatures for the last 1000 years.

This is one version of a temperature trend graph somewhat misrepresenting the original work that Hubert Lamb first presented to the UNESCO-WMO Changes of Climate symposium in 1961.[339] It was often used in climatic change assessments during the mid-1970s to indicate the general trend. At that time, there was much talk that the recent cooling trend might signal a return to the Little Ice Age. The graph is derived from 'winter severity' data for the region of Moscow. When it is compared with Lamb's corresponding charts for Western Europe, the Medieval Warm Period is found truncated and the Little Ice Age deepened. For more on the history of these graphs, see 'Hubert Lamb and the assimilation of legendary ancient Russian winters'.[340]

in the early 1970s is that such a minimum seemed to have passed in the 1960s. By 1975 he was sufficiently sure of this 'climate reversal' to publish the results. While he would continue to speculate about the possibility of a return to the Little Ice Age in the distant future, his new forecast was that the recent spate of mild winters in Western Europe is likely to continue.[341,*]

It was one thing for Hubert Lamb to publicly speculate about the possibilities of a return to the snowy winters of the young Charles Dickens.[344] However, it was another thing altogether when others took this cooling forecast down a level. Just as Lamb and other historical climatologists were indulging in the

---

*Lamb's change of forecast has been misinterpreted by Sanderson[342] and Walker[343] as though he moved from ice age alarm to global warming alarm.

156

prospect of an extended Little Ice Age, there was also speculation that such century-scale cooling might only be the beginning of a cooling trend on an even grander scale. Foreboding the imminent onset of the next (big) ice age has a history almost as long as ice age theory itself. The revival of this speculation in the early 1970s was not only triggered by the recent bad weather. This was but the favourable context for speculation arising from a revolution in the understanding of climatic change during the recent geological past. Once again a movement among European geologists would arise and disrupt the meteorological establishment.

A breakthrough with geological time

In the late 19th century, general agreement settled on the idea that there had been four ice ages already in the current geological period, the 'Quaternary'. This convention persisted well into the 20th century, while the dating of these four events always remained obscure. Cesare Emiliani, an Italian-trained geologist, was one of the first to challenge the textbooks on this teaching. This was mostly through his pioneering studies of extracted 'cores' of deep-sea sediment. After taking up a position at the University of Chicago, Emiliani eventually had the opportunity to investigate new, exceptionally long sedimentary cores cutting through deposits laid down over hundreds of thousands of years, deep into the Quaternary. Isotope analysis of microscopic crustaceans in these samples left no doubt in his mind that there were many more than four major cool/warm oscillations. Later, George Kukla also found clear evidence of many more glaciations in the sedimentary profile of clay quarries in his native Czechoslovakia. Analysis of Greenland ice cores also started to produce similar results. However, what made these findings difficult to dispute were new dating techniques that allowed the timing of these changes to be better framed. By the end of the 1960s, the climatic history of the Quaternary period would be turned on its head.

Indicators were found in the compressed seabed mud marking the Earth's last geomagnetic reversal, which occurred just over 700,000 years ago. This reversal framed eight major oscillations in the intervening years. Other techniques helped to mark out the timeline and reveal that the ice advanced and receded on a distinct 100,000-year cycle. Moreover, the new dating cast aside a presumption of prevailing warmth. The entire Quaternary period—covering the last 2.5 million years—is best described as one big 'ice age' that is only punctuated every 100,000 years by brief 'interglacial' warm epochs, of which the current Holocene is the latest. Not only did the timeline of Quaternary geology need total reconfiguration, but the discovery of these cycles delivered the sobering realisation that human civilisation had developed during a relatively brief and exceptionally warm epoch. And this epoch is about to end.

The realisation that the end is nigh for the Holocene is not in itself alarming. On these geological timescales, a rapid change would be of the order of 1 °C per millennium. Such gradual changes would be hard to detect among all the much more dramatic local, regional and even global fluctuations across decades and centuries, such as the recent multi-century Medieval Warm Period and the Little Ice Age. This is why geological-scale climatic change is generally regarded as irrelevant to forecasting across one or even several human generations. But, *what if short-term jolts and swings accompanied the grander climatic transition?* This possibility could not be ruled out, as most of the climate proxies of the deep past only measure changes on scales of millennia. With such crude 'resolution', even century-scale changes would go unnoticed.

But then some scientists started to claim finer resolutions in their prox-ies, sometimes down to two centuries or even less. It was upon these claims that alarm was raised. These proxies revealed climatic 'spikes', 'leaps' and 'step-changes' in the deep past, including signs of instability at the end of previous interglacial epochs. *Should such changes be expected for the 'breakdown' of the Holocene?* The Danish geologist Willi Dansgaard certainly saw it this way. Ac-cording to Dansgaard, if the deep past is anything to go by, then 'the conditions for a catastrophic event are present today'.[345]

One further step was taken by some of these geologists that would bring their counsel in line with that of meteorologists also warning of imminent cli-matic cooling and instability. Already there was some alignment of their find-ings with meteorological speculation about the cooling effect of aerosol emis-sions: this would only exacerbate the natural cooling predicted by the geologists. But meteorological speculation soon emerged upon presumptions of inherent climatic instability, where the 'delicate balance' of nature is easily upset by hu-man interference. The geological contribution to this story was that the avoid-ance of such interference would be all the more important if the natural balance of climate were particularly unstable at the time of interglacial breakdown.

The nature of the human interference that might trigger changes was not always specified: *any* perturbations of the atmosphere might trigger dramatic changes that otherwise would not have occurred, or not as much, or not so soon. Thus, warming by carbon dioxide emissions was not always excluded as a cooling trigger. It was not new to imagine that the direction of the change might be opposite to the direction of the perturbation. Already in 1956, it had been famously argued by an eminent geophysicist that higher sea levels and a much-anticipated ice-free Arctic, *due to natural warming,* could cause in-creased snowfall on the surrounding lands and thereby bring on *the next ice age.*[†] Logically, inadvertent or deliberate human interference could produce the same result, whether through a general warming or through the planned or inadvertent melting of the Arctic ice.

Edward Lorenz's work on dynamical systems was also brought into play. Anthropogenic disruption might drive the system up and over an attractor 'tipping point' where the climate might settle into an altogether different and hostile regime. Scientists spreading this alarm did not always specify the direction of the fall; it might be cooling, it might be warming, or it might involve wild swings from one to the other. Agreement came with the cautioning that *any* perturbation would be highly risky.[347] For Emiliani, the new geological evidence suggested that the exceptional warmth of the current epoch was in '...a precarious environmental balance, a condition which makes man's interference with the environment...extremely critical'.[348]

One of the American meteorologists who met the geologists on these terms was Reid Bryson. A geologist by training, Bryson also happened to be the founding director of the Center for Climatic Research at the University of Wisconsin, which was one of a kind until Lamb founded the Climatic Research Unit at the University of East Anglia in 1972. By the time the Quaternary geologists were raising alarm about natural cooling, Bryson was already warning that the late-20th century cooling trend might be due in no small part to the rising levels of agricultural dust in the lower atmosphere.[349,350]

Another scientist who took an interest in this work was an oceanographer also trained as a geologist. Wally Broecker was early with impatience for research and action on global *warming*, but his affinity with the Quaternary geologists came through his use of evidence of wild fluctuations in the deep past to warn of the climate system's sensitivity to human interference.[351,352,‡]

There was also the young Steven Schneider. He won early fame after announcing that a simulation of increased aerosol emissions caused a dramatic global cooling in his simple climate model. The aerosol effect turned out to be erroneously overestimated, and he soon switched to the simulation of the en-

---

[†]This idea was proposed by Maurice Ewing and William L. Donn. Ewing was a leading oceanographer, president of the AGU and director of Columbia University's Lamont Geological Observatory.[346]

[‡]Broecker was never consoled by inherent system stability. One of his later concerns was that the ocean current 'conveyor belt' (that he had helped describe) might shut down, and that this could cause dramatic cooling in the northern North Atlantic. In this regard, he would conclude:

> The record of events that transpired during the last glacial period sends us the clear warning that by adding greenhouse gases to the atmosphere, we are poking an angry beast.[353]

hanced greenhouse effect in accord with rising warming alarm.[354,355] But even through this transition, an alignment with the empirical findings of the geologists remained in his view of a climatic system poised in such a fragile balance that any perturbation would risk all sorts of changes.

And so it was: out of the department of science plumbing the inconceivable depths of deep time for unimaginably slow changes, a message of urgency emerged. This message quickly found affinity with receptive meteorologists and amplified their own forecasts of catastrophe in the near climatic future. The message was: *the balmy epoch in which civilisation developed is now ending, with the risk of catastrophic climatic fluctuations.*

This view of the climatic future first consolidated at a small conference in January 1972. The idea for this meeting was hatched by George Kukla and the editor of *Quaternary Research*, Lincoln Washburn, along with Robert Matthews, who would end up playing host at Brown University, Rhode Island. Meteorologists were invited, as well as geologists. Only 22 scientists ended up attending, mostly from North America, but as many again were 'non-attending participants'. Many of these were European geologists who sent over papers or abstracts that could serve in various ways to help those assembled at Brown University answer a single question:

The present interglacial, how and when will it end?

Later that year a complete issue of *Quaternary Research* would be dedicated to articles arising from the conference, including papers by those Europeans unable to attend. A summary of the conference findings also appeared in *Science* around the same time. This includes a statement that had been drafted, circulated and agreed by all 46 participants. It contains the following warning:

In man's quest to utilize global resources, and to produce an adequate supply of food, global climatic change constitutes a first order environmental hazard which must be thoroughly understood well in advance of the first global indications of deteriorating climate.[356]

The need for an urgent response 'well in advance' of the empirical confirmation presents one of the familiar ingredients of a good scare. However, between the conference and these publications, nature herself stepped in to provide stark warnings of what might be in store.

That year there emerged one of the strongest El Niño events in the meteorological record. While today the El Niño/La Niña oscillation is the most familiar and studied of all climatic oscillations, back in 1972 its effects were only vaguely known. The Peruvian anchovy harvest is typically impacted by El

Niño, but overfishing caused its complete collapse late in 1972, and it took some years to recover. Not easily associated with the oscillation were other weather extremes during that year and into the next. In fact, 1972 would go down in meteorological history as one of the most volatile years in the global weather record. Western Europe was unusually cold and parts of Italy and the Balkans were drenched by twice their normal annual rainfall. By contrast, the bread-basket regions of the Soviet Union were gripped by drought. Grain yields were so drastically reduced that famine could only be avoided with massive wheat purchases from the USA and other western countries. The Chinese were also affected, while in India a late monsoon caused widespread famine. Worst of all, a fourth year of drought on the southern fringe of the Sahara Desert caused hundreds of thousands to die of starvation and the mass migration of survivors. This drought was vividly portrayed on Western television as it continued into 1973, amidst renewed speculation that the Sahara is expanding. Indirect global effects of these weather events included an alarming reduction of grain stockpiles and rises in world food prices.[357] This only exacerbated concerns about our ability to feed a booming world population and it led directly to the organisation of the first UN World Food Conference, scheduled for November 1974—at which the possibility of climatic changes would be much discussed.

The publicity around these events marked the beginning of an unprecedented widespread interest in climatic change, and this interest could not have been better timed to highlight the message arising from the Brown University conference. The editorial introducing and summarising the special November 1972 issue of *Quaternary Research* begins thus:

> The year 1972 showed again how vulnerable our society is to large anomalies of the weather...[358]

During the year, Kukla had been in the process of migrating from his troubled homeland while holding a visiting position at the Lamont-Doherty Geological Observatory. Before the year was out, he was sufficiently moved by the urgency of the climatic situation to pen a letter to his new supreme leader, the President of the United States. This letter to Nixon, signed by Kukla and Matthews, cites the conclusions of their conference in support of their warning that...

> ...a global deterioration of climate, by order of magnitude larger than any hitherto experienced by civilized mankind, is a very real possibility and indeed may be due very soon...[359]

With no mention of any possible human influence, the letter went on to explain that 'the present cooling now under way in the Northern Hemisphere could be the start of this expected shift. If this rate of cooling were to continue, then it could bring 'glacial temperatures in about a century'. The consequences could be food shortages and increased frequency and amplitude of extreme weather events such as 'floods, snowstorms and killer frosts'. They recommended coordinated research to 'possibly find an answer to this menace' while in the meantime precautions should be taken, including food stockpiling.[360]

The US government response to the global cooling scare

This letter was taken seriously within the Nixon administration. The convening of an interagency ad-hoc 'Panel on the Present Interglacial' came as a direct response. In 1974 this panel not only reported its findings but produced perhaps the first proposal for a national climate program.[361,§] Over the next few years, the push for such a program was driven by the ongoing policy advice of the Quaternary geologists in concert with others who were linking erratic extremes with forewarnings of a transition to a cooler climatic regime. A related development during 1974 was the White House Environmental Resources Committee establishing a Subcommittee on Climate Change. The committee chairman explained this move thus:

> Changes in climate in recent years have resulted in unanticipated impacts on key national programs and policies. Concern has been expressed that recent changes may presage others.[362]

It is important to note the language used here and elsewhere. It was not *extreme weather events* that had pushed the prospect of climatic change up the policy and planning agenda. The discussion had already moved beyond weather. Rather, the prospect of further climatic change was being raised on the understanding that changes in climate were already underway and were already having profound impacts on society.

Robert White was enlisted to chair the subcommittee, which, before the year was out, produced a report that did no less than outline, as its title says, *A United States Climate Program*. In the meantime, White had arranged for NOAA to organise a series of interagency meetings to develop a National Climate Program proposal. And then, right on cue for Jimmy Carter's inauguration, nature stepped in once again with a reminder of the calamities that might be in train for a cooling world.

---

§The *ad-hoc* panel included representatives from the Department of Agriculture, the Department of Defense, the National Science Foundation and J Murray Mitchell from NOAA. The US government response to the cooling scare at this time also triggered the events at the WMO that would lead to the establishment of the World Climate Programme, as we shall see below.

The Ice Age Cometh...
Or Is It Just Another Lousy Winter?
By Steven S. Ross

"...If the Arctic region continues its cooling trend, it could trigger a new ice age, with glaciers covering a third of the country..."

Just after Christmas, I dropped into my local hardware store in Leonia, New Jersey, and asked for a ten-pound bag of calcium chloride—those little white flakes that help make snow melt away. When the clerk replied that he had only 25-pound bags, I pointed out that it had taken me more than two years to use up the ten-pound bag, and that the stuff is messy to keep around. "I don't think you'll have that problem this year, even with the big bag," he said. ~~rt~~

then listed a dozen states that could suffer fuel shortages if the weather stayed cold. As of mid-January, Ohio, Texas, Missouri, Kansas, Oklahoma, Nebraska, Pennsylvania, New York, Indiana, and Minnesota were in trouble. Only Ohio and Indiana had been on the agency's November list.

Robert Dickson, who's with the Long-Range Prediction Group of the National Weather Service, blames the cold on the unusual behavior of high ~~ude~~

Figure 10.2: Renewed speculation about an ice age.

The severe North American winter of 1976-7 triggered renewed speculation about an Ice Age. This speculation was often linked to the energy crisis. It was no coincidence that this feature article in *New York Magazine* of 31 January 1977 follows directly from another feature on 'How we will beat the energy crisis'.

January 1977 was one of the coldest months on record in many densely populated regions of the North American continent. The adverse weather put enormous pressure on energy infrastructure, a shortage of heating fuel supplies and concerns that some folks might freeze to death without resupply. Deep snowpack across many states crippled transportation and shut down business in some centres for a week and more. There was no escaping to Florida, where the rain turned white and settled on the lawn for the first time in places as far south as Miami. The month ended with a great blizzard blowing for five days across New York State. It shut down entire cities and caused a national disaster. The snowdrifts lingered into February and energy supplies were rationed during what Carter called 'an energy emergency'. This was just two months before his televised call to arms on an energy crisis that was 'the moral equivalent of war'. A subsequent 'climate impact assessment' of the calamities across the entire winter calculated total direct losses at $27 billion.[363]

It was especially in the accounting of the impacts of the weather of 1977 that two issues of great currency came to be firmly linked, namely, the energy crisis and the prospects of a new ice age. Energy and climate were linked through the impact of climate on energy demand and not, as some scientists at the time

Figure 10.3: The ice age scare and the energy crisis.

The linking of energy crisis with the ice age scare through the reporting of the extreme winter of 1977 is demonstrated by the layout on this page of the *Calgary Herald* of 1 February 1977.

so earnestly wished, through the impact of energy production on climate. The warnings of the potential future warming due to an increasing reliance on fossil fuels had a meagre impact on the science-policy debate. Whether emanating from national laboratories or from the NAS, these warnings gained little traction with the public or policymakers. However, as we have seen, this is precisely when the generous funding of the Carbon Dioxide Program began. That it took off at this time is remarkable not only because it came without public or policy-maker engagement, but because the public and policy engagement on climate was otherwise and contrary.

A warning of warming simply could not compete with the fear of more deadly and damaging weather events such as those that were publicised during the 1970s, all of which had been linked (rightly or wrongly) with a cooling climate. This perception was supported by general agreement among climatologists throughout the late 1970s that the records showed the continuance of the cooling trend.[364,365] With the pause in the 20th century warming trend now pushing through to the end of a fourth decade, *surely a little manmade warming could only be for the good*. It was one thing if the burning of fuel for indoor warmth effected an outdoors cooling (as per Bryson, by aerosol emissions), but it was quite another thing if indoor heating might *moderate* the severity of the winter pressing in from outside. Thus, we can see why there was no outcry of climatic change alarm when Carter sought to solving the immediate oil crisis through the mass production of 'synthetic oil' from the plentiful supplies of coal.

Weinberg's cautioning about dependence on coal went largely unheeded. But where such concerns did arise, they only had a dampening effect on climate alarm. In fact, this had been J Murray Mitchell's sobering message at the Brown University conference back in 1972. On the arguments of the geologists, Mitchell rejected anything more than a moderate and gradual century-scale natural cooling trend, while concluding that the likely net manmade effect in future centuries (from aerosols and carbon dioxide) would be warming. If this did eventuate, it would serve to offset a cooling trend that the geologists were, in his view, overstating.[366] So even as scientific agreement consolidated around this idea that any manmade effect emerging in the next century would likely be a net warming, this view tended only to allay climate alarm. In the late 1970s, therefore, the warmers' message was not alarming.

Well, mostly not. Where it was alarming, it did not cause alarm. This was the case with Wally Broecker. His warning against complacency was that manmade warming is being masked by a natural cyclic cooling that will turn around in the 1980s, when we will be in for a rude shock. But this view was marginal and, much to Broecker's frustration, it gained no traction in the science-policy debate during the 1970s (nor even during the early 1980s). Instead, in the reports and in testimonies at the various committee hearings, what was driving the case for climatic change research was the social and economic impacts of the cooling-linked weather extremes. These extremes were themselves interpreted as *changes in the climate* that might *presage others*. This interpretation of current events is what grabbed the attention of government, and especially the new Carter administration. And this is what led Carter to finally sign into law the National Climate Act in 1978.

Figure 10.4: *Climates of Hunger.*

The release in 1977 of Reid Bryson's book-length climatic forewarning was timely. The jacket blurb opens: 'Climate is changing. Parts of our world have been cooling. Rain belts and food-growing areas have shifted. People are starving. And we have been too slow to realize what is happening and why…'

The cooling scare in the rest of the world

Thus, in the United States, it was concern about a cooling trend—and about an inability to cope with it due to the energy crisis—that got traction with the federal government. If tentatively and inadequately, this is what drove the coordination of climatic research as a national priority in the USA.[367,368]

Elsewhere, the scare played out differently. The link was less to any energy crisis and more to impacts on agriculture and food supplies. In Africa the focus was on droughts, especially as the southern fringes of the Sahara Desert continued to receive low rainfall throughout the decade. Already in 1973 most of

166

the affected states had formed a permanent international collaboration (*Comité Permanent Inter-États de Lutte contre la Sécheresse dans le Sahel*') to address what they saw as an ongoing problem. A few years later, UNEP decided to hold a major conference on desertification at its base in Nairobi, Kenya. There, the question of a possible climatic cause rose to prominence in discussions of such questions as 'Was the Sahelian drought evidence of larger changes in the global climate?' and 'Was the Sahara expanding south?'[369]

Meanwhile, up north in Europe, scientists sounded the cooling alarm right through to the end of the 1970s. But in general, old-world governments responded with less enthusiasm and much more scepticism.

After the policy campaign was launched by the geologists in 1972, various scientific conferences and workshops addressing the impacts of global cooling were convened across continental Europe. Many of these meetings (and soon the research informing them) were supported by private charities that were already involved with the issue of global food and resource scarcity. It was with these concerns in mind that the Rockefeller Foundation teamed up with the Nobel Foundation and the Swedish monarch to form the International Federation of Institutes for Advanced Study (IFIAS). Aimed at promoting international collaborations to address humanity's great perils, IFIAS soon made the transition from food scarcity to climate. One of the more influential personalities driving this interest was the inaugural director of NCAR in Boulder, Walter Orr Roberts. We came across Roberts much earlier when, as NCAR director in the 1960s, he raised concern in the press about the climatic impact of subsonic contrails. Now, in the mid-1970s—and indeed through to the end of the decade—he was raising concerns about global cooling, especially through his leadership of IFIAS's climate project.

In 1974 IFIAS began promoting the cooling scare, supporting several scientific meetings in Europe and America on the subject that were influential in the various policy discussions arising around the globe. Perhaps the most influential was a workshop held in Bonn, West Germany, in May 1974. Chaired by the German climatologist Hermann Flohn, this workshop on 'The Impact on Man of Climate Change' produced an agreed statement that IFIAS circulated widely among meteorologists. The 'Bonn Statement' warned:

> The nature of climatic change is such that even the most optimistic experts assign a substantial probability of major crop failures within a decade. If national and international policies do not take such failures into account, they may result in mass deaths by starvation and perhaps in anarchy and violence that could exact a still more terrible toll.[370]

Similar climatic concerns were repeated in Rome at the World Food Conference later that year. In June 1975, the Rockefeller Foundation separately sponsored a conference on 'Climate Change, Food Production and Interstate Conflict' in Bellagio, Italy. Its agreed statement said that...

> ...there is some cause to believe—although it is far from certain—that climatic variability in the remaining years of this century may be even greater than during the 1940–70 period.[371]

This greater variability, it said, 'could cause major crop failures quite beyond the current capability of agricultural science and technology to control or mitigate.'[372] A previous conference on a similar theme was sponsored by Rockefeller Foundation in New York City, in January 1974.

European governments' response to all this activity was mostly subdued. Some political interest was raised by extreme weather, and in particular by a severe drought across Wales, England and other parts of Western Europe during the summer of 1976,[||] the influence of which is evident from the 1978 report of a two-year investigation by the European Commission, which recommended that the EEC establish its own generously funded 'Climatology Research Programme'. 'In announcing the recommendation' so *New Scientist* reported,

> ...the Commission drew attention to the recent remarkable sequence of climatic extremes in Europe—in the space of 15 years we have experienced the coldest winter since 1740, the driest winter since 1743, the mildest winter since 1834, the greatest drought since 1727 and the hottest month (July 1976) since records began 300 years ago.[374]

This warning was not enough to shift European governments from climate complacency. While substantial funds were invested in long-range weather forecasting, no climatic research program was ever implemented.

The government interest (even if usually negative) was in most cases driven by the mainstream media after picking up on any dramatic speculation among scientists. Ever ready to amplify apocalyptic claims, experts were quoted under alarming headlines. An exception to this pattern came in Britain, where perhaps the greatest media-driven public controversy arose. There the scare was bolstered by the carefully orchestrated intervention of a science journalist late in 1974.

[||] Of the British government response to this drought, Malcolm Walker says: 'So serious did the water shortage become in England and Wales in 1976 that a Drought Act was passed by Parliament (on 5 August) and a Minister for Drought was appointed (on 24 August)...Only a few days later, the heavens opened.'[373]

The weather machine and the threat of ice

In November 1974, BBC television screened a two-hour special on weather and climate. Written by the British science journalist Nigel Calder, *The Weather Machine* documentary, with its simultaneously published book, was ground-breaking in the way it introduced to a popular audience the science of weather, climate and the theories of natural climatic change. In doing so, it also introduced the human influence on local climate, although the treatment of possible *global* influences was decidedly cursory and sceptical. But the most powerful impact of *The Weather Machine* was undoubtedly in its promotion of the possibility that an imminent further decline into a cooler weather regime would mark the beginning of the next ice age.[375] The documentary was soon screened in other countries, including the USA, where an adapted version with an American narrator appeared early the following year. However, its greatest controversial impact remained in Britain.

In the documentary itself, 'the threat of a big freeze' is only briefly and simply introduced. It is in the book that the reasoning behind the expected 'return of the ice' is extensively elaborated, while a feature in the *Guardian* on the morning of the first screening provided the most conspicuous platform for Calder's icy forecast.[376] This was how Calder's science lesson hatched a frightening message. Just as with Rachel Carson twelve years earlier, where popular ecology was born foreboding ecological disaster, similarly now with popular climatology. Only Calder's offering was much more fatalistic: while the spread of carcinogenic chemicals might be controlled, there would be no stopping the return of the ice.

Calder bases his ice age speculation mostly on the recent revolution in Quaternary geology, but his proposal for its rapid onset is most extreme and of obscure origins. According to his so-called 'snowblitz' theory, continental ice sheets don't creep slowly outwards from the poles. Instead, they build across vast regions upon one winter's snow cover after it manages to linger right through one cool summer. The snow cover's reflection of the Sun locally reinforces the global cooling trend as the next winter's snowpack builds on the first, and then the next winter the same, and so on; the snow thickens, compresses, and forms into an ice cap like those covering Greenland and Antarctica.

There is nothing new in the proposal of a feedback mechanism promoting ice cap formation through its own cooling effect. This was one of a number of positive feedback loops to feature in early proposals for ice age causation. The novelty in Calder's dispatch to the British public was the frightening speed of the process: *one winter soon, a single snowstorm might be this very 'snowblitz' that buries your farm, never to be seen again*. There would be nothing for it but to

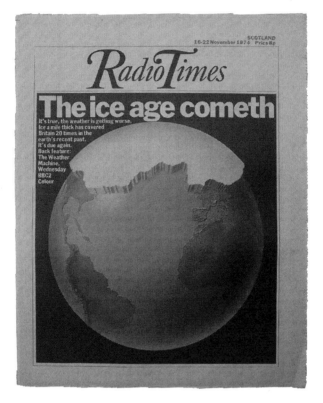

Figure 10.5: *The Weather Machine* at the BBC.

The 14 November 1974 issue of *Radio Times* promoted the first screening of *The Weather Machine* with a cover illustration of the Earth wearing an ice cap that obscures almost the entire British Isles.

walk off the icy wasteland towards the Equator. Not that much respite would be found there, as many regions closer to equator would fall into enduring drought. Calder reckoned 'a very high risk' of an event sometime in the next 100 years 'that could easily kill two thousand million people by starvation and delete a dozen countries from the map'.

> The evidence…for the episode of the sudden cooling and for the mechanism of the snowblitz favours a catastrophic view of the threat of ice.[377]

BBC promotions (see Figure 10.5) and newspaper reviews of the documentary were almost entirely preoccupied with the ice age threat; not that we could blame them for any hyperbole, for this was entirely in accord with Calder's own

170

promotion of the screening. After all, the book claimed that more than half the world's population is likely to starve to death and it arrived wrapped in a dust-cover heralding the peril: 'The Weather Machine and the Threat of Ice'. Similarly, the scary feature in the *Guardian* was published under the banner: 'Snow Blitz'. In that article, Calder claimed that recent scientific discoveries 'imply that the threat of a new ice age must now stand besides nuclear war as a likely source of wholesale death and misery for mankind'.[378]

The reaction of the meteorological establishment

In the British parliament there was little sympathy for Calder's media blitz. The concern was more in that the *British* Broadcasting Corporation had launched a campaign of meteorological alarm without any prior consultation with the *British* Meteorological Office. In response to this complaint, Parliament was assured that the claim of a shorter interglacial is 'incompletely substantiated', as is the claim of rapid glaciation.[379,380] If, in his research, Calder had consulted the nation's chief weatherman, John Mason, then the response would also have been scathing. Mason's scepticism of all claims of non-random climatic change was no secret.

The Met Office's preference for computer modelling of weather over the study of past climates was also well known, and strong differences over this direction of research had been why Lamb had left the Met Office to found CRU. However, relations between the two men did not improve when Lamb returned to Bracknell to discuss collaboration with his old employer, only to have Mason insist that CRU cease its seasonal forecasts, so as to avoid 'possible embarrassment' over any conflict with the Met Office's monthly forecasts.[381]

The Weather Machine book presents all sides of these multi-axial conflicts, which Calder found to be 'debilitating' British climatic research. Yet it displays special sympathy for Lamb. At the time, CRU was threatened with closure due to its failure to win grants: it is 'esteemed throughout the world', explained Calder, but yet it is 'denied government funding owing to opposition from the meteorological establishment'.[382],¶

Mason soon directly entered the controversy, which developed in the opinion pages of *The Guardian*. There, he belittled Calder's credentials as much as

---

¶In 1975 Lamb won substantial funding from the Rockefeller Foundation in response to his public appeal to save CRU. As we shall see below, this charity was promoting research into global cooling and its impacts at this time. Although other private funding followed, it was the Rockefeller grant that Lamb credits with saving CRU.[383] Thus, it could be seen that CRU was saved twice: once in 1975 by funding associated with cooling concerns and then at the end of the decade by funding from the Carbon Dioxide Program associated with global warming concerns.

his snowblitz theory.[384] The controversy soon spread beyond *The Guardian* to other papers and forums. Hubert Lamb stepped in, attempting to moderate, but, in the end, he was only tainted by association.[385,386]

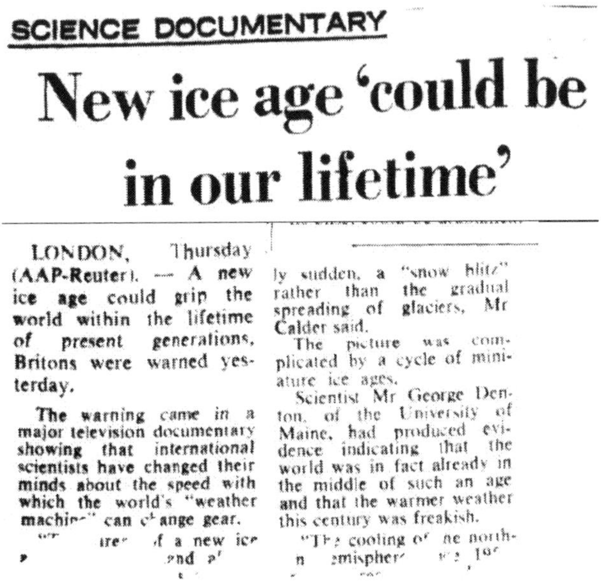

SCIENCE DOCUMENTARY

New ice age 'could be in our lifetime'

LONDON, Thursday (AAP-Reuter). — A new ice age could grip the world within the lifetime of present generations, Britons were warned yesterday.

The warning came in a major television documentary showing that international scientists have changed their minds about the speed with which the world's "weather machine" can change gear.

...re... f a new ice ...nd a'

ly sudden, a "snow blitz" rather than the gradual spreading of glaciers, Mr Calder said.

The picture was complicated by a cycle of miniature ice ages.

Scientist Mr George Denton, of the University of Maine, had produced evidence indicating that the world was in fact already in the middle of such an age and that the warmer weather this century was freakish.

"The cooling of the northern hemisphere...

Figure 10.6: Calder's ice age scare.

Scary syndicated articles reporting on Calder's ice age theory appeared in Australian newspapers prior to any screening of *The Weather Machine* in Australia (*Canberra Times*, 22 Nov 1974).

For the documentary, Calder had interviewed Lamb and other experts in the UK, across the channel and in the USA. Their work and their views were used to support Calder's conclusions. However, for many of them, their own forecasts of cooling did not extend beyond the possible return of the Little Ice Age. This is mostly true for Reid Bryson, but his promotion of alarm was outstanding. While 'ice sheets in our time' was not a big part of his story, throughout the 1970s he was perhaps the most widely quoted meteorologist forecasting an imminent cooling catastrophe. More extreme again were the views of the geologists—Kukla, Dansgaard and Emiliani—and they would be the only scientists supporting something close to Calder's version of the icy apocalypse. An impression of more widespread support among scientists often came through the press, especially where reporters were careless with the distinction between ice ages, big and little. In fact, on the whole, any contribution from meteorologists that was not cautionary was in direct opposition.

It was in accord with this opposition, and despite all the press interest, that European governments failed to respond to the cooling scare with any positive action. In the USA, where the government response was stronger, meteorologists also stepped in to dampen alarm. But an effect of the scare generally was to bring climatic change into the public discourse and to force meteorologists to engage with it. While this engagement was mostly to calm the excitement, meteorologists also tried to leverage the interest in climate that it had aroused.

Among meteorologists, claims of high resolution in climate proxies never gained broad acceptance, and so the supposed discovery in the deep past of rapid changes across centuries (even across decades) remained in dispute.[**] Such claims were fundamental to the instability argument that was behind the predictions of rapid changes. Therefore, the standard response remained pretty much as it had been before the scare:

> Even if the phase of the longer-period changes is such as to contribute to a cooling of the present-day climate, the contribution of such fluctuations to the rate of change of present-day climate would seem to be swamped by the much larger contribution of short-period (if more ephemeral) historical fluctuations.[388]

This quote is from the 1975 US GARP report entitled *Understanding Climate Change*.[††] It gives much the same response to cooling alarm that Mitchell had delivered at Brown University back in 1972, but this time in a widely circulated booklet promoting a program of climatic research. Other sceptical assessments soon followed the US GARP report. One of these, by an Australian Academy of Science committee, was requested after delegates returning from the World Food Conference described the anxiety over climatic change pervading the discussions. This expert panel (including one geologist) not only attempted to allay alarm but also to explain its genesis: it found that 'the more popular accounts', including Calder's, tend to give greater currency than is warranted to 'over-simplified and extreme points of view'.[389]

[**] Such claims remain in dispute today. New proxy data on abrupt changes do indicate rapid shifts *into* the interglacial warm periods, at least on the available scales, which remain no finer than multiple centuries. An unusual reversal of the warming into the Holocene—the so called 'Younger Dryas'—was also rapid at such scales. However, it is the rate and variability of temperature decline *out of* interglacial warm periods that is more relevant to the ice age scare. Variability across decades (or years) still cannot be conclusively determined. On the crude scales of the proxies, past transitions into glaciation do show more variability than the transition into warming, but the rates are moderate and the overall trend gradual across the millennia.[387]

[††] See p. 140.

At the international level, the Executive Committee of WMO afforded little attention to the scare. Previously, and long before the cooling alarm took hold, its Commission for Climatology had taken up an interest in climatic change, but the Executive Committee only got involved after goading from the US government. Its initial response to an urgent call from the USA was to allay cooling concerns. Yet, in doing so, it set in train a series of events that would raise climatic change high on its agenda. Before the decade was out, the WMO would launch its World Climate Programme at the first ever World Climate Conference, at which carbon dioxide emissions would emerge as a key concern among meteorologists for the first time. Thus greenhouse warming may never have attained such a grand platform if not for the US government first promoting interest in climate by raising concerns about global cooling.

11 The WMO response to cooling alarm

The American intervention came in April 1974, not yet a year into the oil crisis and only a few months before Nixon resigned. Addressing a special session of the UN General Assembly, Nixon's secretary of State, Henry Kissinger, called for the world community to come together to solve common problems.

> ...to translate the acknowledgement of our common destiny into a commitment to common action, to inspire developed and developing nations alike to perceive and pursue their national interests by contributing to the global interest.[390]

The United States proposed developing a 'global agenda', with special attention given to the problems of industrially underdeveloped countries, including hunger, unemployment, excessive population growth and the need for greater energy production. Science featured prominently in Kissinger's speech, introduced in its Cold War guise, a power of both darkness and light. Science should be applied 'to the problems which science has helped to create', Kissinger argued. However, to the list of adversities manmade, he added another, which had not been caused by science, industry or otherwise by mankind:

> The poorest nations, already beset by man-made disasters, have been threatened by a natural one: the possibility of climatic changes in the monsoon belt and perhaps throughout the world. The implications for global food and population policies are ominous.[391]

To address the natural climatic change threat, the US proposed 'that the International Council of Scientific Unions and the WMO urgently investigate this problem and offer guidelines for immediate international action'.[392]

It was at this time that Robert White was coordinating efforts to convert the prevailing cooling anxieties into a proposal for a US national program of climatic research.* Now, Kissinger's call for global action was his cue to try to get a similar response from the WMO. As the head of the US delegation, White formally brought the proposal to the attention of the Secretary General, the British meteorologist D. Arthur Davies, so that 'it might be considered in relation to the on-going and planned activities of WMO'. White suggested that 'potentially, the resources of many parts of the United Nations system will be involved' in developing…

> …a better understanding of the meteorological aspects of climate change, together with an appreciation of the impact of such change on the well-being of the world…†

Davies and the WMO Executive Committee would be quick to respond to this request. However, US anxieties over natural changes did sit rather awkwardly within the committee's 'on-going and planned activities'.

The emerging interest in climate

We will remember that back in the 1960s the WMO Commission for Climatology had taken up an interest in natural climatic change, publishing in 1966 a report on statistical methods of detection.‡ In that same year, it had asked the Executive Committee to encourage national meteorological services to distribute any existing climatic data and to begin collecting more.

Around this time the Committee was also launching GARP, the impetus for which had been another pronouncement from another US president.§ GARP had a preoccupation with weather forecasting, and its climatic research would advance slowly, but interest in climate was sufficient for the Executive Committee to promote the issue at the preparatory meetings for the Stockholm 'Human Environment' conference. This effort succeeded to the extent that the conference upheld a recommendation suggesting that climatic research could be extended beyond GARP: after promoting GARP, 'Recommendation 79' also flags

---

*The push for national and international research programs was not the only reason that 1974 marks a turning point in the US policy responses to climatic change. On p. 133, we saw that 1974 was also the year that the ERDA was being established under the guidance of the nuclear scientist Alvin Weinberg, while at the same time Weinberg was beginning his campaign for research into the climatic impacts of fossil fuel production. Later that year there were also the negotiations with the Soviets to control the military use of environmental modifications, which marked a decline of interest in deliberate climatic modification.

†The full text of the letter can be seen in the US archives.[393] This and other extracts are found in WMO records.[394]

‡See p. 150.

§See p. 140.

the establishment of 'new programs to understand better the general circulation of the atmosphere and the causes of climatic changes whether these causes
are natural or the result of man's activities'.[395] With such references to *manmade* climatic change, we should again remember that around this time they
were often referring as much to deliberate as to inadvertent effects. However,
the Executive Committee had already taken an interest in the latter through an
initiative to promote the monitoring of global 'background' pollution. It was in
this context that further discussions of climatic change would be taken up.

The Executive Committee's interest in globally dispersed pollutants began
in the late 1960s when it established two expert panels to coordinate and direct WMO activities concerning the monitoring of 'background' pollution in
the atmosphere and in the oceans. Already by 1970, the atmospheric panel was
giving priority 'to monitoring those constituents which would most affect long-
term changes in the climate' included carbon dioxide and aerosols.[396] After the
Stockholm conference, interest in global pollution intensified and the committee responded by replacing the two pollution panels with a single expert panel
to better coordinate all activities relating to 'environmental pollution'. Special
interest in climate-active atmospheric constituents continued under this new
panel.

This work was the only context in which the Executive Committee was giving climate change any consideration in its 'on-going and planned activities'
when the American request arrived in 1974. This is why the US request for
the WMO to address the threat of *natural* climatic change was considered by
its Executive Committee under the heading of 'Environmental Pollution' (see
Figure 11.1). This incongruity marks the beginning of a jostling for position
between those more interested in promoting research into the physics of climate generally, and those keen to investigate specifically the inadvertent effects
of industrial pollution. This would continue for the next few decades.

The American request had arrived just weeks before the 1974 annual session
of the WMO Executive Committee. To support the discussion at this meeting,
Davies prepared a brief response, in which he talked up the WMO's various
climatic activities, particularly those under GARP, where the WMO and the
ICSU were already collaborating in the coordination of climatic research. But
Davies also acknowledged that this activity was insufficient to match the level
of concern that had arisen, and so his paper concludes by recommending they
convene an 'Expert Panel on Climatic Change' to oversee, coordinate and plan
future investigations of climatic change and its impacts. Not only was this recommendation upheld, but Davies was also directed to follow Kissinger's advice
and seek collaboration with other UN bodies.

When the Seventh World Meteorological Congress convened the following

WORLD METEOROLOGICAL ORGANIZATION

EXECUTIVE COMMITTEE

TWENTY-SIXTH SESSION, GENEVA, 1974

Distr.: RESTRICTED

EC-XXVI/Doc. 70
(23.V.1974)

ITEM 5.6 (3)

Original: ENGLISH

ENVIRONMENTAL POLLUTION AND OTHER ENVIRONMENTAL QUESTIONS

Implications of possible climatic changes

(Presented by the Secretary-General)

Summary

This document conveys to the Executive Committee a request from the Government of the United States of America to consider the problem of the implications of possible climatic changes on the well-being of man. The present WMO activities in this field are reviewed and it is suggested that the Committee may wish to establish a Panel of Experts as the focal point within WMO on the subject of climatic changes.

References:
1. Resolution 7 (CAS-VI) - Working Group on Problems of Climatic Fluctuation

2. Resolution 15 (CoSAMC-VI) - Working Group on Climatic Fluctuations and Man

3. EC-XXVI/Doc. 14, Add. 1 - Report of the president of the Commission for Agricultural Meteorology

4. EC-XXVI/Doc. 51 - United Nations Environment Programme

5. EC-XXVI/Doc. 66 - WMO drought project.

Figure 11.1: The WMO is asked to investigate climatic change.

In this document, the Secretary-General D Arthur Davies presented the Executive Committee of the WMO with the US request to investigate possible natural climatic change and its human impacts. As the heading shows, the request was considered under the only agenda item where climatic change was already being addressed: changes due to industrial 'pollution'. The recommended response to the US request in this document would begin a chain of events at the WMO that led to the investigations of natural climate systems in the World Climate Programme and, separately, to assessments of the influence of industrial carbon dioxide by the IPCC.

year, it seemed that not only the Americans, but also the other heads of national meteorological services had been affected by the cooling scare.[‖] Congress re-solved that the WMO should 'take the lead' in promoting and participating in climatic change studies. In doing so, it should seek collaboration especially with UNEP and the ICSU. Many delegations were keen to see the WMO providing a voice of authority in the public controversy and so Congress asked the Committee to prepare and issue 'authoritative statements' on the threat. The first of these should be issued within a year.[397] The Executive Committee responded by adding the preparations for such a statement to the briefing for its Expert Panel on Climatic Change.[¶,398]

In 1976 the Expert Panel on Climatic Change duly returned a brief assessment, which was then used by the Executive Committee as the basis for its first 'authoritative statement'. Issued on Midsummer's Eve, its preamble opens by explaining its motivation:

> Several controversial statements on climatic change have been issued in recent years by various bodies and individuals, and some governments have expressed concern about the grave implications, for global food and population policies, of possible climatic changes.[399]

Curiously, neither this preamble nor the following 'WMO Statement on Climatic Change' so much as mention an ice age. However, to the contemporary reader, there could be no doubt that the 'grave implications' to which it refers are mostly the consequence of global cooling. And the ice age threat is also suggested when the statement explains how extreme weather has...

> ...led to speculation that a major climatic change is occurring on a global scale which could involve a transition to one or another of the vastly different climates of past ages.[400]

At the time there could be no doubt that the climatic change speculation to which it referred was a rapid and rocky 'transition' out of the warm Holocene. In response to this speculation, the WMO statement goes on to concur with the US GARP** and other similar statements formulated by meteorologists, saying that 'such a change is likely to be so gradual as to be almost imperceptible'.

‖ Every four years, the directors of national meteorological services of the member states and territories of the WMO convene an extended meeting, a 'Congress', to discuss a range of topics and to elect the Executive Council of the WMO.

¶ At this stage, the panel included Hermann Flohn, J Murray Mitchell and Bert Bolin and was under the chairmanship of the Australian, WJ Gibbs.

** See p. 140.

But then, as well as allaying cooling fears, the statement makes another move that would become significant over the next few years. In placing the emphasis on short-term variations for their great impacts on civilization, it discusses 'the possible change of climate resulting from Man's activities', and says that human influences are 'at least of equal concern' as natural changes. The leading candidates are then mentioned, namely carbon dioxide *warming*, direct *heating* and aerosols, although, with the latter, there is no mention of its proposed *cooling* effect.[401]

In short then, the main message of the WMO Climate Statement is to put the emphasis on short-term changes over long-term changes. But the secondary message is to discredit the coolers and favour the warmers. The coolers are found to be mistaken because of a confusion of time scales—in other words, *the geologists have over-reached themselves*. The warmers are promoted when, among the short-term changes discussed, emphasis is given to manmade effects, especially manmade warming. Now, we should remember that, at the time, many meteorologists remained altogether sceptical of threatening natural climatic change, apart from random fluctuations on various scales, and many more remained sceptical of any global human influence. The coolers and warmers were at the extremes. But the effect of the approach taken by the statement would become clear by the end of the decade: it bolstered the warmers.[402]

In London, the *Times* certainly saw it this way. Its interpretation of the statement was supported by quotations of Oliver Ashford, the WMO's planning director. Under the headline 'World's temperature likely to rise', the newspa-

World's temperature likely to rise

From Our Correspondent
Geneva, June 21

A warning that significant rises in global temperatures are probable over the next century has been issued here by the World Meteorological Organization (WMO).

This would be the consequence of a build-up of atmospheric carbon dioxide—which has already risen by 10 per cent in the past 50 years—because of increased use of oil and coal fuels.

In com⁷ years this carbon diox⁷' aas⁷ ⁷⁷⁷ e⁷⁷e⁷

ded also as a considered reply to reports last month predicting catastrophic changes in climate produced by a return to the Little Ice Age, a period of temperatures one or two degrees centigrade lower than today, which lasted from about AD 1550 to 1850 and was part of the neo-glacial cycle.

WMO agrees that this is suggested by knowledge of past natural climatic changes, but emphasizes that such an asessment is completely in⁷alidated b⁷ the additi⁷⁷ ⁷c⁷⁷ on ⁷d⁷ ⁷

Mr Oliver Ashford, the WMO planning director, said the potential increase in consumption of fossil fuels could mean a rise of several hundred per cent in carbon dioxide in the next 100 years or so, "according to our present understanding."

One possible result of the raised temperature would be a reduction in Arctic sea ice. "In some parts of the world", he continued, "the effects would be very beneficial—for example higher agricultur⁷' ⁷duc⁷⁷ ⁷xtre⁷⁷⁷

Figure 11.2: World's temperature likely to rise.

The *Times* of 22 June 1976 was exceptional in drawing attention to the WMO Climate Statement in a report that amplifies the subtle shift away from the prospect of dangerous natural cooling towards dangerous pollution-driven warming.

per claimed that the WMO had come out against those 'predicting catastrophic changes in climate produced by a return to the Little Ice Age'. According to the WMO, it said, such assessments based on historical climatology are 'completely invalidated by the addition of carbon dioxide to the atmosphere and other effects of human activities'.[403] Not two years after *The Weather Machine*, this distortion of the WMO statement appears less an attack on the extreme views of Calder and the Quaternary geologists and more about directing criticism towards historical climatology, especially towards historical climatologists who were more moderate in their cooling concerns. However, this *Times* report is exceptional, and not only for providing an interpretation of the WMO statement as a thinly veiled jab at Hubert Lamb. It is also exceptional in drawing attention to an announcement that otherwise received little press coverage.

In the late 1970s, natural cooling alarm continued to capture the attention of the press and the public, but interest among meteorologists was shifting in the other direction, towards manmade warming. This contrary view would often appear in the press, but rarely (as with this *Times* report) was it the lead story. Mostly it would be a foil, or the 'balance' sought by the reporter from his expert contacts. When contacted, meteorologists would often oppose both sides, as would John Mason, and some occasionally actively sought publicity to pacify the debate.

One of these was J Murray Mitchell. Unlike Mason, who kept his distance, Mitchell was open to the scientific and public discussion on all sides. He was ready with his considered view at all sorts of gatherings, from Earth Day teach-ins to Soviet-bloc conferences. Yet his moderating voice is easy to miss in those historical accounts that tend to emphasise the extremes. Soon enough, such moderate voices would be drowned out in the rising excitement, but not quite yet. In 1976, a story by Walter Sullivan in the *New York Times* appeared under the headline 'Two climate experts decry predictions of disasters'. It opens:

> Two authorities on climate change have termed 'irresponsible' recent predictions of an impending disaster. They also said that any global effects of air pollution on the climate to date remained obscure.[404]

The two authorities were Mitchell and the historical climatologist, Helmut Landsberg. Through Sullivan's reporting, they delivered many reasons not to extrapolate too much from short-term trends or weather extremes, noting that various pollutants of the atmosphere could have all sorts of effects. Extreme effects are possible, but instead of raising fear about the improbable, Mitchell preferred to talk about the more likely outcome, which is that 'long-range climate problems will be ordinary and manageable, rather than extraordinary and unmanageable'. Mitchell was also quoted observing that 'the media are having a lot of fun' with the debate between warmers and coolers:

> Whenever there is a cold wave, they seek out a proponent of the ice-age-is-coming school and put his theories on page one. Whenever there is a heat wave they turn to his opposite number, [someone willing, for example, to predict] a kind of heat death of the Earth.[405]

Actually, it would have been hard for a reporter to find foreboding of warming so apocalyptic as Mitchell suggests, but, if we allow him a certain rhetorical latitude, his point is well made: even a small amount of press 'fun' with a weather story tends to exaggerate the scientific controversy between warmers and coolers.

The other side to this story is that attempts to calm the public excitement—by the likes of Mason, Mitchell and Lamb in their various ways, and by the WMO—were in conflict with the general interest of the meteorological community in attracting research funding. To some extent, both games could be played at the same time. The WMO's response to Kissinger is a case in point. Kissinger's call was firstly to 'urgently investigate' natural climatic change, and secondly to 'offer guidelines for immediate international action'. The WMO's answer to the first request in its climate statement might be interpreted as answering the second: as there is no cooling emergency, *no immediate international action is required*. Instead, the Executive Committee would go on to leverage the interest generated by the cooling scare for the promotion of international action of one particular kind, namely a massive expansion of climatic research.

Support for such a response was also arriving from elsewhere. The growing awareness of the enormous social and economic costs of weather extremes brought with it demands for the weather experts to provide further guidance about what is going on. These demands were coming from the wealthy north, but also directly from the poorer south. UNEP was well positioned to hear the voice of sub-Saharan Africa, having set up headquarters in Nairobi. At its 1977 conference on desertification,[††] discussion continued on the impacts of overgrazing, but also on the possibility of natural climatic shifts. One of the conference recommendations addressed the 'evident' and 'critical role' of climatic change 'in most desertification processes'. It identified...

> ...the need for improving understanding of the causes of climate change and development of improved methods of climate prediction.[406]

[††] See p.167.

But what could be done? The conference could only plead for more research. However, in doing so, it endorsed the plans, only just announced by the WMO, for a coordinated global climatic research program.

The WMO Executive Committee had taken up another of the recommendations of its Expert Panel on Climatic Change, which was to prepare an international conference on climatic change. The Expert Panel was reconvened and it formally proposed what the Americans had been lobbying for, which was a programmatic international response to the threat of climatic change: a 'World Climate Programme'. The conference—now called 'The World Climate Conference'—was soon seen as preparatory to this program, and it was set for early 1979. But even this response would not be sufficient. The Executive Committee also felt duty bound to alert policymakers to any climatic concerns arising in the meantime by asking the Expert Panel to prepare and issue further authoritative statements on climatic change (both natural and manmade). They were also to prepare for a possible ministerial conference to follow the experts' one, 'and to define possible courses of action for [its] consideration...'[407]

Although the WMO Executive Committee was now active on the climate issue and using the public excitement to promote interest in climatic research, it would be a mistake to think that it was ever very much concerned with natural cooling. The cooling alarm that had come through at the Seventh Congress—from whence came the directive for a climate statement—was never much in evidence at the Executive Committee. Instead, their interest was shifting to the carbon dioxide issue. At their 1977 session, the discussion of carbon dioxide was somewhat separate from the general discussion of climatic change and the preparations for the World Climate Programme. It resulted in a separate resolution calling on the Secretary-General 'to formulate a detailed plan of action for future work on the problem of atmospheric CO_2'.[408] Then, the following year, where there was discussion of a climate threat, it was all about the carbon dioxide problem. In that discussion, the Committee was told its Commission on Atmospheric Science was already active on the issue, having set up a 'Working Group on Atmospheric Carbon Dioxide'. The Committee asked this group to draft a statement for its consideration, including the anticipated consequences of various carbon dioxide emissions scenarios. The Committee also decided that following the World Climate Conference they would sponsor another conference specifically addressing the carbon cycle. These moves towards interest in manmade warming at the 1978 Executive Committee (which followed the other initiatives in the USA)[‡‡] provide a small hint of what was to come.

In February 1979, at the first ever World Climate Conference, meteorolo-

---

[‡‡] See pp 162–165.

gists would for the first time raise a chorus of warming concern. These meteorologists were not only Americans. Expert interest in the carbon dioxide threat had arisen during the late 1970s in Western Europe and Russia as well. However, there seemed to be nothing in particular that had triggered this interest. There was no new evidence of particular note. Nor was there any global warming to speak of. Global mean temperatures remained subdued, while in 1978 another severe winter descended over vast regions of North America. The policy environment also remained unsympathetic. Energy supply and global cooling remained the overriding concerns, with heating and transport demand keeping pressure on energy supplies, a pressure that was exacerbated when the Iranian Revolution triggered the second oil crisis, which would unfold as a backdrop to the World Climate Conference.

One factor behind climatologists' shift of focus may have been the US Department of Energy's Carbon Dioxide Program. As described in Chapter 10, this support for warming research arrived in the late 1970s against a background of continuing funding cuts in many other programs. Early in 1978, DoE had decided to double that year's climate research budget of $1.5 million for the US 1979 fiscal year. By the time of the World Climate Conference, hundreds of scientists had a career interest in the topic through involvement in dozens of new research projects that were already underway in the USA and abroad. However, the direct influence of those scientists who were heavily committed to these programs was less obvious at that conference than it would be in later years. Whatever caused this early peak of excitement at the World Climate Conference, it was soon successfully subdued, and it would take some considerable effort throughout the early 1980s for it to be revived.

12 The greenhouse warming scare begins

Call for action at the World Climate Conference

The World Climate Conference was a grand affair, bringing together 450 experts from 60 countries and involving the WMO in collaboration with a number of other UN agencies: principally UNEP, but also the FAO, UNESCO and the WHO.

Background presentations by specialists on the agricultural, social and economic impacts of climatic change reflected this broader involvement, but on the prospects for future climate it was very much a meteorologists' affair. The geologists who were instrumental in bringing global attention to climatic change—Kukla, Matthews, Emiliani and Dansgaard—did not attend. One of the few geologists in attendance was IP Gerasimov from the USSR. His background paper on 'past geological epochs' was very much in line with Mitchell's paper at Brown University in 1972. While arguing that carbon dioxide warming would override any natural long-term cooling trends, he neglected even to mention the work of the geologists claiming otherwise. Mostly, the gathered meteorologists concurred on the warming prognosis, and, by the end of the conference, a new authoritative statement would be issued. The 'Declaration of the World Climate Conference' would make clear which influence on the immediate climatic future should dominate the discussion from then on.[409]

The first step towards this outcome was to emphasise the human influence. Following the opening ceremony, in his keynote address, the chairman Robert White dwelt upon the possible disastrous consequences of our inadvertent interference.[410] Then, throughout the first week, the overview papers on various topics mostly came around to human influences, in particular those due to energy production. In these papers, and in the ensuing discussion, concerns about warming due to carbon dioxide came to dominate concerns about other anthropogenic factors. There were calls by speakers to further the urgent study of this topic, while the contributions of direct heating and aerosol cooling were downplayed. For the second week of the conference, it was planned that selected experts should deliberate in four workshops on four components of the World Climate Programme: climate data, application of knowledge, impacts of climate on human activities, and research into climatic change and variability. However, before these deliberations began, 'the conference noted an additional issue of special importance' and a change was agreed. This 'special issue' was the possible human influence, and it cut across all four components. Thus all the workshops in the second week were asked to give this topic particular consideration.[411]

At the end of the second week, the Conference Declaration was prepared, discussed and agreed. The section on 'Climate and the future' drew attention to the possible human influences, including direct heating from energy production, from nuclear war, and from deliberate weather modification, but the emphasis was placed on the threat of carbon dioxide emissions. Emissions warming 'appears plausible', it said, because carbon dioxide 'plays a fundamental role' in determining temperature. While this effect is not yet detectable, it should be noticeable 'by the end of the century'.[412]

Most prominent in the Declaration is its opening 'Appeal to Nations'. This says that it is now urgently necessary for all nations to apply and improve knowledge of climate, but it is also urgently necessary…

> …to foresee and prevent potential man-made changes in climate that might be adverse to the well-being of humanity.[413]

The appeal to *prevent* foreseen climatic change is profoundly significant. This was the first time that the counsel of any assembled group of meteorologists had included a call for policy action on manmade changes. Moreover, this came as the headline to a declaration agreed at the greatest gathering of climate experts that had ever been assembled on the planet.

Exactly which human influence is of greatest concern is made clear in the Declaration's supporting documentation. This consists of statements from the four workshops and an additional statement on the human influence. This last document tells how the human influence could have a profound effect on mankind in the decades to come, and so there is 'a special sense of urgency' for international research on carbon dioxide accumulation, a subject that 'merits immediate attention'.[414] The statement concludes that, of the several forms of human impact, the impact of carbon dioxide accumulation 'deserves most urgent attention of the world community of nations', and it calls for 'accelerated' research 'on various aspects of the CO_2 problem'.[415]

186

The call for action on carbon dioxide is resisted

With these strong words, the World Climate Conference may have drowned out cooling alarm, but it did not exactly set the warming scare on fire; this unambiguous messaging on the urgency of the carbon dioxide problem attracted only moderate press interest, and even where it did, the meteorological leadership tended to temper alarm. Consider the conference convenor, Robert White. Already during the opening discussions of the first week, the carbon dioxide issue had come to the fore. It was clearly newsworthy. Thus, it is not surprising that White would be found at the midway press conference promoting the urgency of research in this field. Yet he downplayed the urgency of any policy response. It is way too early for that. After all, the impacts, good or bad, are not yet determined.

> We're not sure if the carbon dioxide-induced warming will be beneficial or deleterious. It could be beneficial, on a net basis, by extending the growing season and enabling us to grow more food to feed the increased population of the next century. If it is decided that its impact will be one of benefit, then maybe we just let things go the way they have been going. But if we determine that the impact will be deleterious, then we will have to do something.[416]

This sort of ambivalence did not exactly help the reporters build screaming headlines and, accordingly, public and government interest remained subdued.

Subdued also was the interest of the leadership of national metrological services. One might have expected that all the excitement about manmade climatic change whipped up at the World Climate Conference might have spilled over to Eighth World Meteorological Congress, which followed soon after. Congress might have encouraged public involvement in the climate debate just as they had four years earlier. However, while busy finalising, approving and funding the World Climate Programme, it decided that a ministerial conference on climatic change would be 'premature', and no special attention was given to manmade climatic change.[417] The structure of the program remained substantially as it was before the World Climate Conference. There was no special accommodation for the matter that the world expert review had found deserving of 'most urgent attention' and 'accelerated research'.

The topic could not be avoided at the following Executive Committee session. We will recall that the WMO Commission for Atmospheric Science had already been investigating the grounds for concern, and its carbon dioxide working group had been asked to prepare a report. The report was presented to the committee as a 'draft WMO statement'. Its sobering conclusion was that

much doubt, uncertainty and disagreement continued to plague the field. It recommended intensifying 'efforts to reduce uncertainties', and it urged the main
World Climate Programme partners—the WMO, UNEP and the ICSU—to col-
laborate on further study. After some discussion, the committee decided that no
WMO statement should be issued. Instead, the assessment was annexed to their
meeting report and duly forgotten.[418]

And so was set the pattern for the early 1980s. While the leadership of UNEP
became bullish on the issue, the bear prevailed at the WMO. The responses to
pressure from UNEP to cooperate on the planning and implementation of a
program of activities would be hesitant and lacklustre, with moves to avoid or
suppress the topic and especially to avoid stirring up public excitement. This
attitude to the carbon dioxide question would prevail not only in the WMO's
governing committee, but also in the leadership of the secretariat. The same
Congress that launched the World Climate Programme had also elected a new
Secretary General, the Danish meteorologist Aksel Wiin-Nielsen.* Soon after
arriving at the WMO's Geneva headquarters, he would be telling the 'research'
arm of the World Climate Programme that the carbon dioxide question was one
best left alone.

The component of the World Climate Program established to coordinate
research into 'climatic change and variability' came to be called the World Climate *Research* Programme (WCRP). The plan was for WCRP to evolve out of
the continuing atmospheric research activities under GARP. The transition towards emphasis on climatic matters would be coordinated by the WMO and the
ICSU through a 'Joint Scientific Committee' (which we will hereafter refer to as
the 'WCRP committee'). In fact, this coordinating committee for WCRP was
only a continuation of a previous GARP coordinating committee, and it would
remain under the shared governance of the WMO and the ICSU.

The WCRP committee first met under the new banner in early 1980, with
the new WMO Secretary General in attendance, almost exactly a year after the

---

*Wiin-Nielsen came to the WMO after serving as founding director of the European Centre
for Medium-range Weather Forecasts, which had been established in Bracknell, England, across
town from the UK Met Office. Wiin-Nielsen's doubts about the science behind the alarm over
carbon dioxide emissions eventually led to a clash with the IPCC leadership. His concerns about
the IPCC's Second Assessment were raised in the Danish press during the lead-up to CoP3 in Kyoto. He was even quoted as suggesting that the IPCC should be dissolved. A public rebuke from
the founding chairman of the IPCC, Bert Bolin, marked a rift in a close relationship between the
two Scandinavian meteorologists that had begun when Bolin supervised Wiin-Nielsen's PhD thesis. There is a good summary of Wiin-Nielsen's concerns with the science in one of his papers.[419]
Later, Bolin reflected that he could not understand Wiin-Nielsen's opposition to the IPCC effort.
He had made 'a well-recognised scientific contribution in the past', yet this critique was 'simplistic'
and not supported by 'adequate data'.[420]

World Climate Conference. That conference's overriding concern with manmade global warming was addressed in a paper presented by Secretary General Wiin-Nielsen 'concerning the international framework for effort on the carbon dioxide problem'. It opens by acknowledging the 'widespread publicity' that the topic has received, and also the proliferation of bodies organised to address it. 'Some of these developments', he says, 'have come about in a rather haphazard manner which carries with it the risk of duplication, over-organisation and confusion'. In his view, WCRP providing another level of coordination is no solution.

> Several national programmes which are highly relevant are already under way and it is believed that, in accordance with the principle of minimum interference being advocated here, there is no need for some scientific body to plan and attempt to organise a programme of research into the CO_2 problem.[421]

In other words, the WCRP should not address the carbon dioxide question. This was a big call given that the WCRP brief included investigating the climatic effects of human activities. It was an even bigger call given the importance of this item had been underlined by the World Climate Conference's appeal to all nations to 'foresee and prevent' potential manmade climatic change as *an urgent necessity*. However, it was not that the Secretary General was saying the WMO should ignore the issue entirely; in fact, he *did* see the need for ongoing assessments of the situation. To this end, he recommended the institution of a separate standing board of international experts to inform policymakers on the carbon dioxide question by undertaking assessments and issuing further authoritative statements.

The WCRP committee fully agreed with Wiin-Nielsen. It supported the recommendation for an independent international assessment board, and it undertook to leave the carbon dioxide problem alone.[422] This latter position the committee would maintain for the next five years, even as they came under increasing pressure to relent. Whenever challenged, they came back with the argument that the WCRP could best serve carbon dioxide research by continuing its general investigations of the climate system. They argued that the great gaps in general climatic knowledge mean that there are a great many uncertainties about the carbon dioxide effect. Thus WCRP's 'most effective contribution to the CO_2 question is to continue its fundamental attack on the basic processes of the climate system'.[423] While this view was in complete accord with the view expressed in the Commission for Atmospheric Science assessment report, it was disruptive to the organisation of the World Climate Programme. Most significantly, it brought the WCRP leadership into conflict with the major UN partner in the World Climate Programme that had been excluded from involvement in its research component: UNEP.

UNEP always had a special interest in the topic of pollution-driven climatic change. It was not just that its original mandate—to implement the recommendation of the Stockholm conference—included (albeit in a very long list) the possible establishment of new programs to address manmade climatic change; climate was also a particular interest of its founding executive director, Maurice Strong. In his address to the opening session of UNEP's Governing Council in 1973, Strong announced his intention to engage the world scientific community to discover the 'outer limits' of manmade changes, beyond which adverse impacts would be set in train. He gave specific examples of the kinds of human influences that needed to be monitored: pollution affecting the oceans and atmospheric ozone levels, and also pollution that might cause direct atmospheric warming, in particular carbon dioxide.[424] The following year, the second session of the Governing Council affirmed these 'outer limits' investigations and mentioned in particular biological tolerance and climatic change.[425] The new executive director, Mustafa Tolba, took up the specific issue of carbon dioxide following the strong recommendations to do so from the World Climate Conference. Soon after that conference, the Governing Council asked Tolba to develop an action plan for carbon dioxide, including reviews and research, and to do this in collaboration with its World Climate Programme partners, the WMO and the ICSU.[426]

This was not the only reason UNEP would badger the WMO on the carbon dioxide issue. Another reason relates to its assignment of responsibility for one of the four components of the World Climate Programme: climatic change impacts on human activities. This fitted nicely with its ongoing involvement with climatic change impacts, especially desertification. However, its study of potential future impacts would require outside support. In order to anticipate the impacts on agriculture and other human activities, it needed some idea of the climatic changes that might be in store. Predicting natural climate trends was difficult enough, but it was this component of the World Climate Programme that had been most affected by the emergence of the carbon dioxide issue at the World Climate Conference. According to the view that the influence of carbon dioxide might soon override natural changes, the emphasis of the program's global impact studies would be 'on the impact of CO_2-induced changes'.[427] The trouble was that while carbon dioxide causation might be *global*, the climate impacts needed to be studied *regionally* so that the regional environmental and social impacts could be assessed. Thus to begin to address human impacts of emissions-driven climatic change, various future emissions scenarios needed modelling, and the modelling needed to provide climatic forecasts at a regional level. The plan was that such scenario forecasts would be provided by WCRP. When UNEP's request for this scenario modelling duly arrived with the WCRP committee, they baulked at the idea: computer modelling remained too primitive and, especially at the regional level, no meaningful results could be obtained. Proceeding with the development of climate scenarios would only risk the development of misleading impact assessments. The committee did at least agree to meet with UNEP representatives, but only to ensure that they avoided 'unwarranted conclusions'.[428] Thus a stalemate developed: the World Climate Programme's impact studies component could not proceed without reliable regional climate forecasts, but WCRP, the research program assigned to make the forecasts, said that they were not yet possible. This led UNEP Executive Director Mustafa Tolba to put pressure on the meteorological researchers through WCRP and the WMO, badgering them to advance the modelling and empirical science of the carbon dioxide effect.[†]

WCRP's resistance to Tolba's pressure to fast track carbon dioxide research was supported by the WMO Executive Committee. It agreed with Wiin-Nielsen and the WCRP committee that 'there was no need for additional organization of research on CO_2 at this time'. This resistance does not seem to have been about money or the prioritisation of the research program's scare resources. On the contrary, UNEP always seemed to come up with generous support for any joint activities relating to environmental pollution. Indeed, it had become an important benefactor of WMO activities. It was precisely for this reason that the WMO Executive Council 'stressed the importance of this collaboration'.[‡] Nevertheless, through the early 1980s they remained obstinate in the face of increasing UNEP enthusiasm for progress on the carbon dioxide issue.[430]

While this conflict over research planning and coordination was never entirely resolved, there was general agreement that further assessments were required. Proposals abounded. Wiin-Nielsen's standing board of assessment[§] was never exactly realised, but various *ad-hoc* assessments would be arranged, all of which were actually, or effectively, separate from WCRP. The first of these

---

[†]The personnel on one side of this controversy are of interest. During the transition from GARP, WCRP's Joint Scientific Committee was chaired by a climate modeller, Joseph Smagorinsky. As founding director of the climate modelling program instigated by von Neumann at the Geophysical Fluid Dynamics Laboratory, at Princeton University, Smagorinsky was sceptical of using models as predictive tools. His vice-chairman was John Houghton of the UK, who took up the chair in 1982, after which the committee's position on the role of climate modelling remained unchanged.

[‡]At its 1984 session 'the Executive Council stressed the importance of this collaboration' with UNEP, 'which had resulted since 1974 in an amount close to US $5.25 million approved funds by UNEP'.[429]

[§]See p. 189.

was initiated independently by UNEP in its attempt to get things moving on the issue. Convened as a collaboration of the three main partners in the World Climate Programme, it would be presented as a World Climate Programme activity, even though it was not conceived as part of that organisation's work, and remained entirely outside its four-component structure.

At the invitation of its Austrian delegation, the meeting was held in the alpine town of Villach in the autumn of 1980.[431] This first-ever international conference to assess the climatic impacts of carbon dioxide was attended by representatives of the convening partners as well as just eleven country representatives, mostly from those countries that were already undertaking research on the topic. After a week of workshops, this small group produced an assessment like so many others that had gone before, highlighting all the gaps in the science. The report explains that the great many uncertainties make 'premature' any development of 'a management plan for control of CO_2 levels in the atmosphere or of the consequent impacts on society.'[432]

For those wishing to maintain the momentum of the World Climate Conference, the Villach meeting was an all-round failure: it did not come up with anything very new or interesting; it called for further investigations but found no mechanism for progressing them and it failed to draw any attention to the carbon dioxide question. Moreover, its view on human impacts delivered a dismal prospect for the World Climate Programme's impact studies even getting off first base. However, for those on the other side of the debate, it could be seen as a successful assessment that confirmed that studies of impacts and possible policy responses would be premature while there was so much uncertainty about the climate system and its response to external influences generally.

This is how the WMO Executive Committee viewed the results from Villach. After a discussion of the brief Villach conference report, the Executive Committee reiterated its conclusion that 'more research in this field was necessary'. Indeed, it went further, cautioning that care should be taken about raising alarm on the issue:

> In connexion with the general matter of public pronouncements on the CO_2 issue, the Committee re-emphasized that conflicting opinions based on inadequate study could prove confusing to decision-makers and urged continued caution in making public statements.[433]

These concerns were reiterated and re-emphasized by the WMO Executive Committee when carbon dioxide was raised again in 1982.[434] This explains why the committee issued no more 'authoritative statements' on (manmade) climatic change. It also helps explain the committee's continued support for WCRP in avoiding the carbon dioxide question.

Their support remained strong until after another meeting in Villach; a meeting that would change everything. Just before that second Villach conference in 1985, the Executive *Council* (note the name change) agreed with the WCRP leadership...

> ...that resolving the present uncertainties on global and regional climate impacts would require considerable progress in most, if not all, aspects of the WCRP and agreed that the current scientific plans laid out for the implementation of that programme constituted the best approach for further refinements of the quantitative assessment of the climatic impacts of greenhouse gases, including carbon dioxide.[435]

Thus, we can see that at least until 1985, the body representing the heads of national meteorological services agreed with the scientists coordinating its global research program that the development of climate science was inadequate to a proper assessment of the carbon dioxide threat. However, by 1985 this was not the view of the UNEP Executive Director. By then, Tolba was already convinced that the science had established a level of threat so high that it required an urgent and drastic globally coordinated policy response. If the meteorologists were subdued, then not so the temperature record. At last, during the early 1980s, Nature gave some clear signals that it was coming out on the side of the warmers.

The late 20th century warming trend begins

In the early 1980s it started to become clear that the four-decade general cooling trend was over. Weather station records in the northern mid-latitudes began again to show an upward trend, which was traceable back to a turnaround during the 1970s.[||]

James Hansen was early in announcing this shift, and in doing so he also excited a foreboding of manmade warming. In a 1981 *Science* article, he had ex-

---

[||] This turnaround had been forecast regionally before a global trend was entirely evident. We will recall Hubert Lamb at the UK's Climatic Research Unit in 1975 starting to predict the continuation of milder winters, at least for Britain. Also note John Mason at the Met Office quieting the cooling scare late in the winter of 1976 with this:

> There is no real basis for the alarming predictions of an imminent ice age, which have been largely based on extrapolation of the 30-year trend of falling temperatures in the northern hemisphere between 1940 and 1965. Apart from the strong dubiety of making a forecast from such a short-period trend, there is now evidence that the trend has been arrested.[436]

plained the importance of this trend reversal to the promotion of global warming:

> The major difficulty in accepting this theory has been the absence of observed warming coincident with the historic CO_2 increase.[437]

Right back to Revelle in 1965,¶ warmers had been embarrassed by the observation that the great bulk of the industrial emissions supposedly causing a general warming had occurred during an extended period when there was no such warming. Now Hansen's article not only showed that a warming trend had returned to the well-monitored north, he also brought into question the extent and depth of (what could now be seen as) the mid-century cooling. *Perhaps this dip was not so deep and not so global.* Better data for the southern hemisphere told a different story, of a more consistent and more gradual warming right across the last 100 years (See Figure 12.1). Moreover, the deeper cooling in the north could be explained by high levels of volcanic dust and (the global effect of) solar variability. When all of this was taken into account, Hansen found that the global mean temperature trend since the end of the 19th century came into line with what the greenhouse modelling had been indicating all along.

Hansen could not quite declare that carbon dioxide emissions had *caused* the 20th century warming. There was still a chance that it might only be natural variability. This natural 'noise' meant that the carbon dioxide 'signal' would probably not become clear until the end of the century (a projection he shared with many others at the time). He also had some other reservations. One was that the expected warming effect was based on models that 'do not yet accurately simulate many parts of the climate system, especially the ocean, clouds, polar sea ice, and ice sheets'. Indeed, it was a similar assessment of the models by the WCRP committee that caused it to recommend against their use in 'emissions scenario' forecasting of the climatic impacts.

For Hansen, these limitations did not stop him from contributing to the speculation on the environmental impacts of projected runaway warming. Yet, even here, he had doubts. It had been known for a long time that slight moderations in the extreme cold over and around ice caps could bring *more* snowfall and thus contribute to their growth. By the early 1980s this was widely acknowledged by the warmers. Because of this, Hansen explained, 'it is not certain whether CO_2 warming will cause the ice sheets to shrink or grow'.[439] If warming did cause them to grow then they would no longer threaten human settlement with rising seas. Not that Hansen or any others ever much discussed this possibility. As doubts arose about this cause of sea level rises, there was a

---

¶See p. 138.

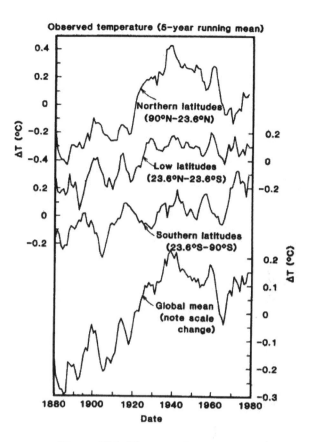

Figure 12.1: The warming returns.

In 1981, James Hansen published a set of graphs that showed the lack of a mid-century cooling trend in southern mid-latitudes and a return to global warming from the late 1960s.[438]

shift towards explaining the same outcome by a different cause: the previously neglected contribution of the oceans' thermal expansion.

Besides, there was a much grander diluvian story that continued to gain currency: the semi-submerged West Antarctic ice sheet might detach and slide into the sea. This was for some an irresistible image of terrible beauty: displacement on a monumental scale, humanity unintentionally applying the lever of industrial emissions to cast off this inconceivably large body of ice. As if imagining some giant icy Archimedes slowly settling into his overflowing bath, Hansen calculated the consequential displacement to give a sea-level rise of 5 or 6 metres within a century. At the low end of this estimate, the USA would lose '25

Study Finds Warming Trend
That Could Raise Sea Levels

By WALTER SULLIVAN

A team of Federal scientists says it has detected an overall warming trend in the earth's atmosphere extending back to the year 1880. They regard this as evidence of the validity of the "greenhouse" effect, in which increasing amounts of carbon dioxide cause steady temperature increases.

The seven at~ spheric scientists predict a gl~ .in~ ~~' .ost un~

precedented magnitude" in the next century. It might even be sufficient to melt and dislodge the ice cover of West Antarctica, they say, eventually leading to a worldwide rise of 15 to 20 feet in the sea level. In that case, they say, it would "flood 25 percent of Louisiana and Florida, 10 percent of New Jersey and mar other lowlan~, througho~' .e worl ~i~~'n~~ ~ry or'~~

Figure 12.2: Hansen wins his first *New York Times* front page on 22 August 1981.

percent of Louisiana and Florida, 10 percent of New Jersey'. Droughts were also threatening, but talk of the seas rising up to swallow lands and cities soon proved the greatest attractor of press interest in this new climatic change story.[440]

That the advance of the waves was about to replace the advance of the ice is demonstrated by the reception for Hansen's paper. His re-evaluation of the southern hemisphere temperature trend and his claim of a global temperature turnaround were scientifically interesting, but it was his extrapolation towards sea-level rises of around 15 to 20 feet that won him a front page headline at the *New York Times*.[441] And we should remember that this time Walter Sullivan's story was not the result of some telephone conversation about some vague back-of-the-envelope calculations circulated to nobody; this was peer-reviewed and published science. Moreover, it had the imprimatur of the American Association for the Advancement of Science; the AAAS journal, *Science,* was esteemed in the USA above all others. Thus we can forgive Sullivan his credulity of this string of claims: that the new discovery of 'clear evidence' shows that emissions have 'already warmed the climate', that this supports a prediction of warming in the next century of 'almost unprecedented magnitude', and that this warming might be sufficient to 'melt and dislodge the ice cover of West Antarctica'. The cooling scare was barely in the grave, but the warmers had been rehearsing in the wings. Now their most daring member jumped out and stole the show.

Hansen had taken a reputational risk with his speculations, but his success in drawing interest to the science had come at a critical time in his career. As an expert in the atmosphere of Venus, he had taken up a position at NASA's

196

Institute for Space Studies, a New York satellite of the Goddard Space Flight Center in Maryland. It had been created in 1961 to accommodate Robert Jastrow, its founding director. Jastrow happened to be as sceptical as anyone about the whole warming scare. When in 1981 he resigned, NASA tried to close down the operations at this New York satellite. A few staff resisted, and it was after Hansen won his front-page story that NASA cut the funding. But then the EPA stepped in. Soon enough, NASA funding was revived, Hansen fully shifted his interest from Venus to Earth, and global warming research thrived in New York under his leadership. His 1981 paper and Sullivan's story would set the trend for climate research. Interest in global warming continued to rise as the warming trend continued. And it did continue; slowly through the 1980s the global mean temperature line crept closer and closer to the level of its prewar peak.

Meanwhile, a second attempt was underway to kick-start the global warming movement at the UN.

13 UNEP and the push for a climate treaty

The failure of the 1980 Villach conference was most disappointing for its chairman, the Swedish meteorologist, Bert Bolin. A leading figure in atmospheric research, especially through his involvement in establishing GARP, Bolin had been investigating the carbon dioxide cycle since the late 1950s. His view was that the Villach group had insufficient time to complete a proper assessment, which meant that its report could not be very penetrating. On the train ride home from the conference, Bolin discussed his misgivings with some of the participants. Later he recalled how he raised the possibility of a more substantial analysis with a wider scope, greater depth and involving the world's leading scientists.[442]

The idea of a more considered assessment had also gained currency elsewhere, including at the WMO. However, it was Mustafa Tolba who again got things moving. Two years after that first Villach meeting, the Executive Director of UNEP approached Bolin, telling him of his plans for a new assessment and asking him to again take the lead.

The 1985 'SCOPE 29' assessment

Bolin remembers that it was Midsummer's Day, auspicious for meteorologists, but doubly so for one in Scandinavia: a day full of light, a night full of day, the Sun encircling Stockholm, city on top of the world. That day Bolin was invited to Tolba's hotel to discuss Tolba's plan for a comprehensive international assessment of the carbon dioxide problem. The UNEP Executive Director had secured funding and was asking Bolin to take up the project. This was a venture Bolin had long considered necessary, and he would now be charged with its realisation. He would select the other editors, commission chapters and organise peer-review while working out of the International Meteorological Institute at the University of Stockholm.[443]

For the most part, this would be a UNEP project, but it was soon acknowledged that other formal institutional associations should be maintained. In his memoirs, Bolin admits that engaging the WMO was an afterthought. Their involvement 'soon became obvious', he reflects, because they were, after all, the UN agency responsible 'for meteorology and hydrology'. It might have been thought that 'climate' was also within their remit. However, the pretence of their involvement would be useful for another reason: to maintain associations with the other partners in the World Climate Programme. Involvement of the other main sponsor of the World Climate Programme, the ICSU, would also overcome a problem that then plagued all UN agencies, and a problem that is easily discounted in the internet age, namely the problem of report distribution. Commercial publication improved the chances of a report ending up in bookshops and libraries and therefore making an impact. The ICSU could provide this guarantee through their 'SCOPE' series.[444]

The Scientific Committee on Problems of the Environment (SCOPE) began with the ICSU's contribution to the Stockholm 'Human Environment' conference. Since the early 1970s, SCOPE has produced a highly respected series of environmental assessment reports, each published by John Wiley & Sons. Previously Bolin had led a review of research into the carbon cycle that was published by SCOPE in 1979.[445] When the SCOPE committee was approached with Tolba's proposal this time around, they were preparing a detailed report on the sulphur cycle. This was based on papers from a 1979 conference that was evidently motivated by the cooling scare, a motivation reaffirmed by the SCOPE chairman in a foreword that harks back to those former times. Such a study 'has become especially urgent', he explained, due to substantial 'man-made disturbances' of the sulphur cycle at a time when our demand for 'food, fuel, and fibre is increasing more and more rapidly'.[446] Now, in agreeing to publish Bolin's report, SCOPE adopted the new vogue: an opposing manmade disturbance, a warming, the modelling for which gave scant regard to the cooling effect of sulphur emissions.

SCOPE 29: The Greenhouse Effect, Climatic Change and Ecosystems would be the first comprehensive international assessment of that subject. As its title suggests, it extended the assessment to environmental (and so human) impacts, both negative and positive. One positive impact given considerable attention is that upon agriculture. Studies of the direct stimulus to plant growth of higher concentrations of atmospheric carbon dioxide had shown that crop yields would increase more or less depending on the crop, but by as much as 50% with a concentration doubling. The most surprising finding of the experiments discussed in the report was that substantial yield increases would come even under stress caused by other limiting factors, such as insufficient nutrients and drought.[447]

As for the main concern of the report, namely the climatic effect and its impacts, there was nothing particularly new or surprising. It concludes that at current trends, a doubling of carbon dioxide concentrations should be expected around the end of the 21st century. By that time, it estimates a general warming will have reached somewhere in the range of 1.5–5.5 °C. This extraordinarily broad range contains all but the lowest estimates published since the earliest calculations late in the 19th century.[448] At its high end, the rate of warming would be the cause of great alarm, but the report places the emphasis on the plausibility of the lower half of the range (i.e. 1.5–3.5°C). As for the environmental impacts, regionally these cannot yet be established. For sea level, at the low end, the estimated general rise would be barely noticeable above natural variability or above the natural background rising trend. At the (less plausible) extreme, levels might rise 1.65 metres, mostly due to thermal expansion. The contribution of ice-cap melt is completely discounted. As for the scare that had won global warming its first *New York Times* front page, the collapse of the West Antarctic icesheet is nothing to worry about according to this expert review.[449]

Of course, these speculations, are all based on the modelling of the climatic effect. Consideration of how this effect might become evident is addressed in the 'Empirical studies' section, which Bolin had commissioned from the UK's Climatic Research Unit, now under its second director, Tom Wigley. Wigley's assessment of the state of the empirical science is based on investigations that CRU and others were conducting, mostly under the DoE Carbon Dioxide Program.

These 'detection' studies were a field beset by obstacles. Wigley reported that confirmation of the carbon dioxide effect on climate was going to be difficult and is not yet even nearly possible when applying current techniques of analysis to currently available data. The problems begin with identifying any natural variability that otherwise might be confused with the predicted manmade effect. The first step towards distinguishing any natural variability in the recent (and future) global trend is to find it in the past. To do this, historical data is required at timescales relevant to that of the expected manmade effect. In other words, the resolution would need to be fine enough to show changes across decades, or at least across centuries. Wigley did find, as Lamb had, that 'many warming and cooling episodes occurred on the 100-year timescale'. The trouble was that the quality of the data did not permit extrapolation to a *global* trend.

> Because these events are only observable through indirect or proxy evidence, which is both local and often poorly dated, we do not know how global mean climate changed, or even *if* global mean temperature has changed since, say, 6,000 [years ago].[450]

The doubts Wigley raised about using proxy data to find past global trends were much the same as those raised with the 'coolers' at Brown University by Mitchell all those years ago.* The crudeness of their resolution was not the only reason for the proxy data being inadequate for the purpose of speculation about the future global trend. Other difficulties arise when attempting to extrapolate a *global* trend from local proxies of many different types found across the land, sea, and ice surfaces of the globe. Lamb had used documentary evidence as well as climate proxies to show that the recent trend varies greatly across the European continent. But then the US GARP report of 1975 had taken a regional segment of these data and used it to suggest a generalised trend. Other reports followed suit, some of these obscuring the locational origin of the original data.† This was no misrepresentation that Wigley was prepared to maintain. He was sceptical enough about Lamb's charting of local trends (he had looked at the data collection and was not impressed[454]) so he was not going to have any locational proxy charting writ large in some pretence to telling the global story. No temperature trend charts appeared in the SCOPE report.

There is another emerging convention from which Wigley retreats in his section of the SCOPE report: forecasting when detection of the carbon dioxide effect might be achieved. By the early 1980s, as 'detection' studies took off, it was often speculated that the emissions 'signal' would be detectable above natural 'noise' by around the end of the century. Indeed, in a 1981 *Nature* article, Wigley had proffered this very prediction. But since then his doubts about the ability to distinguish 'signal' from 'noise' seem to have increased. Given that the best efforts of historical climatologists had failed to pick up any amplitude or frequency characteristics of the natural 'noise', it is not surprising that in the SCOPE 29 report he offers no prospect of when the 'signal' might be identified.

In all, the SCOPE 29 assessment found that any alarm over carbon dioxide could only be based on the vague and varied predictions of modelling, which continued to present many shortcomings. This three-year study had delivered a sobering assessment of the science behind the scary speculations that were already attracting press coverage. Yet, despite this awkward handicap, a conference called to address the implications of its conclusions came to mark the beginning of the movement for global climate action. This conference would be

---

*See p. 160.
†A version of the 1975 US GARP graph is presented in Figure 10.1. This distortion of Lamb's original findings continued right through to the first IPCC Assessment. Figure 7.1c of the Working Group 1 report,[451] which shows a pronounced hump of medieval warming, is loosely derived from another millennial temperature trend graph specific to Central England.[452] For the full story on the use of these graphs, see the author's blog post[453]

convened in Villach again, five years on from the last climate conference there. The second Villach conference would succeed on all counts where the first had failed. This was due in no small part to the enthusiasm that arrived in the Austrian Alps with the appearance of Mustafa Tolba.

Tolba at the 1985 Villach conference

Tolba came to the second Villach conference in October 1985 primed and ready. Early that year in Vienna he had successfully negotiated the Framework Convention for the Protection of the Ozone Layer, and he was already openly advocating the commencement of negotiations for a similar climate treaty.[455] In his memoirs, he enumerates some of the lessons learned while leading the world to one of the greatest multilateral treaty achievements of all time. One important contribution to the ozone success was that the UNEP representatives had chosen to abandon 'the traditional mediator's role of noncommittal neutrality'. Instead, Tolba and his colleagues 'actively sought solutions'.[456] Opening the new treaty campaign at Villach, his speech was strident and rousing by all accounts, and without so much as a hint of noncommittal neutrality (emphasis added).

> ...we have now *laid aside most of the doubts* as to the effect of the build-up of CO_2, and other trace gases on global climate. [The SCOPE 29] assessment has *confirmed* that there is *almost unanimous agreement* that the global average surface temperature would increase in response to a doubling of the greenhouse effect. Differences in the amount of increase [in the modelling] are modest, in fact *insignificant for our current purposes*. It is clear now that scientists are *reasonably confident* that at current rates of build-up a global mean temperature increase of several degrees will probably occur over a period of half a century or so.[457]

With the SCOPE 29 assessment in the bag, Tolba is telling the gathered scientists they have already delivered the basis for policy action. All they need do now is clearly enunciate the policy implications so that their findings can direct policymakers. Many would later feel that Tolba had leapt over all the vagaries and uncertainties elaborated by the scientists in the SCOPE 29 report and elsewhere; his speech at least suggests that he had exaggerated scientists' confidence and dismissed their doubts. But to leave it at that would be to overlook much of what was going on.

If we wish to understand the events at the science-policy interface, we need to pay more attention to the treatment of doubt by Tolba and the other policy advocates who would soon come into play. For Tolba, commitment to actively seeking solutions does not require the dismissal of substantial uncertainty about the reality of the problem being solved. On the contrary, Tolba would highlight uncertainty, just as he did later in this speech: 'the picture is clouded with uncertainty', he said. The level of uncertainty in the science behind the warming scare was nothing new to him, nor was it for the avoiding. Rather, he would muse over doubt, recast it and use it for rhetorical advantage, before, as in the passage quoted above, dropping it in a flourish of unbridled confidence and an imperative to action.

To understand Tolba is to understand that there is no incongruity, no disgrace, but, instead, evident pride, in moving forward in the face of formidable doubt. We should remember Richard Benedick, the US chief negotiator, describing the state of ozone science at the start of the talks that would result in the Vienna Convention. It was all speculative projections, contradictory modelling, no observed depletion and no evidence that harm would arise if the depletion began.[‡] This is not far from Tolba's own assessment of a threat 'still distant' and an issue 'shrouded in scientific uncertainties'. Back then, in 1982, Tolba told the gathered delegates 'we cannot even say with certainty that the ozone layer is being depleted'. He was almost boasting when he said that they are dealing with 'a problem that has yet to be proven conclusively', a problem compounded by an 'imperfect understanding' of natural variation. All of this and more comes in a speech proudly reprinted in his memoirs, a speech in which he is imploring the country delegates that 'we have to act now' to solve a problem that 'will not be experienced until well into the future, perhaps beyond our own lifetimes'. The Doubting Thomas in his story is the public. Tolba explains in his memoirs how the public required the immediacy and 'scientific certainty' of the Antarctic ozone hole before it would come onside; but not he and not delegates like Benedick, although some other delegates for industrialised countries did use 'the lack of scientific evidence' for ozone depletion as a 'pretext' to tactically avoid legally-binding controls. It helps if such confirmation eventually arrives—as it later did in the shape of the Antarctic ozone hole and all the frightening discoveries announced soon after—but for Tolba, confirmation was a luxury unavailable to those forging that first framework agreement in Vienna.[458] Doubt has a completely different appearance to these heroes on the other side of the science-policy divide. To say that the policy advocates did not falter in the face of crippling doubts is only to amplify their achievement of the Vienna Convention.

So then we find Tolba in Villach, a few months after Vienna, and he is telling the gathered scientists that 'we have laid aside most of the doubts'. In the spirit of Vienna, this is to say: *we have laid aside the reservations and fears that might*

[‡]See p. 97.

paralyse the timid. Consider, even if the danger were to arrive tomorrow, even if it were here today, we would not be able to recognise it; that is basically what Wigley had reported in the assessment that Tolba had requested, that he had paid for. Yet, to this way of thinking, Wigley provided no reason for hesitation in confronting the hidden danger. Not that Tolba would dwell on this or other aspects of the assessment that might undermine confidence in the reality of the hidden danger; better instead to concentrate on speculation about the terrifying consequences of inaction if it were real. When, at Villach, Tolba explained that the West Antarctic icesheet could 'break away from its continental moorings' and slip into the sea, it might have been that he simply could not resist the spectacular imagery. As it happened, this risk was specifically dismissed in the SCOPE 29 report, and this dismissal was emphasised in the executive summary,[459] and would be repeated again in the brief consensus statement of the meeting he was addressing.[460] The contradiction scarcely mattered, and the fact that it scarcely mattered is revealing about the policy movement.

At Villach, at the beginning of the climate treaty movement, when the science was supposed to be informing and directing the policy, the rhetoric of the policy movement was already breaking away from its moorings in the science. Typically, in this rhetoric, the uncertainties are re-envisaged as the challenges of the heroic policy mission. As such, these uncertainties will no longer be about whether danger lies ahead, but how much, when, and in what form. Tolba's clouds of uncertainty are not about whether carbon dioxide would cause dangerous runaway warming, but exactly how much warming, how fast, and with what precise environmental and social impacts. This recasting of the uncertainties disarms the doubter of the whole show: the sceptic. Recast the doubts in this way, and they can even be used in rhetorical offensives against the sceptic.[§] Thus doubts raised over the wildest speculation can be turned around, in a rhetoric of precautionary action: *we should act anyway, just in case*. After all, no one could say *for certain* that the West Antarctic icesheet would *not* break its moorings. With the onus of proof reversed, the research can continue while the question remains (ever so slightly) open. This approach operates on a more general level too: so long as no-one can say for certain that the climate system is robust, that our effects on it are puny, and that they will be damped by negative feedback, *the imperative for policy action remains*. After facing up to all these

---

[§] Uncertainty is treated in this form: objections to the scare are captured and recast as disagreements between the scared, then any ongoing objections to the scare are recast as irrational or immoral denial of the scare—attempts to pollute the debate between the scared—which should therefore be dismissed. Such treatment of uncertainty is exemplified by Mike Hulme, especially in his aptly titled, *Why we Disagree about Climate Change*, where all genuine disagreement is cast as only that between the scared.[461,462]

fearful possibilities and to our doubts about them, the final step is to lay aside the doubts, to redact the caveats about uncertainties and erase the broad lanes of error from the graph lines. This allays the fear that acting on our fears might be a frightful overreaction. Removing the caveats smooths the way for the policy decision, often with the excuse that policymakers require certainty for action (as if the political, economic and social basis for all their other policy decisions are grounded in unassailable certainties!). What policymakers might prefer is greater confidence about the level of danger than any panel of experts would be prepared to give. No doubt Tolba and other policy advocates advanced their counsel in the hope that the science would catch up, that it would all turn out right, just as it had with ozone; when confirmation arrived, the effect was detected and the doubters were silenced. Alas, it would not be so easy with global warming.

Tolba's opening address to the 1985 Villach conference concludes with a celebration of the momentous historical import of their scientific achievements. 'The greenhouse problem brings science to a new dawn', he proclaimed. This new dawn is nothing less than an inadvertent realisation of the old vision of climate modification. Tolba would not be the first, nor the last, to introduce the idea that to control emissions is to control the climate. In other words, *we have found a global thermostat.*‖ Thus, a climate-policy response could bring an empowering turnaround: in rising to meet the threat of self-destruction, we realise self-determination. Through discovering our destructive power, these scientists have also discovered our power for good. In taking up Kissinger's call to apply science 'to the problems which science has helped to create', these scientists have surpassed their mission, and in doing so they have transformed von Neumann's bleak prospect of technological self-annihilation. In Tolba's vision, von Neumann's climatic-control *Übermensch* —complete with its 'forms of climatic warfare as yet unimagined'—has morphed into the American Superman, who uses his powers, in spite of his doubts, to impede destruction and act for the good of all. Tolba wraps it up thus:

In spite of all uncertainties, it remains clear that the world's climate is now

---

‖Mitigation discussions presupposing climate control include the calculations in the EPA report two years earlier (see p. 230) and those by the economist William Nordhaus in the mid-1970s.[463] The 1977 Foreword to the NAS Energy and Climate report (see p. 141) finishes with the musing that we might soon address the question: 'What should the atmospheric carbon dioxide content be over the next century or two to achieve an optimum global climate?'[464] The idea of climate control by intervening in the carbon cycle arrived soon after the carbon dioxide ice age theory was first proposed. In 1901 the Swede Nils Ekholm speculated that one day we might control carbon dioxide levels to 'regulate the future climate of the earth and consequently prevent the arrival of a new Ice Age'.[465]

subject to the intervention of human beings. Debate—I believe—must focus on how best to handle this intervention... If we have the power to alter the climate, why shouldn't we harness that power for the good of humankind? It is an exciting prospect.[466]

The Villach consensus

This exciting prospectus for greenhouse science arrived with the considerable institutional weight afforded the head of UNEP. By comparison, the principal editor of the assessment, Bert Bolin, carried less weight, but, anyway, his speech avoided both the rhetoric of disaster and the rhetoric of triumph. Emphasising the work required to reduce the uncertainties, he made no call for policy action.[467] If this were a letdown after Tolba, then more so would have been the address by Donald Smith, the WMO Deputy Secretary-General. In agreement with Bolin, he found that the main challenge ahead would be for scientists to identify the nature of 'the most important uncertainties' and 'to find objective ways to deal with them'. For Smith, this attention to the uncertainties is important because only when the uncertainties in the current evaluation are sufficiently narrow will there be the prospect of global consensus on social and economic action.[468] In other words, the science behind the scare is not yet sufficiently developed as the basis for a policy push. Place Smith's reluctance besides Tolba's enthusiasm, and one might think that the WMO would again drag its heels and again end up playing second fiddle to UNEP. Not this time. This time around, the WMO Executive Council would take a leading role from the beginning, mostly thanks to the initiative of one Canadian delegate.

We have seen how the US government and the Seventh World Meteorological Congress had to push the WMO Executive Council into the debate over global cooling. We have also seen how the Executive Council actively resisted taking up the issue of carbon dioxide warming, despite an unambiguous call for action at the World Climate Conference. Earlier, we also saw how UNEP took over the lead on the ozone issue. By the mid-1980s a minority view had developed within the Executive Council that the WMO had been too slow to act on ozone and that it should now take the lead on carbon dioxide.[469] One of those who saw it this way was the director of the Canadian Atmospheric Environment Service, Jim Bruce. As an enthusiast for action on global warming, he was largely responsible for WMO leadership on the issue, including through his chairing of the Villach conference.¶ Bruce's own opening address at Villach

¶Following the Villach 1985 meeting, Bruce would retire from his position as director of Canada's meteorological service and take up employment in the WMO Secretariat in Geneva, where, in 1986, he became acting Deputy Secretary-General of the WMO. In that role, he would

closed on a personal note:

> I feel a real sense of urgency on this issue—it is not a matter for decisions in the next century but for major decisions in the next few years.[470]

Thus, it was Tolba and Bruce who were ready to move towards policy action, and at Villach they were encouraging the scientists to help pave the way.

The gathering of scientists at Villach this second time around was much bigger than the first, and it was of a very different character. This time 76 invited participants arrived from 29 countries (together with 13 representatives from the organising bodies), but most were not attending as representatives of their country. Instead, they had been invited because of their expertise in the relevant fields. This difference is often seen as significant because it freed the scientists to express their scientific views less constrained by national policy perspectives.** What is perhaps more significant is that it allowed attendance to be more representative of interest. One quarter of the experts at Villach were from the USA, mostly from institutions currently researching the question under DoE funding. They included one from Weinberg's Institute for Energy Analysis, another from the Oak Ridge Laboratory, two from the DoE, as well as Mike MacCracken from Lawrence Livermore Laboratory. Also invited were Wigley and two colleagues from UK's CRU, which since 1979 had been entirely dependent on funding from the US DoE Carbon Dioxide Program. Other participants included a number of European and Canadian scientists who would soon become active in promoting policy action.

Jim Bruce set the gathered experts two important tasks. The first was to develop 'a consensus statement' on the present state of knowledge about global warming, including the physical and socio-economic impacts. The second was nothing less than to use this consensus to develop policy-response recommendations for implementation by countries and international agencies.[474] While the scientists would complete the first task, they made only limited preparations for the second. Nevertheless, these would be the basis upon which other groups of scientists, policy experts and advocates would soon build a variety of policy options to meet the threat of greenhouse warming.

---

lead the WMO in negotiating the development of the UN process for addressing the global warming emergency.

** In the report of the 1984 WMO Executive Council it is explained that: 'Participants at the conference will be representatives of governments having active CO_2 programmes and individual scientists engaged in CO_2 research to be nominated by WMO/UNEP/ICSU'.[471] (36th EC P36) No particular distinction was made between each of these classes of scientists during the discussions, nor in the meeting report. On the importance of scientists operating in their 'personal capacity', see the PhD thesis by Hirst[472] and the paper by Franz.[473] However, this may be overstated as an explanation of a strong consensus statement that can be otherwise explained.

The original plan for the Villach 1985 conference was that the consensus on the science and the recommendations for policy action would be based on the report before them, namely SCOPE 29, which had been recently completed but not yet published. However, the report was barely even mentioned in the opening speeches. In fact, in the deliberations during the following week, the conference would come to conclusions markedly different from those found in the executive summary of SCOPE 29. This was partly due to the incorporation into their predicted warming rate of new calculations on the additional impact of greenhouse gases other than carbon dioxide. Where SCOPE 29 looked toward the time required for a doubling of the atmospheric concentration of *carbon dioxide*, at Villach the policy recommendation would be based on new calculations for the equivalent effect when *all emitted greenhouse gases* were taken into account. The impact of the new calculations was to greatly accelerate the rate of the predicted warming. According to SCOPE 29, on current rates of emissions, doubling of the *carbon dioxide* concentration would be expected in 2100. At Villach, the equivalent warming effect of *all greenhouse gases* was expected as early as 2030.

This new doubling date slipped under a psychological threshold: the potential lifetime of the younger scientists in the group. Subsequently, these computations were generally rejected and the agreed date for 'the equivalent of CO_2 doubling' was pushed out at least 20 years; indeed, never again would there be a doubling estimate so proximate with the time in which it was made.[††] However, at Villach, they were promoted by both Tolba and Smith in their opening addresses, and they seemed to have had the desired effect. Reports from participants say that the news of an accelerated climatic effect instilled a greater sense of urgency at the Villach workshops.[‡‡]

This goes some way to explaining the unprecedented strength of the concluding statement agreed by all the scientists. Not that the scientists were quite so strident as Tolba, but the consensus statement does display some impressive confidence. It concludes:

> Based on evidence of effects of past climatic changes, there is little doubt that a future change in climate of the order of magnitude obtained from climate models for a doubling of the atmospheric carbon dioxide concentration could have profound effects on global ecosystems, agriculture, water resources and sea ice.[478]

[††] This would be the first of many cases in the global warming treaty push where an extraordinary scientific claim was rushed into service before proper scientific scrutiny could assess and qualify it.

[‡‡] The new calculations were published in a paper by Ramanathan et al.[475] For the impact of this work on the conference attendees see Hecht[476] and Hirst.[477]

Like so many of the consensus statements from this time on, this one is twisted so that it gives the appearance of saying more than it actually does. In this way, those pushing for dramatic effect and those concerned not to overstate the case can come to agreement. In fact, this passage of the statement brings the case for alarm down to the reliability of the modelling, which is pretty much true of SCOPE 29. The only claim made with confidence—the 'little doubt'—is from the evidence of past climates. However, this does not sit well with SCOPE 29. We recall how Wigley had reported his failure to establish past global trends at the relevant scales. At Villach, Wigley had been elected to join the all-important 'Impact on Climate Change' working group. Its report sits much better with his 'Empirical studies' section of the SCOPE 29 assessment, except where it descends into a greater scepticism.

The brief report of the 1985 Villach 'Impact on Climate Change' working group does not even discuss the evidence of past climatic change. On the empirical science problem of detecting the signal in the actual climate record, there is only a vague reference to Wigley's entire 'Empirical studies' section of SCOPE 29, which we have seen is sceptical about the very possibility of such evidence ever supporting a detection claim.[479] Instead, the group's report expresses a great deal of concern about the reliability of the models. One particular concern was that some of the calculations of the average temperature change seem to give undue weight to the large changes expected at high latitudes. The report even goes so far as to say that 'model imperfections' are such that 'for the time being'—until they improve—the temperature range given in the SCOPE 29 report should be interpreted only as 'a measure for agreement among the different model approaches'. This is referring to the 1.5–5.5 °C range of warming for a doubling of carbon dioxide concentration given in SCOPE 29. Of this range the report says that it…

> …should not be interpreted as a prediction for the range of the eventual impact on climate.[480]

In other words, the Impact on Climate Change working group concluded that the models are not yet ready to make predictions (however vaguely) about the impact of greenhouse gas emissions on the global climate.[*]

When SCOPE 29 was published the following year, the Villach consensus statement would appear at its head, before the executive summary of the 520-

[*]The scepticism of this report is even more striking when we consider that this working group contained a mix of empirical climatologists and modellers. It even included Syukuro Manabe. Perhaps the most senior and respected climate modeller at the time, in 1983 he had succeeded the retiring Smagorinsky as leader of the modelling group at Princeton's Geophysical Fluid Dynamics Laboratory.

page peer-reviewed report. As for the Villach working group reports, these only appear at the back of a subsequent WMO publication, report number 661.[481] Thus, only in that obscure report from Villach can be found the incongruences between the Villach Impact on Climate Change working group findings and the Villach consensus statement. Much more accessible are the contradictions between the Villach consensus statement and the executive summary of Bolin's comprehensive assessment. These would be plain for all to read in the commercially published and widely distributed book. But even there, they were hardly noticed. No one saw fit to point them out, and they caused no palpable disruption to the campaign for policy action.

The silence of scientists on such gross contradictions is an early indication of an unwritten protocol for those who would come to work under the umbrella of the warming scare: they would not speak out publicly against contradictions, exaggerations and misrepresentations of the science while it was being interpreted and reinterpreted on its journey to the policy interface. The public airing of such criticism would come to define the boundary over which the climate change sceptic, or 'denier', has stepped. This boundary would have little to do with the level of scepticism expressed within the scientific discourse. As we have already seen, and as we shall see, many scientists working in the field would deliver results counter to the alarmist narrative, and even expressed profound scepticism about the science behind predictions of catastrophe. While exercising this freedom they would certainly not win the fame of those scientists like James Hansen who were prepared to raise alarm to levels only topped by the most extreme political activists. Yet, these scientists could continue to practice their science with relative impunity so long as they, and all those involved, were careful to ensure that any sceptical conclusions remained obscure within the scientific discourse, just as the scepticism expressed at the Villach conferences has remained obscure to this day.

In the published SCOPE 29 report, directly following the Villach consensus statement on the science (and before the SCOPE 29 executive summary), also appear the policy recommendations agreed by the scientists at Villach. Several of these ask for more monitoring and further assessments of impacts—just as so many assessments had before. This was not exactly what Bruce had in mind when he asked delegates for recommendations for government policy responses. But although specific policy responses are not elaborated, the 'Recommended actions' section of the statement does go on to say that these should now be developed. 'While major uncertainties remain', the statement concedes, 'nevertheless the understanding of the greenhouse question is sufficiently developed' to move forward with policy development:

...scientists and policymakers should begin an active collaboration to explore the effectiveness of alternative policies and adjustments.[482]

This is the first time since the World Climate Conference that a council of scientists had made this recommendation. While the World Climate Conference had appealed to nations that they should *foresee and prevent*, six years later the experts at Villach agreed that the science behind the *foreseeing* has been *sufficiently developed* for nations to proceed towards developing prevention strategies and other responses. The recommendation is for assessments of 'the widest possible range of social responses aimed at preventing or adapting to climate change', and that 'these assessments should be initiated immediately.'[483]

The final recommendation of the 1985 Villach conference is a call for the sponsoring organisations to establish a small taskforce, whose responsibilities would include periodic assessments and advice to governments on action. The final responsibility of this group would be...

...to initiate, if deemed necessary, consideration of a global convention.[484]

With this international scientific consensus in place, the movement for international policy action could now begin.

The Advisory Group on Greenhouse Gases

This final recommendation of the Villach 1985 conference was something that Tolba himself specifically requested when, in his opening address, he called for the creation of an 'international co-ordinating committee on greenhouse gases.'[485] The model for this committee was evidently the Co-ordinating Committee on the Ozone Layer, which was chaired by Robert Watson. This small group had been working closely with UNEP in providing assessments and expert advice to the governmental negotiators ever since negotiations began in 1977.

Following Villach, a similar group was established as the 'Advisory Group on Greenhouse Gases' (AGGG). Elected to the AGGG were just seven scientists, including Bert Bolin, with Jim Bruce serving as secretary. The chairman would be his fellow Canadian, Ken Hare, a climatologist from the University of Toronto. It was to this group that the Villach conference entrusted the responsibility for taking the greenhouse gas issue forward. Alas, the AGGG turned out to be a failure, mostly because it did not win the backing of governments, and in particular the USA. It also had few supporters among scientist–activists. Bert Bolin considered that such a small group would not be up to the task of engaging governments on such an important policy issue.[486,487] But anyway, no process of negotiations had even been established for it to advise.

A general realisation of the failure of the AGGG was already developing around the time of its first meeting in the summer of 1986, when Tolba triggered the negotiations that led to the formation of the IPCC. How that came about, we will see shortly. But while these negotiations were evolving, and while the IPCC was being brought into being, they were influenced by another series of events that saw the greenhouse gas issue suddenly breaking out onto the world stage. This would even lead to a rush of international meetings, some at ministerial level, where collective responses to the crisis were discussed and planned. These talks would threaten to pre-empt the process that was being mapped out through the establishment of the IPCC. Indeed, it would take some careful negotiations to rein in all this excited activity and bring the global response strategy back under the control of the two responsible UN bodies.

The background to all this excitement in the late 1980s is the building of global environmentalism's second wave.[488]

The influence of the sustainable development movement

We will recall that in the summer of 1986, the ozone protection campaign had emerged from its 'dark years', when publicity about the Antarctic ozone hole brought environmental advocacy groups back to the cause. However much Tolba would later celebrate the achievement of the Vienna Convention, the reality was that it only produced a *framework* for commitment. But from 1986, progress accelerated tremendously, so that the following year real commitment could arrive through the Montreal Protocol. This unplanned course of events soon became the model for the climate treaty process. Meanwhile, both processes were invigorated as they were swept up into the rising environment movement under the banner of 'sustainable development'.

The idea of 'sustainable development' had its beginnings even before the 'Human Environment' conference in Stockholm. This idea was to bring together the apparently conflicting goals of industrial development and environmental protection.[489] To some extent this was already the project of UNEP when it was formed to implement the resolutions agreed at Stockholm. However, the concept was given greater clarity and impetus late in 1983 when the UN General Assembly established a special commission to propose 'long term environmental strategies for achieving sustainable development by the year 2000 and beyond'. Elected to the chair was the former prime minister of Norway and leader of their Labour Party, Gro Brundtland. The activities of the 'Brundtland Commission', from 1984 to 1986, were notable for high-profile public hearings in impoverished regions, shining a spotlight on the environmental perils affecting communities underrepresented in the normal processes of intergovernmental policy development. Also brought into the conversation with an unprecedented level of legitimacy were environmental non-governmental organisations (NGOs). Brundtland submitted her report, *Our Common Future*,[490] in a blaze of publicity early in 1987.

For the sustainable development movement, greenhouse warming presents a classic case of apparent conflict, with energy production for economic development threating the global environment. It also has the future-orientation of the sustainability movement: the environment is to be protected, not just for the current inhabitants of the Earth but for generations yet unborn. Global warming is thus treated with some prominence in *Our Common Future*. In the energy section, it is discussed ahead of the nuclear issue. But its prominence should not be overstated; it was no headline issue, and it rated well below poverty, which was the primary concern. In promoting the report, Brundtland herself gave the greenhouse effect little emphasis.[491]

This is not to say that the Brundtland Commission played no part in the eventual triumph of the global warming scare: it had a tremendous influence, but this influence was otherwise. It came through the links that formed between people in the two camps, in particular those facilitated by Jim Bruce through his countryman Jim McNeill. While Bruce had moved directly from his role as the 1985 Villach conference chairman to become secretary of the AGGG, McNeill, as Secretary General of the Brundtland Commission, had taken on responsibility for drafting its report.[492] The affinity between the two groups first becomes apparent in the coverage of the carbon dioxide threat in *Our Common Future*, which relies heavily on the conclusions at Villach.

Following the release of the report, the Commission lacked a flagship cause; its great priority issue (poverty) was so large and so multifaceted that it had no clear course of resolution. What the movement required to sustain momentum after the publication of its report in 1987 was a discrete campaign issue that could lead to a new global call to action. After the triumph of the ozone treaty, there was no obvious candidate cause until, the following year, global warming turned up in the right place at the right time.[493]

And the influence worked in the other direction too. After the sustainable development movement had joined the global warming cause, its own goals would soon be taken up by global warming advocates and incorporated into the climate treaty negotiations. This was most especially to address poverty and to aid development by helping to reverse the net transfer of resources from poor to rich. The path to glory for the greenhouse cause began soon after the first meeting of the AGGG in July 1986.

214

The 1987 Villach–Bellagio policy workshops

One of the dissatisfied members of the AGGG was Gordon Goodman, a Welsh ecologist who was founding director of the Beijer Institute in Stockholm. Privately funded, yet under the auspices of the Royal Swedish Academy of Sciences, the Beijer Institute was established in 1977 to investigate issues around energy and environment, which, as we have seen, were much in vogue at the time. Following his participation at the 1985 Villach conference, Goodman began to apply his considerable entrepreneurial skills and the resources of the Beijer Institute to promoting policy action on climatic change. He won specific funding from the Rockefeller brothers for two workshops that would further assess the impacts of warming and develop policy responses. These would be convened late in 1987: one again in Villach and the other, one month later, south across the Alps in Bellagio, Italy.

The 1987 Villach–Bellagio workshops, 'Developing Policies for Responding to Climatic Change', were organised in direct response to the recommendation of the 1985 Villach conference that consideration of policy options 'should be initiated immediately'. They were nominally part of the process authorised at that conference, with a draft report of both workshops submitted to an AGGG meeting in December 1987, and the final version later published by the WMO under the banner of the World Climate Programme. However, that brand was thereby stretched to the limit of toleration. If the 1980 Villach conference, the SCOPE 29 assessment and its 1985 Villach review were all on the periphery of the organisational structure of World Climate Programme activities, then these 1987 workshops were surely over the line. A special note at the head of the published report does explain that it 'is not an official WMO publication'. What it does not explain is that the Villach–Bellagio workshops were undertaken without the guidance of any WMO, ICSU or UNEP committees, commissions or panels.[494] We should also note that neither were they under the guidance of any government. As such, they were an early example of global environmental policy entrepreneurialism, and so it pays to consider their organisation.

As the chief policy entrepreneur behind the workshops, Goodman first enlisted the help of Jill Jäger, who was also at the Beijer Institute and who had been one of the editors of SCOPE 29 under Bolin. He also brought in Michael Oppenheimer, an atmospheric scientist working for the US Environmental Defense Fund (EDF), the group that first won fame for its legal fight to ban DDT. One of EDF's founders, George Woodwell, was also involved, as the founding director of Woods Hole Research Center, a newly established and privately funded environmental research institute. None of these main organisers was employed by a government or held office in a UN body.[495,496] The Steering Committee of

215

the workshops included two other AGGG members: Bert Bolin and Ken Hare. There was also Howard Ferguson from the Canadian Atmospheric Environment Service, who had followed Bruce and McNeill into the theatre of international science policy.† Also notable on the Steering Committee was William Clark, who had previously worked at Weinberg's Institute for Energy Analysis, where he edited its *Carbon Dioxide Review, 1982.*

If we next consider those invited to participate in the policy deliberations, the organisers followed the Villach recommendation that scientists should 'begin an active collaboration' with policymakers.[497] They also came in line with the broader community engagement strategy of the Brundtland Commission. Scientists and technical experts were still dominant, especially in the first workshop where the focus was on assessing the environmental and social impacts of warming, but on the policy-development side there were many representatives from government agencies as well as environmental NGOs.

The strategy that seems to have underpinned the success of the Villach–Bellagio workshops was to first explicitly bed down the scientific foundation of the policy discussion. This foundation was established directly on the authority of the Villach 1985 'scientific consensus' about the threat.

> It is now generally agreed that if present trends continue 'in the first half of the next century a rise of global mean temperature could occur which is greater than any in man's history'.[498]

In the report of the workshops, this quotation from the Villach consensus statement is followed by other excerpts and paraphrases from the same document, after which it is explained that these 'conclusions of the 1985 Villach conference were used as a starting point' for the workshops. In this way the Villach–Bellagio workshops followed the stated intent of Tolba and Bruce at the start of the 1985 Villach conference, namely that the scientific consensus achieved at that meeting should provide a firm basis upon which policy could be developed.[499] In this way, questions and doubts about *whether* we are causing dangerous climatic change would not delay discussion of its specific environmental and social impacts, nor would such doubts hinder the development of specific policy responses.

The report of the workshops attests to the success of this strategy, with some significant policy proposals evident. In particular, it promotes two possible legal

---

†Canadians had a strong record of leadership in global environmentalism since the time of Maurice Strong, who had led the organisation of the Stockholm 'Human Environment' conference before becoming the founding executive director of the Environment Programme. By the late 1980s, Canadian interest was consolidating around this one specific issue.

mechanisms to mitigate global warming: a general 'law of the atmosphere' and a procedure for developing emissions targets.

A law of the atmosphere

The idea of a law of the atmosphere, or 'law of the air', is to develop a legal and regulatory framework to support the management of the atmosphere as a 'global common'. Drawing from the idea of the village common, an international common is conceived of as a part of the world that falls outside national jurisdictions and that is taken to be the common heritage of mankind. The first such common to come under consideration for an international agreement was the continent of Antarctica. The need for an agreement arose soon after the International Geophysical Year, when twelve participating countries established research stations on that continent; the ensuing Antarctic Treaty of 1959 rejected all claims of national sovereignty. Subsequently, the international commons idea was used in the Outer Space Treaty of 1967, in the failed Moon Treaty and in the protracted negotiations for a Law of the Sea, which began in the 1970s, but would continue throughout the 1980s.

With all of the interest in atmospheric pollution during the 1970s, it is not surprising that consideration turned to our common atmosphere and so to a 'law of the air'. Indeed, the possibility of just such an international agreement was the impetus for a conference convened by anthropologist Margaret Mead back in 1975. Mead's anthropology would later be discredited, but at this time she was riding high on its fame as president of the American Association for the Advancement of Science.

In her opening address to the 1975 conference, Mead explained that she had invited the gathered atmospheric specialists 'to consider how the very real threats to humankind and life on this planet can be stated with credibility and persuasiveness' so as to inform an anticipated international agreement like a 'law of the air'.[500] A high concentration of science-advocates were among the 35 participants, including Wally Broecker, Stephen Schneider, George Woodwell and William Kellogg, all of whom were soon active on global warming. While this conference is notable for an early emphasis on carbon dioxide, other pollutants considered potentially subject to a 'law of the air' included those causing ozone depletion, acid rain and desertification.

A decade later, all these pollutants and more came under consideration by the Brundtland Commission.[‡] Indeed, the tally of threats was so enlarged that

---

[‡] The 'global commons' theme runs right through the Brundtland report, which even extends it from dimensions of space into the dimension of time; in other words, a common *future*. The

any comprehensive 'law of the air' was starting to look decidedly complex beside the singular target of the ozone treaty. Nevertheless, in the section on climatic change in the Commission's report, it called for some form of agreement on management policies for all environmentally reactive chemicals. Specific responses to climatic change were only a fallback, and then only to adapt to it, not to stop it:

> If a convention on chemical containment policies cannot be implemented rapidly, governments should develop contingency strategies and plans for adaptation to climatic change.[501]

Turning now to the Villach–Bellagio report, we find that it promotes both a general and specific responses, and it even refers both proposals to the 'intergovernmental mechanism' that would soon be the IPCC. Such an organisation should…

> …examine the need for an agreement on a law of the atmosphere as a global commons, as was developed in the Law of the Sea, or the need to move towards a convention along the lines of that developed for ozone.[502]

A law of the air was never given much consideration by the IPCC and the proposal would soon fall by the wayside as too difficult to negotiate. However, one Villach–Bellagio recommendation for environmental regulation would be taken up.

Emissions targets

This was the idea of emissions targets. The process by which they could be established would begin by determining a tolerable *rate* of warming. Elsewhere and later, interest in targets was about determining a tolerable *upper limit* of warmth: a point beyond which the effect is 'dangerous'.§ But at the Villach–Bellagio workshops the first consideration was to find a rate of change that would be sufficiently slow to allow the environment and society to adapt. This was mostly about the adaptation of flora and fauna, and the adaptation of agriculture, although in the determination of tolerable warming, consideration was also given to a slight acceleration in the continuing natural rate of sea level rise.

---

three parts of *Our Common Future* are: 'Common concerns', 'Common challenges' and 'Common endeavours'.

§ For an early example see the paper by Nordhaus.[503] See also the 'dangerous levels' controversy between the IPCC (Bolin) and the secretariat of the UNFCCC (Estrada), on p. 268.

Once an acceptable rate of warming has been determined, the total global *emissions* rates that would affect such a *warming* rate could be calculated as a total global emissions target. This target could then be used as the basis for agreement on national emissions reduction targets.

The efforts to develop formulae for emissions targets around this time represented considerable advances in policy thinking, and it is worth considering just how advanced they were. Today, emissions targets dominate discussions of the policy response to global warming, and total emissions rates are tacitly assumed to be locked to a climatic response of one, two or so many degrees of warming. Today's discussions sits on top of a solid foundation of dogma established across several decades and supposedly supported by a scientific consensus, namely that there is a direct cause–effect temperature response to emissions. Back in the 1980s this dogma was by no means established. Scientific assessment of the climatic effect of emissions was stuck on the unsolved and complex problems of determining natural climate variability. Projections of the emissions impact on this variability were only coming from climate modelling that the modellers admitted was deficient for that purpose. The failure to establish the emissions effect on climate in turn inhibited progress towards establishing the climatic effect on human activities, whether good or bad. Yet at the Villach–Bellagio workshops, a line had been drawn under these concerns, and this allowed the participants to deliver something close to Tolba's envisaged *climate control*; not yet a *thermostat* but an emissions *throttle*, a way to determine just how much the world needed to ease off the accelerator pedal of energy usage so as to keep the warming rate below dangerous levels. The Villach–Bellagio report even suggests a target warming rate of 0.1°C per decade, which is 0.2°C slower than the middling expected rate taken from the modelling.[504] The report does not take the calculations much further, but there is at least the recognition that sufficiently slowing the warming rate 'could only be accomplished with significant reductions in fossil fuel use'.[505]

The 1988 'Changing Atmosphere' conference in Toronto

The Villach–Bellagio report called for its recommendations to be considered by a number of institutions and by some conferences already in the planning. One of these was a Second World Climate Conference that had already been planned for the spring of 1990. Members of the Villach–Bellagio steering committee hoped that negotiations for a climate convention would begin there.[506] One of them, Howard Ferguson, had been appointed co-ordinator of preparations, and he would end up chairing the conference. In the end, other plans for international negotiations would come into play before that grand event, but not before another conference already announced by the Canadian government for Toronto in the spring of 1988.

The Canadians had little to lose with a shift away from fossil fuels: most of their electricity production was already nuclear, they had a significant renewable sector with their hydropower, and there was the potential to sell excess power to their southern neighbour.[507] Their Toronto 'World Conference on the Changing Atmosphere' would generate the most excitement about the global warming issue of any meeting ever. Steven Schneider later dubbed it 'the Woodstock of CO_2'.[508]

The Toronto conference was pitched as a response to the Brundtland Commission's call to improve awareness among policymakers of pollution-driven changes in the atmosphere. It was structured to first address the scientific basis of concerns about these changes, before considering response strategies and the socio-economic implications. There were three headline concerns: the ozone layer, acid rain and global warming. This trio would be addressed across the four days of the conference, but the emphasis invariably inclined towards the latter.

Among the 300 guests from 46 countries who were invited to participate, there were many scientists and experts in the relevant fields. On the policy side, the contingent of environmental policy advocates was most significant, alongside UN and government agency officials. New with this conference would be government representation at the ministerial level. The Brundtland Commission was well represented, with Brundtland herself giving an opening address right after the Prime Minister of Canada, Brian Mulroney. This meant that the conference opened with the endorsement of two prime ministers, as Brundtland herself had been returned to that office in Norway even before submitting her report.

For Mulroney, the atmospheric pollution that most concerned him was acid rain; otherwise his address generally promoted the aims of sustainable development.[509] As for Brundtland, she again urged the need for international regulation to cover all atmospheric pollutants, but this time the emphasis came down hard on global warming:

> ...it is established beyond doubt that we *will* experience a global change in climate...The impact of climate change *may* be greater and more drastic than any other challenges that mankind has faced with the exception of the threat of nuclear war. [510]

The reference to the nuclear war threat was very much in keeping with the global security theme of the conference as expressed by its subtitle:

220

Chapter 13. UNEP and the push for a climate treaty

The changing atmosphere: implications for global security

This theme continued through to the 'conference statement', which famously opens:

> Humanity is conducting an unintended, uncontrolled, globally perva-
> sive experiment whose ultimate consequences could be second only to
> a global nuclear war.[511]

The 'experiment' in question is mostly explained as the 'inefficient and wasteful use of fossil fuel' by a rapidly growing population. This is reminiscent of the 1970s' energy crisis, of Jimmy Carter and of the predicted catastrophic conse-quences arising from the wasteful consumption of energy. With that crisis over, humanity is now faced with another crisis, where the same inefficient and waste-ful use of fossil fuel is causing changes to the atmosphere that will have even more drastic consequences; so drastic, in fact, that the imperative to action is again the moral equivalent of war. The conference statement does not use those terms, but instead declares a security threat so high that only the most catas-trophic war imaginable is worse. Further down, it continues with the Cold War arms race theme by saying that 'the growing peril of climate change and other atmospheric changes' threatens not only the global economy but 'may well be-come the major non-military threat to international security'. It may be a *non-military* threat, but it threatens actual hot warfare, which itself constitutes an imperative to action on global warming and ozone depletion:

> The best predictions available indicate potentially severe economic and
> social dislocation for present and future generations, which will worsen
> international tensions and increase risk of conflicts between and within
> nations. It is imperative to act now.[512]

One of the main recommendations for mitigating these dire consequences is a comprehensive global treaty to protect the atmosphere. On the specific issue of global warming, the conference statement calls for the stabilisation of atmo-spheric concentrations of one greenhouse gas, namely carbon dioxide. It esti-mates that this would require a reduction of current global emissions by more than 50%. However, it suggests an initial goal for nations to reduce their current rates of carbon dioxide emission by 20% by 2005. This rather arbitrary objective would become the headline story: 'Targets agreed to save climate'.

And it stuck. In the emissions-reduction policy debate that followed, this 'Toronto target' became the benchmark. For many years to come—indeed, until the Kyoto Protocol of 1997—it would be a key objective of sustainable develop-ment's newly launched flagship.

North American enthusiasm

The Toronto conference statement effectively launched global warming as the preeminent global environmental issue, a position that was unexpectedly enhanced even further as the threat of nuclear war receded following the fall of the Berlin Wall the following year. Its claims were sufficiently shocking to raise a high level of public alarm. However, these claims sat uncomfortably beside all the uncertainties acknowledged in the various scientific assessments. Thus, it might be asked how the atmospheric scientists at Toronto could have agreed with them.

The statement had been drafted by a committee led by the conference chairman, the UN ambassador for Canada, Stephen Lewis, and it had blown out to over 6000 words before being revealed for a brief review at a large plenary session on the final day. At this meeting of policymakers, policy advocates and scientists, three meteorologists *had* raised concerns about the statement, saying that they were being asked to endorse, without discussion or debate, very strong assertions of alarm that they did not believe had a sufficiently demonstrated scientific basis. Two of those who raised this concern did so with some authority. One was the first vice-president of the WMO and head of the Australian Bureau of Meteorology, John Zillman, while the other was the third vice-president of the WMO and head of the UK Met Office, John Houghton.[513] However, their protests were to no avail and the statement was not toned down. Everyone else, it seems, was swept along with the enthusiasm of moment.

In the Australian delegation to Toronto, Zillman's concerns about the premature spread of warming alarm were in contrast with the views of one of his two compatriots, Graeme Pearman, who had already been most effective in promoting the issue back in Australia.|| Despite their differences, their agreed delegation report reveals why Zillman and Houghton found so little support for their protests. The Australians explained that it was 'impossible not to be affected by the suddenness with which climate change issues are moving centre stage in world affairs', and at Toronto they sometimes felt...

> ...a sense of being swept along in a tide driven more by the Canadian political agenda and emotions born of the contemporary North American drought than by strongly objective assessment of the scientific evidence and arguments...[515]

|| Pearman, of the CSIRO Division of Atmospheric Research, was instrumental in winning early and widespread public awareness for the greenhouse warming threat. It was in no small part due to his advocacy that polling around the time of the Toronto conference indicated not only that 80% of the Australian public were aware of the issue but also that nearly all of these expected a policy response.[514]

As the Australian suggested, politics north of the 49th parallel was not the only force sweeping along this tide. The warming scare was also being used to funnel emotions born of the heat and drought of that North America summer into the US presidential election campaign. In fact, the competition on environmental policy between the vice-president George Bush and the Democratic Party candidate Michael Dukakis was just as effective in driving the climate change issue towards centre stage—if not in world affairs, then certainly in the USA.

Global warming was launched into the US presidential campaign through a spectacular performance by Democrats on Capitol Hill only days before the Toronto conference. The setting for this performance was a congressional hearing called to discuss an ambitious and multifaceted climate change bill that the Democrats had just introduced. As Democratic Party Senator Tim Wirth later explained, the supporters of the bill contacted James Hansen and arranged for him to speak about his new research on the global temperature trend and its possible link with manmade warming. They then contacted the National Weather Service to establish what date historical records suggested would be the hottest of that Washington summer. They then arranged for Hansen to speak on that day. Furthermore, to help the audience make the heatwave connection, Wirth tells of how it was arranged that all the windows of the hearing room would be opened the night before Hansen was due to speak, so that the air conditioning could not cope. The day turned out to be a record-breaker,[516] and with all the camera lights, it was especially hot in the crowded room as Hansen spoke to a written submission entitled:

> The greenhouse effect: impacts on current global temperature and regional heat waves[517]

Hansen's testimony referred to a recently published article, which included a new global temperature graph. This showed that global mean temperature had already surpassed the late-1930s peak just *before* 1980. This was a surprising new finding, given that the graph he had produced back in 1981 showed the trend still significantly below that peak at its last data point in 1980. Moreover, on the version of the new graph used for the hearing, Hansen included monthly data for the first part of 1988, right up to the (exceptionally warm) May, just prior to the hearing. This monthly (upward) trend for 1988 had not been statistically smoothed and so its inclusion had a striking visual effect. It looked like, *the way things are going*, the near vertical trend line would shoot out the top of the chart by the end of year (see Figure 13.1). But Hansen went beyond this graph and beyond the conclusion of his published paper to firstly make a

223

strong claim of causation, and then, secondly, to relate this cause to the heat being experienced that year (indeed, the heat being experienced in the hearing room *even as he spoke!*). He explained that 'the Earth is warmer in 1988 than at any time in the history of instrumental measurements'. He had calculated that 'there is only a 1 percent chance of an accidental warming of this magnitude...' This could only mean that 'the greenhouse effect has been detected, and it is changing our climate now'.[518]

Hansen's detection claim was covered by all the main television network news services and it won for him another *New York Times* front page headline:

Global warming has begun, expert tells Senate[519]

The article paraphrases Hansen saying to the press after the hearing that he is...

...99 percent certain that the warming trend was not a natural variation but was caused by a build-up of carbon dioxide and other artificial gases in the atmosphere.[520]

Hansen's testimony had a huge impact across the USA. Indeed, Americans often see it as marking the beginning of the global warming scare in their country. With Senators Wirth, Gore and other Democrats maintaining the profile of the warming scare in the Dukakis campaign, warming mitigation soon also rated an appearance on the Republican agenda. While Ronald Reagan had been notoriously sceptical of environmentalist claims, especially where they came with recommendations for new industrial regulation, George Bush was using his interest in environmental issues as a way to define his difference with Reagan. On the campaign trail that autumn, he famously declared:

Those who think we're powerless to do anything about the greenhouse effect are forgetting about the White House effect. As president, I intend to do something about it.[522]

Global political enthusiasm

Before winter closed in at the end of 1988, North America was brimming with warming enthusiasm. In the USA, global warming was promised attention no matter who won the presidential election. In Canada, after the overwhelming success of the Toronto conference, the government continued to promote the cause, most enthusiastically through its environment minister Tom McMillan. Elsewhere among world leaders, enthusiasm was also building. The German chancellor, Helmut Kohl, had been a long-time campaigner against fossil fuels

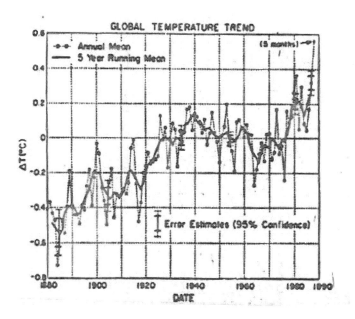

Figure 13.1: Hansen's alarming temperature graph.

This is the graph submitted in Hansen's written testimony to the US Senate, Committee on Energy and Natural Resources, June 23, 1988. Notice that this time the mean trend line rises above the 1939 peak before 1980. Also note a visual effect where it seems to end with a high data point for the first 5 months of 1988.[521]

and had followed his predecessor, Helmut Schmidt,¶ as an early promoter of this new peril. Other governments in Western Europe were also coming on board, not least Norway, with Brundtland back in power and now giving special attention to the issue. In Australia, the Labor Party had been swept to power in 1983 on the back of a promise to meet the demands of Australia's greatest ever environmental campaign (to stop the flooding of the Franklin River in Tasmania). That government went on to promote the new environmentalism by publishing a special edition of the Brundtland Report, complete with a foreword by the Prime Minister, followed by an Australian supplement, and all under the imprimatur of its Commission for the Future.[523] From 1987, this commission, in collaboration with Pearman at the Commonwealth Scientific and Industrial Research Organisation, was an early and effective promoter of global warming alarm, not least through the support of its director, Australia's first ever Minister for Science, Barry Jones.

¶ See p. 63.

Perhaps the most surprising political recruit to the cause that year was the first scientist ever to be elected Prime Minister of Great Britain. Mrs Thatcher had come under the influence of the British diplomat Crispin Tickell, an environmentalist who had been lobbying for international legislation to protect the atmosphere.** In a speech to the Royal Society in September 1988, she professed her conversion to environmentalism by raising familiar concerns about the Toronto trio of atmospheric troubles: acid rain, ozone-layer depletion and especially global warming. Her previous reluctance to address acid rain and ozone protection meant that her sudden conversion to the cause was treated with suspicion. In particular, her advocacy on the new issue of global warming was seen as a way to leverage her scientific credentials and the new environmentalism as justification for closing coal mines.[525] Nevertheless, her conversion that autumn added great authority to the cause, the momentum of which was maintained into 1989, a year remarkable for the rush to board the global warming bandwagon.

The 1989 global warming bandwagon

In February 1989, the year got off to a flying start with a conference in Delhi organised by India's Tata Energy Research Institute and the Woods Hole Research Center, which convened to consider global warming from the perspective of developing countries. The report of the conference produced an early apportionment of blame and a call for reparations. It proclaimed that the global warming problem had been caused by the industrially developed countries and therefore its remediation should be financed by them, including by way of aid to underdeveloped countries. This call was made after presenting the problem in the most alarming terms:

> Global warming is the greatest crisis ever faced collectively by humankind, unlike other earlier crises, it is global in nature, threatens the very survival of civilisation, and promises to throw up only losers over the entire international socio-economic fabric. The reason for such a potential apocalyptic scenario is simple: climate change of geological proportions are occurring over time-spans as short as a single human lifetime. ††

** Tickell outlined his case for a law of the atmosphere in the 1986 edition of his book, *Climatic Change and World Affairs*.[524] His influence extended beyond Thatcher and Britain, especially after he was appointed ambassador to the UN in 1987.

†† Quoted by Bert Bolin.[526] On the significance of this conference see also Hecht and Tirpak.[527]

While poor countries began calling for aid to help them avert the crisis, a cluster of western European governments were urging the immediate commencement of negotiations on emissions targets and timetables to do just the same. However, their campaigning met with opposition from the UK and the USA. It was not that the new US president reneged on his election promise; on the contrary, the Bush administration moved quickly with preparations to stop the warming. Rather, it was that senior policy advisers were firmly for delaying discussion of binding emissions targets. The British position was similar, and so Thatcher and Bush united with calls for deferral until the conclusion of the process undertaken by WMO and UNEP through their IPCC.[528]

The contrast with other European countries, especially Germany and the Netherlands, could not be more striking. Their impatience for legally binding commitments is indicated by the outcome of a ministerial-level environment conference in March 1989. Jointly organized by the governments of the Netherlands, France and Norway, this was convened in The Hague, with representation from 24 countries, including a selection of southern poor countries like Brazil and Indonesia and wealthier others like Australia and Japan. But this was very much a European show. Most impressive was the arrival of four European heads of government: Helmut Kohl of Germany, Francois Mitterand of France, Gro Brundtland of Norway and Felipe Gonzalez of Spain. The UK and the USA were notable absentees.

This 1989 Hague conference produced the so-called 'Hague Declaration', the main subject of which is the means for addressing the threat of imminent climate change. The Hague Declaration first establishes the scientific basis of alarm by noting that, 'according to present scientific knowledge', the consequences 'may well jeopardize ecological systems as well as the most vital interests of mankind at large'. It then recognises that 'industrialized nations have special obligations to assist developing countries' who will be 'very negatively affected' by the changes, but who may only be marginally responsible for them. The solution is 'vital, urgent and global'. It requires 'a new approach through the development of new principles of international law including new and more effective decision-making and enforcement mechanisms'. Decision-making requires 'a new institutional authority' within the framework of the UN but one that permits non-unanimous decision-making. Enforcement requires a mechanism controlled by the International Court of Justice.[529]

In July 1989, the G7 economic summit in Paris was not quite so enthusiastic about the role of supranational institutions of governance, law and enforcement. Nevertheless, Kohl, Thatcher and Bush successfully promoted environmental issues, and all seven leaders of the world's great economic powers agreed to 'strongly advocate common efforts to limit emissions of carbon dioxide and other greenhouse gases'. According to their communiqué, one common effort would be towards the development of a framework convention, and they even promoted 'specific protocols containing concrete commitments' that could be 'fitted into the framework as scientific evidence requires and permits'.[530] For Bush and Thatcher, this was all about following the ozone-protection process and so awaiting assessments of the 'scientific evidence' by the WMO/UNEP panel. But this was not the case for Kohl; Germany and the Netherlands continued to push for emissions action without delay. Already at the Toronto conference, the Dutch environment minister had announced plans for another ministerial conference specific to climate protection.[531] Germany would soon come on board to assist with preparatory negotiations.

The polarisation of the two groups of countries only increased during the preparatory talks, with the hosts impatient and the UK and USA reticent. In October 1989, the stream of VIPs who arrived for this climate change conference was most impressive. Delegations from 67 countries, including all the key players, arrived in the small Dutch city of Noordwijk, which is nestled in the dunes protecting the Low Countries from North Sea storms. They proceeded to negotiate their agreement while the Iron Curtain around the Soviet Union was collapsing, and with it Cold War arms race fears. The resulting 'Noordwijk Declaration' declares that 'the stabilising of the atmospheric concentration of greenhouse gases is an imperative goal'. But in the end the USA, USSR and Japan were successful in blocking moves towards binding emissions targets to achieve this goal. Negotiations on such matters would be deferred until after the negotiation of a framework convention, which would follow the completion of the UNEP/WMO panel assessment.[532]

The year 1989 ended with the meeting in Malta of the Soviet leader Mikhail Gorbachev with George Bush. At the end of his first year in the US presidency, Bush celebrated the end of the Cold War by passing around souvenir chunks of the Berlin Wall. Climate change was not the main topic on the agenda, but Bush did promise to host the first climate convention negotiations in Washington, just as soon as the IPCC had completed its assessment. Thus, by the end of 1989, all the enthusiasm for climate action had been successfully funnelled into the UNEP/WMO processes. As it turned out, Bush would keep his promise to host the first treaty talks, although by then WMO/UNEP were no longer running the process. By then, these two bodies, and their celebrated climate panel, had been pushed to the sidelines.

14 Origins of the IPCC

While the political movement was sweeping towards the inevitability of a climate treaty, out of the perceived failure of Tolba's Advisory Group on Greenhouse Gases there also arose the process that resulted in the establishment of the IPCC. One key impetus to this process was the publication in 1986 of the 1985 Villach report. Mustafa Tolba proceeded to spread the bad news of that meeting's scientific consensus by forwarding copies of the freshly printed report to those influential with government policy. He sent one copy to the US Secretary of State, George Schultz, with a letter urging policy action. Schultz in turn forwarded it to the National Climate Program Office at NOAA. From there the matter was soon raised at the National Climate Program Policy Board.[533]

This was not the first time that this interagency board had addressed warming concerns. In fact, it was not long after the establishment of the various climate institutions in the last days of the Carter administration that their attention had begun to turn from cooling to warming. Already in the early 1980s the National Climate Program Policy Board had come to represent widely varying views on the urgency for policy action on the warming threat. These differences were accentuated in a most embarrassing and public way in 1983 with the almost simultaneous publication of two differently styled assessments of the very same issue.

US government assessments of the carbon dioxide problem

Formal US government assessments of the warming scare began even before the 1980 Villach conference. The first of these followed a 1979 congressional symposium on the implications of the greenhouse effect for energy policy, at which concern was raised that plans to resolve the energy crisis through a much greater reliance on coal could mean that the atmospheric concentration of carbon dioxide would skyrocket, doubling by as early as the first decade of the 21st century. The airing of this speculation led to the first NAS assessment specifically addressing the carbon dioxide emissions question. A study group was convened for a one-week review during the summer of 1979, a few months after the World Climate Conference. Under the leadership of the meteorologist Jule Charney, this group mostly considered speculative modelling, with little regard to empirical validation. Charney found that the modelling predicts carbon dioxide doubling would drive up temperatures by around 3°C by sometime in the first half of the 21st century. This warming rate was based on an assumption of strong positive feedback from increased water vapour, while the 'cloud effect' was only noted as 'a difficult question to answer'. The report's foreword, by the chairman of the Climate Research Board, Verner Suomi, warns that 'a wait-and-see policy might mean waiting until it is too late'.[534]

The following year, the 1980 Energy and Security Act was passed by the US Congress with a request that NAS prepare another, more comprehensive assessment. To undertake this task, the Climate Research Board established a Carbon Dioxide Assessment Committee chaired by the director of Scripps Institution of Oceanography, William Nierenberg. Their assessment drew heavily on work funded through the DoE's Carbon Dioxide Program. It was around this time that keen interest in the empirical evidence arose in the carbon dioxide research community and this interest is reflected in the Nierenberg report, which was released in October 1983. After highlighting the great problems and gaps remaining in the science, it recommends against policy action.

> We do not believe…that the evidence at hand about CO_2-induced climate change would support steps to change current fuel-use patterns away from fossil fuels. Such steps may be necessary or desirable at some time in the future, and we should certainly think carefully about costs and benefits of such steps; but the very near future would be better spent improving our knowledge…than in changing fuel mix or use.[535]

Three days before the release of this NAS report, the EPA released its own briefer internal assessment. Its presumptive title, 'Can we delay a greenhouse warming?' reflects its content. The evidence that there is in fact something to 'delay' is derived entirely from modelling. While there is little discussion of any empirical evidence of the effect, there is much discussion of how to delay it. The executive summary is especially significant for highlighting just how disruptive mitigative action would need to be. The counter-effects of various emissions reduction options are given in carbon dioxide 'thermostat' calculations (which were earlier and independent from those used later at the Villach–Bellagio workshops). These calculations determined that severe taxation on fossil fuels would only delay the warming trend by a few years, while a complete ban on coal by the year 2000 would only put off a 2°C warming until 2055.[536] Accompanying the release of the report came a warning that major changes to the

E.P.A. Report Says Earth Will Heat Up Beginning in 1990's

By PHILIP SHABECOFF
Special to The New York Times

WASHINGTON, Oct. 17 — The Environmental Protection Agency warned in a report made available today that the warming of the earth known as the "greenhouse effect" will begin in the 1990's.

John S. Hoffman, director of strategic studies for the agency, said in an interview today: "We are trying to get people to realize that changes are coming sooner than they expected. Major changes will be here by the years 1990 to 2000, and we have to learn how to live with them."

The report, which was completed last month by Mr. Hoffman's office, said the warming trend, the result of a buildup of carbon dioxide in the atmosphere, is both imminent ʳ d inevitable. In the next century, iᵗ   ns, the world will ' .ve tᵒ learn ʳ   il ⸱ .th majᵒr ch·   ×ᵉ   ⸱⸱ nᵉ

Figure 14.1: The EPA promotes alarm ahead of the NAS report release.

This *New York Times* front-page story of 18 October 1983 was met three days later with another front-page story reporting on the NAS finding under the headline: 'Haste of global warming trend opposed'.

climate would arrive in the 1990s if no action were taken, and this won the report front-page headlines. For example, the *New York Times* declared that 'EPA report says Earth will heat up beginning in 1990s'. However, the same front page countered 'Haste of global warming trend opposed' on the release of the NAS report just three days later.

231

Divergent views in the US lead to the 'intergovernmental' compromise

Three years later, in 1986, when Tolba's call-to-action arrived via Secretary of State Schultz with the National Climate Policy Board, the EPA remained prominent on this interagency board for its alarming assessment of the scale and urgency of the problem. Among the more moderate views around the table, that of the DoE held considerable force. As the principal funder of all the research now underway, its view concurred with the Nierenberg report that the scientific case for action had not yet been established. With the exception of the Department of State, this was also the general view among senior officials across the Reagan administration. Reagan himself was the least concerned, but he was coming under political pressure from senators who were using congressional hearings to promote the issue. In June 1986, seven senators—including one persistent campaigner, Al Gore—were so 'deeply disturbed' by the situation that they wrote to the president of NAS requesting another review.[537] Thus, by 1986, the strong differences of opinion within US agencies were reflected in the politics, bringing these differences into public view. Further debate ensued within the government until a compromise emerged.

This compromise was to hold off from support for any policy action until completion of yet another assessment, this time by an international panel convened by the UN bodies. The recommendation that this assessment should be *intergovernmental* arose in part due to concerns about the policy entrepreneurialism of Tolba, Goodman and others in response to the 1985 Villach conference. At least two members of the Villach–Bellagio steering committee later acknowledged that they came under pressure from the US State Department to rein in their activities: Michael Oppenheimer recalled how during the preparations for the workshops he received…

> …a letter from the Deputy Undersecretary of State warning him that the US Government was concerned about the activities of the private group of policy entrepreneurs that he was part of.[538]

Jill Jäger gave her view that one reason the USA came out in active support for an intergovernmental panel was that the US Department of State thought the situation was 'getting out of hand', with 'loose cannons' out 'potentially setting the agenda', when governments should be doing so.[538] An intergovernmental panel, so this thinking goes, would bring the policy discussion back under the control of governments. It would also bring the science closer to the policymakers, unmediated by policy entrepreneurs. After an intergovernmental panel agreed on the science, so this thinking goes, they could proceed to a discussion of any policy implications.

While this did indeed seem to be the thinking behind the compromise position, not everyone agreed with it. Some thought that even this course of action would be premature and risked politicisation. One of those was no less than the head of the US delegation at the WMO. Richard Hallgren, director of the National Weather Service, had taken over from Robert White as the US 'permanent representative' after White's chairing of the World Climate Conference. In a 1993 speech, Hallgren recalled the 'extensive discussions' in the USA six years earlier, and how he had been against the proposal for what became the IPCC. He thought he had been wrong at the time to believe that such a panel 'would be politicised by individual government positions'.*

Despite Hallgren's opposition, in May 1987 he took the US policy position in support of an intergovernmental review to the Tenth World Meteorological Congress. It was at this meeting, while considering how best to respond to the 1985 Villach report, that the whole discussion came to a head. Not only had the Americans arrived with a view on how to proceed; several other delegations had briefs calling for the WMO to play a stronger role than seemed likely to be played by the AGGG. Indeed, much to Tolba's disappointment, many delegations to the Congress at least agreed that the AGGG was inadequate to the purpose.[540] In the lively debate that ensued, the Australian delegate John Zillman recalls one defining moment. This was...

> ...the impassioned plea by the Principal Delegate of Botswana for WMO to establish some sort of mechanism that could provide her with an authoritative assessment of what was known about human-induced climate change that would enable her to answer the difficult questions she was being asked by her government about whether the greenhouse effect was real and what, if anything, they should be doing about it.[541]

After this plea, it was clear that some formal mechanism needed to be established and that it would be appropriate to involve UNEP 'because of its clear environmental policy dimensions which were beyond the mandate of WMO'.[542] Congress then referred the request for an 'appropriate mechanism' to the Executive Council, which was due to meet the following month. In June 1987, the Council called for an 'intergovernmental mechanism' to carry out the assessment, and, accordingly, a similar decision came soon after at the Governing

---

*In this address to an Australian audience, Hallgren went on to say: 'In WMO, John Zillman, myself and others debated long and hard before we took the step of setting up the IPCC. I was wrong, and I hope that we can keep politics out of the next [second] IPCC scientific assessment now underway.'[539]

Council of UNEP.[†] In this way, the IPCC was instituted *before* the Villach–Bellagio policy workshops, and *well before* Hansen's Congressional testimony and the Toronto conference. Thus when the IPCC was conceived, the political situation of the greenhouse warming threat was a long way from where it would be by the time of the panel's inaugural meeting, which came *after* all the excitement of that 1988 summer had propelled the issue to the top of the sustainable development agenda.

Given what the IPCC has become, it is difficult to imagine what was envisaged for it when it was conceived in 1987. In this regard, it is helpful to recall previous assessments conducted by WMO panels. Consider, for example, the expert panel that conducted the assessment of the cooling scare back in 1976. It had provided an 'authoritative assessment', much as the delegate from Botswana later requested for the warming scare. It was two pages long. In 1987 the warming scare was not much bigger in the public mind than the cooling scare back in 1976. Both had enormous policy implications, although the latter was bigger in terms of institutional developments. We will remember that the main proponents of the cooling scare were marginal to the meteorological establishment, and we will remember that meteorologists had attempted to downplay the scare before the expert panel met. One decade later, the global warming scare had also been downplayed in some previous authoritative assessments, but not all, and there was significant support in the assessments and among meteorologists for continued investigations. Such investigations were well funded, with the US DoE alone pouring $10 million every year into its Carbon Dioxide Program throughout the early 1980s. Not that this necessarily supported the push for policy action, which, since the 1979 World Climate Conference, had been hard going. Yet significant institutional attention had been won, importantly, but not only, in the USA. This was sufficient to bring on a serious discussion of the 1985 Villach 'consensus' at no less a forum than a World Meteorological Congress. So, by 1987 the assessment to be undertaken by the newly established IPCC was undoubtedly regarded as more important, more complex and a much bigger task than the 1976 cooling assessment. *But how much bigger?*

This is a question upon which John Zillman would later reflect after playing a central role in the development of the IPCC from its beginnings at that 1987 Congress. There were certainly a range of views, as he recalls, including on whether the IPCC would continue in an ongoing role after completing its (first) assessment. But, in terms of that assessment, the task appeared contained.

What most of those who drew up the concept of the IPCC had in mind at

---

[†] The ICSU was excluded because it is neither a UN agency, nor is it made up of government delegates.

the time was a group of, say, 40–50 climate experts nominated by 20–30 governments to produce a 30–40 page document which could be passed to all governments by the WMO and UNEP Secretariats to assist them (governments) in addressing the climate change issue in their national contexts, as well as for use in preparations for the already planned 1990 Second World Climate Conference and in support of discussions in other parts of the UN System on a possible 'law of the air'...[543]

This vision was not to be. Within one year of the first IPCC session, its assessment process would transform from one that would produce a pamphlet-sized country representatives' report into one that would produce three large volumes written by independent scientists and experts at the end of the most complex and expensive process ever undertaken by a UN body on a single me-teorological issue. The expansion of the assessment, and the shift of power back towards scientists, came about at the very same time that a tide of political en-thusiasm was being successfully channelled towards investment in the UN pro-cess, with this intergovernmental panel at its core. This transformation of the IPCC assessment process at this time was achieved principally through the lead-ership of one man: this was not the elected chairman of the IPCC, but the chair-man of one of its working groups.

The intergovernmental assessment begins in 1988

The first session of the IPCC convened in Geneva over three days in November 1988. Despite all the excitement over the summer, only 28 delegations arrived to hear the latest plan for the progression from a climate change assessment to a climate change convention. This had already been mapped out in a draft UN General Assembly resolution. The draft title placed the work of the IPCC squarely within the greater realm of sustainable development:

The conservation of climate as part of the common heritage of mankind[544]

The IPCC's role in the strategy to 'conserve the climate' would be to sub-mit their completed assessment, in the first place, to the second World Climate Conference. The conference had been planned for June 1990 as a review of the World Climate Programme as it reached its tenth year. Now, in addition, it was being asked to consider policy action on greenhouse warming. It was not that the World Climate Programme was finally taking up the cause, nor was there any need to keep up a pretence that it had. With the formation of the IPCC, ex-pert involvement in the greenhouse problem was now institutionally indepen-dent from that program, and yet its conference provided a convenient forum for the transition from greenhouse assessment to greenhouse policy action, even if the timing presented the IPCC with a tight 18-month schedule. Calls by Tolba and others to use the conference for the policy transition of the greenhouse issue came simultaneously with requests to postpone it until late in 1990. Soon enough these requests would be honoured and in the planning for the Second World Climate Conference emphasis began to shift away from general climate science and towards addressing the greenhouse problem.

This course of events was already envisaged at the first session of the IPCC, when, following the opening speeches, Bert Bolin was elected by acclamation to the chair. Although at first reluctant to take up the role, Bolin would go on to become the chairman of the IPCC across its first decade. The work of the assessment was then divided between three working groups:

- Working Group 1: The scientific basis of climate change

- Working Group 2: The environmental and socio-economic impacts

- Working Group 3: Formulating response strategies.[545]

Membership of the three working groups was open to all delegations. However, it was decided that each group not only required a chairman but also a designated 'core membership', which together would take on the responsibility of completing the group's tasks in a timely manner.[546] This provided the basic structure of the IPCC, to which was only added a small executive 'bureau' led by Bolin and constituting the leadership of the working groups. The plan was for the working groups to submit their completed reports to the bureau, which would integrate them into a single assessment statement. This would include an executive summary for policymakers with recommendations for co-ordinated international action.

The assignment of the chairmanships and selection of core membership of the working groups was only achieved through careful negotiations. Working Group 3 was the most popular because there was already the view that the terms of a climate treaty would start to take shape in this group. Consequently, it was assigned 17 core members, while the other groups were designated only 12. The USA won the chairmanship of this 'policy' working group, as it came to be called. The Soviets were assigned Working Group 2, on the impacts of climate change. Working Group 1, covering the underlying science, was assigned to the UK and already at that first meeting John Houghton was specifically named to take up the chair.[547] This turned out to be a most significant appointment because within a few years interest would shift away from Working Group 3 and towards Working Group 1. This shift was partly due to external events constraining the policy group's policy work, and partly due to the difficulties Working Group 1 had in establishing the scientific grounds for policy action. However, another reason for the rise to prominence of Working Group 1 was Houghton's exceptional leadership in driving and shaping the assessment process.

The Houghton-led shift away from intergovernmental involvement

John Houghton's career began in Oxford in the 1950s, where he investigated the ozone layer with Dobson and Brewer. He had taken up with Dobson's pioneering work on remote sensing, participating in the development of a temperature-sensing device for satellites. Much later, when John Mason retired, he took over leadership of the UK Met Office. While in that role, he had many other responsibilities, including an involvement with the WCRP as it took over from GARP. He was elected vice-chairman of its joint governing committee, and then chairman in 1982. When he was assigned the responsibility of leading the IPCC scientific working group, he took to this role with alacrity.

The British had arrived at the first IPCC meeting with an outstanding bid for Working Group 1 already prepared. While Houghton saw the management of a thorough and comprehensive scientific assessment as a complex and difficult job, he would not be working alone: the UK Department of the Environment had pitched in with an annual grant of a quarter of a million pounds, which would be used to establish a 'Technical Support Unit' at the UK Met Office, complete with three or four seconded staff. This support made possible the development of a more complex assessment process.[548]

From the first session of his working group, Houghton began introducing a process of assessment very different from anything that had been outlined previously. While the other working groups were scheduling workshops to support panel members in writing their report, Houghton moved towards a model more along the lines of an expert-driven review: he nominated one or two scientific experts—'lead authors'—to draft individual chapters and he established a process through which these would be reviewed at lead-author meetings. He also incorporated a formal step of review by a wider expert peer group, thereby maximising engagement with the global research community. Late in the process, the expert lead authors would come together and draft a summary for policymakers specifically for the working group report. In this model, the involvement of the intergovernmental panel is to review the report and to come to agreement on a final version of the summary. But that is all.

Houghton moved so fast in planning and implementing this assessment process that he was already far ahead of the other working groups by the second IPCC session in June 1989. Bolin and others supported Houghton's model, and it would be gradually implemented across all the working groups as the assessment proceeded. Another change that came at the second IPCC meeting was where the 'core members' idea was abandoned, so that membership of each working group became equal and open to all delegations. In effect this meant that the original intergovernmental panel repeated itself three times in three subsidiary panels. Not that this gave the country delegations more control over the assessment; it only meant that the responsibility for delivering the assessment that had previously been assigned to the core membership now shifted towards the working group chairman.

This expansion and transformation of the assessment process deserves some consideration due to its later implications for both the assessment's success and its reception. The main change was that it shifted responsibility away from government delegates and towards practising scientists. The selection of lead authors, mostly on account of their expertise, made for meritocratic participation in the assessment process. This was a long way from equitable intergovernmental representation. It also had the immediate effect of shifting power towards residents of rich countries, where most of the science was being conducted.

Houghton's expert-led approach differs in one important respect from many comparable scientific assessments on controversial topics, whether by national academies or by international bodies: it lacked the usual protocols to prevent conflicts of interest in its assessors. One common way to avoid such conflicts is to have the assessment undertaken by scientists who are not directly involved in the science (and so in the controversy). Most of the assessments we have discussed so far have followed this policy to some degree. Sometimes, for highly specialised topics, such an approach proves difficult, and sometimes, anyway, panellists are found to be involved in some aspect of the science under review. In such cases, there are often protocols requiring them to declare their interest and then to ensure it does not conflict with the impartiality of the assessment. Such protocols are not evident in Houghton's model or practice, and he deliberately chose chapter lead authors who were among the leading researchers in that specialist field.

The decision to recruit assessors who were leaders in the science being assessed also opened up another problem, namely the tendency for them to cite their own current work, even where unpublished. Unpublished work is usually avoided to ensure that the assessment is based on findings that are open to, *and have survived*, the normal processes of scientific scrutiny. What normally happens in science is that all sorts of advances are proposed, yet the great majority do not survive beyond peer-review, publication and reply, a process usually taking two or more years to unfold. However, under Houghton and Bolin, a culture developed in which *currency* was prioritised over *scrutiny*. From the beginning, at the IPCC, it was more important to use the most recent research, including yet-to-be-published work, to ensure that the assessment was made upon the very latest science. This attitude arose from the pervasive and overwhelming concern that many of the fields under assessment were new, underdeveloped and suffered from a dearth of evidence. This concern could be allayed through an optimism that the research was advancing more rapidly and accurately than prior experience would ever suggest.

In his memoir, Houghton describes how a leading climate modeller arrived at a lead-author meeting with early results of a new atmosphere–ocean coupled computer model. The report had already been drafted, reviewed, and redrafted. However, at the eleventh hour, it seemed that a great advance had just been achieved. One common criticism of the atmospheric modelling of carbon dioxide's climatic effect was that it did not incorporate the dynamics of the giant heat bank that is the oceans below. This new model did, and did so much better than any previous model. There was some hesitation in accepting these new results in the report but, after considerable debate, their inclusion was agreed. This is how there came to be substantial passages of that first IPCC report supported by reference to little more than a 'private communication'.[549,550]

In this and other cases that were to follow, acceptance of unpublished new work was justified by claiming that the lead-author discussion—even *the debate about its very inclusion*—is an adequate substitute for the normal processes of scientific review.[551,552] Whether or not this is true, the IPCC fell into a heavy reliance on work that had not undergone the normal processes of scientific scrutiny and which was also difficult for its own review processes to handle. At the stage of expert review, many of the key findings in the draft reports would be supported by papers that had not been, or not yet been, published in scientific journals. Many would be awaiting publication, and their contents therefore under embargo by the publisher. Some had not even been accepted for publication, and others were not yet written. This practice would only increase during the second assessment, where early criticism by one of its reviewers (see Figure 14.2) would be vindicated when three major controversies that subsequently arose all involved findings based on unpublished work by the lead-authors.[‡]

How could Houghton and Bolin have allowed the authors such liberties? One

---

[‡] These controversies were: firstly, the Working Group 3 'price of life' controversy, where the relevant findings were mostly based on unpublished doctoral theses by two of the young lead authors (see pp. 269 and 285); secondly, the controversy over the release of the latitudinal breakdown of the UK Met Office modelling results (see p. 273); and thirdly, the 'Chapter 8 controversy', where the 'detection' claim was mostly based on the coordinating lead authors own unpublished work (see p. 281).

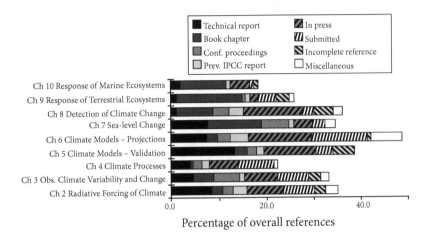

Figure 14.2: Publication status of references cited by the IPCC.

Pat Michaels surveyed the status of references cited in the review draft of the IPCC Working Group 1 second assessment report. Michaels was early in using the World Wide Web to broadcast criticism of global warming science when his newsletter *World Climate Report* went online in 1995. The original caption reads: 'Total percentage of citations contained in the IPCC manuscript that do not meet the standards of top-ranking scientific journals. Acceptable references make up the remainder.'[553] The issue emerged again with the IPCC Fourth Assessment when a similar review was undertaken through a crowd-sourcing initiative.[554]

explanation is that they were naïve, unaccustomed to, or otherwise unprepared for the controversial environment that soon developed around the assessments. After all, their experience of reviews and assessments had been mostly in the very different environment of GARP and WCRP. It is true that they already knew that the topic was controversial. However, we should consider that even with the SCOPE 29 assessment, no problems of this kind had arisen, despite the fact that lead authors were often entirely interested in the work they were evaluating. In the case of Tom Wigley, not only was he involved in the research under review, he was also characteristically very critical of the work of others, demolishing positive claims with a heavy round of scepticism. That such an 'interested' approach to assessment was favoured within the meteorological milieu of Houghton and Bolin is also evident much earlier.

In 1980, a WMO Secretary General had made a specific recommendation to engage scientists who were actively involved in the science under review. Moreover, this recommendation was made in one of the earliest proposals for an international assessment on the carbon dioxide question. We will recall Wiin-

Nielsen visiting the first meeting of the WCRP committee with a proposal for how to address the carbon dioxide question after it had been raised so prominently at the World Climate Conference.[§] He recommended against WCRP involvement, proposing instead 'an international board for the assessment of research into CO_2 effects on the climate'. In this proposal, he made the following recommendation on the composition of the board:

> Furthermore, to be effective, a majority of the expert members should be practicing scientists of high standing in their field. In this way, apart from having the competence to perform the task properly, the members would be able to do much to guide and promote the right kind of research through direct contacts with the working scientists, many of whom would be their colleagues. This kind of approach proved most effective in the organization of GARP.[555]

Recruiting practising scientists who would guide and promote the right kind of research among their colleagues does make sense for GARP and WCRP reviews, where the promotion and development of new areas of science are primary objectives. On the carbon dioxide question, these twin objectives might also hold. However, it might be argued that it is better to pursue them separately on such a highly controversial assessment on a topic with known and substantial political dimensions.

Houghton's decision to shift towards an assessment by practising scientists was one thing, but his recruitment of lead authors revealed a further preference. Nearly one third were UK scientists. If we add to this the highly competent and expert staff in the Technical Support Unit—who were deeply involved in coordinating activities out of the UK Met Office—then we have a very anglocentric affair. It was even more anglophone: 21 of the 34 lead-authors were from either the UK or the USA. With the USA winning the chair of the 'policy' working group (Working Group 3), it is easy to see how the poor-country delegates (and middle-eastern oil producers) came to view the IPCC as heavily biased towards the views of a small group of rich Westerners. This became even more of a concern as the IPCC became further embroiled in the process leading towards a possible climate treaty, which was looking all the more likely as the fateful year of 1989 progressed.

The IPCC embroiled in preparations for a climate treaty

In 1989, George Bush, the new resident in the White House, proved true to his word, not only using the 'White House effect' to bring special attention to the

---

[§] See p. 189.

climate issue, but also reflecting some of its glory down upon the IPCC. The chairmanship of Working Group 3 had been eventually assigned to Frederick Bernthal, who was no less than an Assistant Secretary of State. Then, in January 1989, the Secretary of State himself, James Baker, chose *for his first engagement in that role* to open a session of this 'policy' working group.

Throughout 1989, the IPCC working groups conducted a busy schedule of meetings and workshops at venues around the northern hemisphere. Meanwhile, the outpouring of political excitement that had been channelled into the process brought world attention to the IPCC. By the time of its second full session in June 1989, its treaty development mandate had become clearer: the final version of the resolution that had passed at the UN General Assembly the previous December—now called 'Protection of global climate for present and future generations of mankind'—requested that the IPCC make recommendations on strengthening relevant existing international legal instruments and on 'elements for inclusion in a possible future international convention on climate.'[556] In this second session, it also became evident just how much more attention was being paid to the IPCC: the number of delegations nearly doubled, the total number of attendees more than doubled, and there was now a small contingent of environmental NGOs alongside a business lobby.

In his welcoming address to the second session, Tolba described the IPCC's partnership with the WMO in terms of the goal of climate action as:

> ...helping to unify the scientific and policymaking communities of the world to lay the foundation for effective, realistic and equitable action on climate change.'[557]

Patrick Obasi (who had taken over from Wiin-Nielsen as WMO Secretary General in 1984) then explained how the Executive Council had just approved resolutions so that negotiations for a convention could begin after the IPCC assessment was received.[558] Bernthal added that the conclusion of the 'policy' group report would be the starting point for formal negotiations on a framework convention.[559]

But then Jim Bruce announced that Working Group 3 was not the only forum in which preparations were being made for a convention. He explained how the WMO and UNEP executives had formed another small group, outside the IPCC process, that was deciding how best the two UN bodies could contribute to the treaty development and also preparing a timetable for negotiations.[560] This group would soon evolve into a convention-preparation 'task force'. To ensure continuity with the IPCC process, this eight-member advisory group would include three experts from the 'Legal Measures' subgroup of Working Group 3.[561]

The importance of the sub-group, as Bernthal would later explain, comes with an anticipation that the legal section of the report...

> ...would ultimately provide a road-map for negotiation of the framework climate convention by laying out clearly those issues that require negotiation.[562]

The convention-preparation task force was first convened by Tolba around the time of the Noordwijk ministerial conference. It included Howard Ferguson, who was coordinating preparations for the Second World Climate Conference, where convention negotiations were still expected to begin.[561,563]

Given that this task force was not a designated part of the treaty process, the question begs: *Why was it convened?* One way to understand this move is through noting Tolba's preference for working with smaller, more manageable groups, just as he had in the ozone convention development. In an insightful summary of these developments, Shardul Agrawala suggests that a politically important goal of this task force was...

> ...to keep the deliberations for the climate convention low key and to prevent them from becoming enmeshed in the much more political UN General Assembly.

Agrawala sees this as...

> ...Tolba's attempt...to duplicate the informal 'ad-hoc group on legal and technical experts' which had led to the signing of the Vienna Convention on Ozone.[564]

Tolba's attempt to replicate his Coordination Committee on the Ozone Layer in the AGGG may have failed,[ll] but this new body seems to have been a successful attempt to replicate his small ozone legal advisory group in an initiative largely independent of the IPCC. This initiative was already underway when, in December 1989, the UN General Assembly gave assent to Tolba's request for it to direct Tolba and Obasi to 'begin preparations for negotiations on a framework convention on climate', which the Assembly also resolved should 'begin as soon as possible after the adoption of the interim report of the Intergovernmental Panel'.[565]

The third session of the IPCC in February 1990 was opened in Washington by none other than the President of the United States. In his address, George Bush remained enthusiastic, pledging his commitment to the IPCC process and

---

ll See p. 212.

extending his invitation to host 'the first negotiating session for a framework convention, once the IPCC completes its work'.[566] His enthusiasm came despite some cooling on the issue among his advisers. Concerns had been raised about the economic costs of emissions cuts and also the reliance on climate models, given all their known shortcomings, in particular their failure to adequately simulate the movement of heat through the oceans.

These concerns reflected those widely raised in the scientific press; a backlash of scepticism was starting to develop (more on that below). One of the administration's responses to these concerns was to pour money into improving the models in an effort to reduce the uncertainties.[567] Generally, federal funding for climate research, and specifically for global warming research, was expanded massively at this time, way beyond the wildest dreams of those scratching for funds in the late 1970s. In his address to the IPCC, Bush boasted a budget that included 'over $2 billion in new spending to protect the environment'. The spending on the Global Change Research Program, he said, 'will increase by nearly 60% to over $1 billion'. Bush presented the provision of this funding as assurance of support for the cause in the face of rising scepticism and in the face of continuing concerns that the shift from the science to policy had been premature:

> Where politics and opinion have outpaced the science, we are accelerating our support of the technology to bridge the gap.[568]

The sudden movement of the climate change issue to centre stage in world affairs and the way it came about produced some powerful and rapid flow-on effects for those in the world of environmental science and policy. For one, it was an undoubted boon for the atmospheric sciences. Indeed, with the broad range of impacts to consider, both natural and social, it started to provide opportunities for experts across a range of natural and social sciences. Another effect was that the IPCC's dominance of the policy development process reflected glory back onto its parent bodies, drawing funding and attention to the WMO's efforts to coordinate work on the underlying science, and involving UNEP (through the IPCC and separately) in the coordination of the policy development of a global policy issue much grander than ozone layer protection. The assessment panel itself achieved a renown that challenged even that of its parent bodies, and as it became more prominent, it became more independent than any scientific assessment panel that had gone before.

But the most remarkable effect of the sudden rise of the climate change issue in world affairs came partly in response to the transformation of the intergovernmental assessment into an expert assessment. The expected constraints of an 'intergovernmental mechanism' had been overcome in various ways:

244

- the executive 'IPCC bureau' consisted of scientists and policy advocates acting independent of country representation

- the Houghton-led changes to the assessment process meant that the assessment development was largely independent of the intergovernmental panel, which was only brought in at the end to review and accept the report

- Tolba's convention development 'task force' was outside the process and entirely unrepresentative.

However much these distortions of the original intent may have displeased those who had first insisted on an 'intergovernmental mechanism', many poor-country delegations became dissatisfied with their perceived disempowerment in what had now become a major new forum for international negotiations. Some soon sought to remedy the situation by forcing a return to a properly intergovernmental mechanism independent from the IPCC and its parent bodies. This discontent and the push for its remedy would soon explode into a full-blown revolt within the IPCC, jeopardising the delivery of its assessment right at the end of the process.

The poor countries' aid campaign in UN negotiations

From the early days of the greenhouse policy debate, the poor-country governments had been lobbying for additional aid, using a variety of arguments. In the first place, they argued that the problem was/is caused by the rich in becoming rich and in maintaining their wealth. Rich countries therefore had a responsibility to address all the consequences, including those suffered by people in poorer regions. The governments of the poor countries need aid to conduct research into the impacts. They need aid to mitigate the damage. They need aid to progress their own industrial development in ways that would still preserve the climate for all. They also need help overcoming the barrier of western-owned patents and intellectual property rights rendering unaffordable the development of technologies for all these purposes. This so-called 'technological transfer' is required so that they can address the problem of climate change without hindering their modernisation.

The terms of this lobbying has a history stretching back to long before the greenhouse policy debate began, even to the early days of the UN trade talks, when the poor countries consolidated into a formidable force. The G7 might have the economic power, but, as postwar decolonisation progressed, the poor countries found that they had the strength in numbers through a collaboration known as the G77. This voting bloc came to serve its members well, not only in the trade talks, but also when voting on issues of shared interest at the UN General Assembly. Just as some semblance of this group was making demands at the climate talks, its members were also stalling the Law of the Sea negotiations with demands for money and technological assistance to develop their underwater resources. The Brundtland Commission had only given succour to this activism, while reparations to the poor featured prominently in the 1989 Hague Declaration.

In the ozone-protection negotiations, governments of poor countries demanded support for moving to substitute chemicals. They achieved some success; in June 1990, an amendment to the Montreal Protocol established an interim fund of $240 million for 1990–93. Once this financial incentive was in place, China and India, who had previously held out, moved to become parties to the protocol.[569]

With the climate treaty, the campaign for aid was taken up earlier in the process, which meant that it took place in parallel with the more advanced ozone negotiations. From the beginning, the IPCC had been raising concerns about participation in their processes, which 'developing' countries could ill afford, and for which they could rarely provide suitably qualified representatives. In its first session, the IPCC moved quickly to make funds available to support delegation attendance. At the second session, a special fund was established and a committee formed to address the participation problem. From that time on, various other methods of affirmative action remained under consideration or in the process of implementation.[570,571,572]

Where these rich/poor sensitivities mostly emerged was in the 'policy' working group. Cracks along the wealth divide had already been exposed at the plenary session convened for final acceptance of its assessment (this was in June 1990, around the same time of poor countries' success in winning aid under the Montreal Protocol). These cracks were then split wide open a few months later at the IPCC session where all three working group reports were up for acceptance and the top-level summary finalised.

The 1990 poor countries' revolt at the IPCC

The finalisation of the IPCC assessment was scheduled for its fourth session, at the end of August 1990, in Sundsvall, Sweden. The session started well, especially for Houghton, who arrived with boxes full of completed and commercially published copies of the Working Group 1 assessment report. For Bolin, it also might have been triumphal. Each line of the three working group summaries had already been given intergovernmental approval, and he had summarised their content into a single overview, which the meeting was now to approve. Once this had been done, this summary-of-summaries would be ready for delivery to the Second World Climate Conference and then to the convention negotiations that would follow.

However, instead of a triumph, the meeting degenerated into a shambles. As one UK delegate later reflected:

> ...having started in a very organized fashion with songs about the future from children's choirs...the meeting came close to a breakdown. It finished at four o'clock in the morning, one day late, with most of the delegates having abandoned their chairs in the conference hall to gather on the front podium and shout at each other.[573,¶]

At these meetings, considerable goodwill has to prevail among the delegations, as every single line requires unanimous agreement that its wording fairly represents the underlying report. But in Sundsvall, from the beginning, goodwill was in short supply among the Brazilian delegates. They led a campaign for interpolations advancing the poor countries' view, and it was mostly this campaign that pushed the final session into the early hours of the following morning with no prospect of completing the approval. Way past midnight, the meeting had ground to a halt, with a dangerous polarisation of positions. The USA had chaired the fractious final meeting of Working Group 3, and in Sundsvall they continued to resist amendments to the text requested by Brazil, which were intended to advance the poor-country view. The antagonism was distressing for all involved, because Brazil featured prominently in new plans for the treaty development process. The framework convention was to be made ready for signing at a grand 'Earth Summit' in 1992. This meeting, marking the 20th anniversary of the Stockholm Human Environment Conference, would be hosted by Brazil in Rio.

In the official report of the Sundsvall meeting, there is barely a hint of what went on, but in his memoirs written 17 years later, Bert Bolin does discuss the struggle and his disappointment with the outcome, if only politely and diplomatically.[575] The Australian Delegation Report, released within two weeks of

---

¶ Tony Brenton's recollection that 'most of the delegates' were 'on the podium' shouting at each other, is an exaggeration according to John Zillman, who was also at the meeting. In a private email, Zillman recalls instead that: '...about half a dozen delegates went up on the podium at one stage for informal consultation on the best way to proceed. It was very tense but it was entirely civil. In the end, Bert [Bolin] phoned the Brazilian Minister for Science, Jose Goldemberg (himself a distinguished climate scientist), to get him to overrule the Brazilian delegation'.[574] Even given this account, the tense informal conversations and the early morning international call for an over-ruling by the Brazilian Science Minister does not entirely diminish the impression of desperation.

the meeting, is much more descriptive and also explanatory. In the foreword, it notes that tension between the industrialised and developing countries...

> ...had been increasing during the IPCC process as the time draws closer for the start of negotiations for a climate convention. This session was seen as a final chance to influence the IPCC towards a more political stance in the report and to reflect the desires of the developing countries for free transfer of technology and capital from industrialised countries. As a result, relatively little time was devoted to the discussion of science...[576]

The Australians were uncommonly explicit about the motives they saw behind the obstructionism that almost caused the collapse of the meeting:

> It became clear as the session developed that a number of developing countries led by Brazil were anxious to prevent Panel endorsement of the proposed Overview and Conclusions in order to ensure that the main international action on the climate change issue moves out of the WMO-UNEP framework and under the umbrella of the UN General Assembly.[577]

Already, in the IPCC Bureau meeting that directly preceded the full meeting, the Brazilian representative had 'made clear that Brazil saw the further action, both on negotiation of a climate convention and the future of the IPCC, as being in the hands of the UN General Assembly'.[578]

In the face of consolidating opposition, the chairman, Bert Bolin, remained determined that the meeting should not be abandoned or reconvened (as had been suggested), but should produce some agreed text for delivery to the Second World Climate Conference, only three months away. In the end he succeeded, but only after his partly approved summary-of-summaries had been discarded.

In the early hours of the morning, what became the policymakers' summary of the entire assessment was approved. This top-level document was as shambolic as the agreement that produced it. It consisted mostly of pasted-together slabs of the working group summaries with a disclaimer on the front and a concise affirmation of the poor countries' view inserted in the 'Response strategies' section. The disclaimer explains that the report does not reflect the positions of all the participating governments, which is to say, in fact, it had not been approved unanimously by the group. This was an admission of failure to meet the main object of the entire process, which was to achieve intergovernmental agreement on the scientific basis for policy action. The other insertion was a declaration that rapid technological transfer is urgently required. This declaration closes thus:

248

> Narrowing the gap between the industrialized and developing world would provide a basis for a full partnership of all nations in the world and would assist developing countries in dealing with climate issues.[579]

Whether or not this were true, it was not a finding of the climate change assessment that could be found anywhere in its chapters. Perhaps for the best, this document was never properly published, nor widely distributed, and it was quickly forgotten. Not so easy to forget was the desperate battle that got it over the line, which had only served to consolidate a bloc of opposition behind Brazil.

Aftermath

Despite the near collapse of the assessment process, and in the face of Brazil's expressed intention to ensure that any coordination policy action moved out of the WMO-UNEP framework, Tolba went ahead with an *ad-hoc* meeting of government delegates to advance the policy action process. Authorised by his Governing Council (and by the WMO Executive Council) just prior to the IPCC session in Sundsvall, this meeting was an 'open-ended working group of government representatives' to develop the 'means and modalities' for negotiation of the convention beyond what had already been achieved in his *ad-hoc* convention task force. As such, it was part of Tolba's final preparations for the first formal convention negotiation session, which was scheduled for Washington early in the new year.[580,581,582]

After the IPCC assessment approval meeting in August 1990, and then this *ad-hoc* policy meeting that September, next on the tight schedule was the Second World Climate Conference in November. As if nothing had happened, UNEP, the WMO and their IPCC remained centre stage in a theatre that brought massive publicity not to the World Climate Programme (a review of which had been its original purpose) but to the new climate emergency. To meet this emergency, the conference reaffirmed the plan for a climate convention to be prepared and made ready for signing at the Rio 'Earth Summit' in two years' time. Margaret Thatcher forged ahead on the eve of her own political demise with a marvellously evocative keynote address that celebrated the WMO-UNEP process, seemingly unaware of the impending doom. 'We must not waste time and energy disputing the IPCC's report', she said; rather it should be taken as 'our sign post', directing UNEP and the WMO, 'the principal vehicles' taking us to our 'destination', a global convention.[583]

Alas, by the end of the year, Thatcher was no longer prime minister and the UN bodies she had extolled were no longer the vehicles of the convention. That December the Brazilians had their way: the treaty negotiations would be placed under the auspices of the UN General Assembly, where the poorer countries could continue to assert their majority. A new and independent Intergovernmental Negotiating Committee (INC) would be established to draft the climate convention.[584] It would first convene near Washington in February 1991, just as President Bush had requested, but by then UNEP and the WMO were little more than interested spectators. Bolin watched on as 'more than 100 people from the UN in New York attended the transformation of the climate change issue into a central UN effort.' In his memoirs, he says that he 'really wondered what they were doing, since their participation in the work so far had been minimal.'[585]

If Bolin was disappointed that diplomats and bureaucrats had taken over from the scientists, then Tolba was just as upset that UNEP had been excluded. In his memoirs, his dissatisfaction with the outcome remains palpable. After carefully recounting the stages by which protection was brought to the ozone layer, there is a brief section on protection of the climate that quickly descends into a catalogue of just how he had again and again made the poor countries' case for funding and how he might have come good with it, just as he had with the ozone negotiations.[586] As it turned out, after UNEP was sidelined—around the time that Tolba resigned from its leadership, and just four months out from the Rio summit—brinkmanship by the poor countries delivered a result. At the fifth session of the INC in New York, while still holding out on binding emissions targets, the USA conceded financial assistance for poor countries, announcing that it would commit $50 million to a core fund and $25 million during the next two years for emissions evaluations and impact assessments.[587]

As for the IPCC, from this time onwards it would steer well clear of policy proposals, restricting itself to 'policy-relevant but policy-neutral' advice. However, before it was able to produce another full report, the struggle for control of the policy agenda threatened its very survival. The convention drafted by the INC invoked a Subsidiary Body for Scientific and Technological Advice (SBSTA), which would provide the parties to the Convention with 'information and advice on scientific and technological matters'. SBSTA's provision of 'assessments of the state of scientific knowledge' would include the preparation of 'scientific assessments' as well as responses to any technological questions of the parties to the convention.[588] The design of SBSTA was also intergovernmental, but its members could only include representatives of countries who had signed-up to the convention. The IPCC could never become this SBSTA, so it was told, because IPCC membership was open to all countries. Nevertheless, the IPCC did still rate a mention in the text of the convention, if only as one provider among others of advice in the 'interim'; that is, it could provide advice until the convention came into force and SBSTA was inaugurated at the first Conference of Parties (CoP1). The wording could not be clearer: at CoP1, SBSTA would

250

replace the IPCC.[589]

The WMO and UNEP were being pushed out into the cold. With the treaty process now run by career diplomats, and likely to be dominated by unfriendly southern political agitators, the scientists were looking at the very real prospect that their climate panel would be disbanded and replaced when the Framework Convention came into force.

The policy gap and the need to find a signal

While Bolin was facing up to the exclusion of the IPCC from the climate treaty process, other scientists in the climate field had very different views on the situation. Many remained back where the assessments were before the 1985 Villach conference, viewing the evidence-base as too weak to support a path of policy action, as with Robert White and his 'inverted pyramid' of knowledge.** Indeed, with the realisation that there was an inexorable movement towards a treaty, there was an outpouring of scepticism from the scientific community. This chorus of concern was barely audible above the clamour of the rush to a treaty and it is now largely forgotten.

At the time, John Zillman presented a paper to a policy forum that tried to provide those engaged with the policy debate some insight into just how different was the view from inside the research community. At a workshop organised by Sustainable Development Australia in August 1991, Zillman tried to strip away any illusions that 'the climate science community' is united behind the moves towards a climate convention. The then head of the Australian Bureau of Meteorology and First Vice-President of WMO, began by explaining how the 1985 Villach statement was 'very low on the consensus scale'. As such, it produced a wide range of reaction in the climate science community. Some saw it as...

> ...quite literally 'little more than a lot of hot air' and, to the extent that it was something more, it was essentially a crude bid for increased research funding from one particular subset of the climate research community.[590]

Zillman went on to explain how the initial response of the mainstream climate science community to the emergence of the scare 'was a measure of cautious optimism that governments were about to take the need for increased climate research seriously' and that this might allow 'climate prediction before the end of the century'. In other words, their objections to the premature raising of public alarm were subdued in the hope of increased funding to advance the

---

**See p. 143.

251

understanding of natural climatic variation and change. But when 'the policy process stripped away the scientists' careful caveats', scientists began to speak out about the uncertainties in the predictions. Meanwhile, the poor countries soon found in the greenhouse debate 'a useful surrogate for a range of other political agendas'. When their action affected the removal of the proposed convention from the auspices of the WMO/UNEP framework, this was...

...widely, though perhaps not completely accurately, interpreted as conveying two messages:

- science has served its purpose in the process and can now be dis-
pensed with;
- the international agenda isn't climate change, it's 'technology transfer' and aid, and it must be pursued in the fora where the developing
world has the numbers.[591]

Among climate scientists, Zillman felt that the initial *modus operandi* at the INC served to reinforce the conclusion...

...that the greenhouse debate has now become decoupled from the scientific considerations that had triggered it; that there are many agendas
but that they do not include, except peripherally, finding out whether and
how climate might change as a result of enhanced greenhouse forcing and
whether such changes will be good or bad for the world.

To give some measure of the frustration rife among climate researchers at the time, Zillman quoted the director of WCRP. It was Pierre Morel, he explained, who had 'driven the international climate research effort over the past decade'. A few months before Zillman's presentation, Morel had submitted a report to the WCRP committee in which he assessed the situation thus:

The increasing direct involvement of the United Nations...in the issues
of global climate change, environment and development bears witness to
the success of those scientists who have vied for 'political visibility' and
'public recognition' of the problems associated with the earth's climate.
The consideration of climate change has now reached the level where it is
the concern of professional foreign-affairs negotiators and has therefore
escaped the bounds of scientific knowledge (and uncertainty).

The negotiators, said Morel, had little use for further input from scientific agencies including the IPCC 'and even less use for the complicated statements...put forth by the scientific community'. This suggested that 'efforts to

communicate scientific views' to the negotiators 'have reached the point of van-
ishing returns and are a waste of time'. Morel then provided his recommendation
as director of the WCRP to its committee, which was...

> ...to curtail further involvement... in the preparation of pronouncements
> which can only be viewed, from the scientific perspective, as surrealistic.
> With the concurrence of [the committee], abstention is proposed from
> further participation in such exercises, such as the [Rio Earth Summit] in
> 1992.[592]

In fact, Morel's proposal to curtail cooperation in the policy discussions did
not carry. Not that there was much to curtail. As we have seen, WCRP had repeatedly and successfully rejected direct involvement in global warming science.
Indeed, partly as a consequence, the direct engagement in the policy process had
been assigned to an entirely new and independent organisation, the IPCC.[††]

The general feeling in the research community that the policy process had
surged ahead of the science often had a different effect on those scientists engaged with the global warming issue through its expanded funding. For them,
the situation was more as President Bush had intimated when promising more
funding: the fact that 'politics and opinion have outpaced the science' brought
the scientists under pressure 'to bridge the gap'.

The existence of this gap was often left unspoken at the time, and remained
largely unspoken until its bridging could be celebrated. Previously, when the
treaty process was only a vague possibility, this gap had been much discussed.
Indeed, in the early 1980s, at least within DoE Carbon Dioxide Program, the
'conventional wisdom' was that policy action required going beyond the models
to deliver empirical evidence.[594] This perceived requirement was the impetus
for a specific program, as MacCracken reported in his 1979 progress report:

> Because it seems unlikely that policy decisions related to fossil fuel energy
> could be made without addressing forthrightly the issue of identifying
> signals of a changing climate, DoE expects to initiate a modest program
> during the next several years [to find these signals].[595]

[††] For an overview of this outpouring of scepticism from the scientific community around this
time, see the documentary 'The Greenhouse Conspiracy', which screened in Britain in August
1990, and remains available on the internet. The scientific basis of the scare is systematically criticised through interviews with scientists, including: David Aubrey (Woods Hole Oceanographic
Institute), Robert Balling (Arizona University), John Houghton (UK Met Office), Sherwood Idso
(US Water Conservation Laboratory), Peter Jonas (University of Manchester), Richard Lindzen
(MIT), Patrick Michaels (University of Virginia), John Mitchell (Meteorological Office), Regi-
nald Newell (MIT), Julian Paren (British Antarctic Survey), Stephen Schneider (US Center for
Atmospheric Research), Roy Spencer (NASA Space Flight Center), and Tom Wigley (University
of East Anglia).[593]

This is what became known as the 'first detection' program. With funding from DoE and elsewhere, the race was soon on to find ways to achieve early detection of the climate catastrophe signal. More than 10 years later, this search was still ongoing as the framework convention to mitigate the catastrophe was being put in place. It was not so much that the 'conventional wisdom' was proved wrong; in other words, that policy action did not in fact require empirical confirmation of the emissions effect. It was more that the policy action was operating on the presumption that this confirmation had already been achieved.

For the scientists at the IPCC to have any hope of reengaging with the policy process, a rear-guard action against scepticism was now required. Scientific evidence needed to be found to support the policy action that was already underway. The only trouble was that the problem of detecting greenhouse gas emissions' effect on the climate was not getting any easier to solve, and hopes of success were diminishing as further complications came to light. We now take up with the history of the search for the climate catastrophe signal and follow this search right through to where its eventual success saved the IPCC.

15 Searching for the catastrophe signal

Early claims of detection

> Few of those familiar with the natural heat exchange of the atmosphere, which go into the making of our climates and weather, would be prepared to admit that the activities of man could have any influence upon phenomena of so vast a scale. In the following paper I hope to show that such influence is not only possible, but is actually occurring at the present time.[596]

So Guy Callendar began his presentation one Wednesday afternoon in the winter of 1938. He had been cordially invited to present his findings to a meeting of the Royal Meteorological Society. Although an amateur in this field, he proceeded to impress the experts with extraordinary learning and research into every aspect of the topic, bringing it all together in support of the conclusion that more than half of the warming over the previous 30 years could be attributed to increasing 'sky radiation' due to industrial emissions of carbon dioxide.

Early speculation that industrial emissions of carbon dioxide might one day contribute some (welcome) warming had faded at the turn of the 19th century with the rejection of the theory that variations in volcanic carbon dioxide emissions could trigger positive water vapour feedback sufficient to explain geological-scale temperature changes; it was established that most of the wavelengths of radiated heat that might be blocked by carbon dioxide were already blocked by (almost ubiquitous) water vapour.[597] These results were announced at the beginning of the most sustained and widespread warming trend in the temperature record of northern climes. Now, in 1938 at the other end of this warming period, a British steam engineer, Guy Callendar, had gathered some new data, revived the theory and announced that he had found a human influence detectable in it.

255

Impressed though they were by the depth and breadth of research, the assembled experts rejected Callendar's conclusion. There was no question 'that there had been a real climatic change during the past thirty or forty years', but this recent warming was easily explained as but a phase in the natural variability of climate, similar to those in the not-so-distant past. And anyway, if emissions were having an impact, it would not be as straight forward as Callendar's calculations suggested. 'An increase in the absorbing power of the atmosphere would not be a simple change in temperature'. Instead, it would have a more complicated impact on energy transfer in the general circulation, including on vertical heat dissipation into the upper atmosphere and so to space.[598]

While climatic change on a geological scale had been generally accepted for decades, British meteorology was also already familiar with natural climatic change on smaller timescales. In fact, one Meteorological Society fellow at Callendar's presentation, Charles Brooks, had been investigating climatic changes and their impacts on civilisation for many years.[599]

Following his presentation to the meteorologists, Callendar continued to develop his theory and collect more evidence, even as the warming trend looked to be subsiding. Before the mid-century cooling became entirely evident, he found an ally across the Atlantic in Gilbert Plass. Plass had revived the 19th century argument that ice ages might be triggered by a lack of volcanic activity, and, as an extension, he also attributed recent warming to industrial emissions. At the 1953 annual meeting of the AGU he declared that:

> The large increase in industrial activity during the present century is discharging so much carbon dioxide into the atmosphere that the average temperature is rising at the rate of 1.5 degrees per century.[600]

Thus, he seems to have attributed all the 20th century warming to this cause, and this claim drew attention from the science press and even some newspapers.

Three years later, Plass gained further publicity for an early application of computer modelling, in which he calculated the atmospheric sensitivity to a doubling of carbon dioxide at 3.6°C. However, this was without the cooling feedback of clouds. When he added middle-level and high clouds, the warming result reduced to 2.5°C. Not long before Keeling commenced his benchmark monitoring of 'background' atmospheric carbon dioxide, Plass estimated that industrial emissions contributed a 30% increase per century and that this was causing a 1.1°C warming per century. Plass published five papers on the topic in 1956, including a popular account in *American Scientist*. Again, the press were interested, although no great alarm was raised. Neither Callendar nor Plass questioned that warming was a good thing. The most alarming statement is found at the end of the *American Scientist* article, where Plass suggests that there may be a problem after several centuries:

> ...the temperature rise from this cause may be so large in several centuries that it will present a serious problem to future generations. The removal of vast quantities of carbon dioxide from the atmosphere would be an extremely costly operation.[601]

Roger Revelle also joined the discussion, downplaying the suggestion of any 'catastrophe'. Around this time, Plass consulted Callendar and they corresponded. In December 1957, Plass reported that Revelle and others had 'attacked the carbon dioxide climatic theory quite vigorously at a meeting earlier this year'.[602] (As we have seen above, by 1965, Revelle himself was speculating about a possible catastrophe.) Others soon published attacks on Plass's version of carbon dioxide driven warming. These included Lewis Kaplan, who found that multiple errors in the calculations had caused Plass to grossly overestimate the effect,[603] while another critical analysis concluded that 'the chain of reasoning appears to miss so many middle terms that few meteorologists would follow him with confidence'.[604] When Fritz Möller attempted to replicate Plass's results, modelling gave exponential increases until he discovered that slight changes in cloudiness and water vapour feedback could cancel the effect. These results led him to conclude that 'the theory that climatic variations are affected by variations in the CO_2 content becomes very questionable'.[605]

Back in Britain, unperturbed by this polite rejection of his attribution theory —or perhaps driven by it—Callendar continued to gather evidence to illustrate this mechanism of warming. To have any chance of overcoming the scepticism of the entire meteorological establishment, what he really needed was to show some evidence of this particular causation in the climate data itself. Firstly, he needed to find evidence in the records that at least some of the recent warming was not entirely due to the natural 'internal' variability of climate, and then he needed to somehow find a way of showing that the trend was caused by emissions and not by any of the other likely candidates for external climate forcing. Callendar did find a way, but not until the 1960s, and by then it was too late.

Others had already noticed, as Callendar had, that the pattern of the recent warming was uneven across the latitudinal zones. So now he began to look more closely at these zonal patterns for signs of the cause. He did this by modelling the pattern of warming that might be expected from the various proposed causes, and then comparing these with observations. What he found was that the solar cycle and volcanic dust were indeed necessary to explain fluctuations of a few years' duration. As for the overall trend, greater warming in high latitudes had been observed, especially in the northern hemisphere, while the tropics were not only experiencing less warming but also less rain. Callendar determined that a decline in stratospheric volcanic dust over this time may have influenced the trend, but the main features of the observed pattern were incompatible with this cause. However, they were 'not incompatible with the hypothesis of increased carbon dioxide radiation.' With this argument, he had taken the first steps towards the technique of warming attribution by 'pattern analysis', a technique later refined and re-pitched as climate change 'fingerprinting'.[606]

Why this pattern analysis failed to gain traction in the atmospheric research community is unclear, although when Callendar was noticed by professional meteorologists it was often with derision. Even after sending Keeling off to set up the carbon dioxide monitoring station on the top of a volcano in Hawaii, Roger Revelle would not let Callendar off first base, refusing to believe that carbon dioxide could have increased as much as Callendar claimed (and as is now confirmed).[607] Callendar's book *Climate and Carbon Dioxide* was never published. Just how embittered Callendar was towards his critics is disputed, but perhaps it was the climate itself that had the last word. While the data in Callendar's warming-pattern analysis cut off around 1950, the article was not published until 1961, after which time it was hard to find any warming to analyse: England was hit with a series of famously harsh winters, and in Fleming's biography there is a marvellous photo, crisp with symbolism, of Callendar shovelling snow after a blizzard in 1962.[608] The following winter of 1963 is remembered as The Big Freeze. One more winter and Callendar and his global warming theory would be dead. By this time all the talk was about *What's causing the cooling?*

Targeting 'first detection' in the Carbon Dioxide Program

During the peak of the early 20th century warming phase, neither Callendar in the UK nor Plass in the USA had succeeded in persuading meteorologists that they had in fact shown any of it to be attributable to carbon dioxide emissions. Modelling of the effect continued sporadically through to the late 1970s, when there began a concerted effort to develop an empirical science basis for the revived greenhouse warming speculation. Much of this arose under the US DoE Carbon Dioxide Program and its attempts to develop the scientific basis for 'first detection'.* The idea was at least to make ready for the earliest possible detection of the effect when it emerged above the noise of natural variability.

The importance of this matter rose to prominence at the National Carbon Dioxide Program conference in Washington, 1980.† As the conference pro-

---

*See p. 254.
†See p. 63.

ceeded, the need for empirical validation emerged as a key theme. It was evident in the introductory speech of the program director, who picked up on this gap in the science even though he was not expert in the field.

> Skeptics and critics can make a good case against unverified model predictions. I would like to know whether verification studies are now being pursued on a global basis in a sufficiently aggressive manner. By verification, I am talking about studies of past climate and past CO_2 concentrations. This seems to me to be terribly important...[A]re there leads that we are not pursuing, leads that might give some evidence of verification, a surrogate kind of verification, in the next 5 or 10 years?[609]

Some of these 'skeptics and critics' had been invited to the conference, including George Kukla, who remained more concerned about the descent into the next ice age than the prospect of warming. At the conference he argued from pattern analysis against any suggestion that the recent uptick in the warming might be caused by carbon dioxide.

At least since Callendar's time, there has been one feature of the carbon dioxide warming effect that has remained relatively uncontroversial: that the warming should be amplified in high latitudes and should thus appear first in the polar regions. When Callendar investigated the early 20th century warming, he claimed to have found this expected polar amplification. However, no such amplification was evident as the cooling started to reverse in the late 1970s. In fact, as sceptics such as Lamb and Kukla would often point out, there appeared to be a continued Arctic cooling. In Washington, Kukla claimed that the carbon dioxide modelling is contradicted by the available data, which indicates that during the previous year the high North had descended to the coldest yet recorded.[610] In a later discussion at the 1980 Washington conference, a modeller makes the same point:

> The amplification is predicted at high latitudes in the atmospheric models. One thing that really bothers me is not that we haven't seen the CO_2 signal in mid-latitudes, but that we haven't apparently seen it at high latitudes where we probably should be seeing it.[611]

Reid Bryson, also at the conference, was concerned that the aerosol emissions effect was not also included in the modelling. He claimed that he had used an 'empirical approach' to get a negligible sensitivity to carbon dioxide, and he also presented modelling results that included 'a term for the emissivity of aerosols'. His results strongly suggested, he said,...

> ...that models ought to be tested against the past. If one can simulate the past, then at least I have more confidence that one might be able to predict

the future. Those models that can simulate the past have much less sensitivity to CO_2 than those that cannot simulate the past... The aerosol effect should also be included in other models, and it will need to be included to simulate the climate of the recent past.[612]

In fact, the aerosol effect would not be factored into the main carbon dioxide warming models for more than a decade, but, when finally introduced, it had a profound impact, as we shall see.

Meanwhile, back at the Washington conference, Jerome Namias also joined the discussion. Namias was a renowned climate forecaster who had led the Climate Research Group at Scripps Institution of Oceanography. He was one of those who came out as sceptical when the climate treaty process began. Namias emphasised potential negative feedbacks, which would reduce the warming effect of carbon dioxide, and he also stressed the importance of historical climatology for verifying the models. In the ensuing discussion, the argument was repeated that an understanding of natural climate change was needed to provide a reference against which to measure the carbon dioxide effect. As another scientist put it:

> ...it is baseline data for DoE. DoE has to have it. I don't see how you can operate without it. I don't see what all the models can do if we don't know [what is caused by natural variability]...[613]

Such concerns were raised with few objections and so it is not surprising to find that the conference workshop on 'Climate Effects' concluded that:

> Unfortunately...there exists no technique or set of past or recent climate conditions that will allow completely satisfactory verification of any numerical climate model.[614]

The following year, the DoE Carbon Dioxide Program held a workshop dedicated specifically to the problem of first detection. At this workshop, the bleak outlook for model verification was generally acknowledged, as shown in these excerpts from the workshop summary:

> For more than a decade, numerical models have predicted that increasing CO_2 concentrations would cause warming of the global climate, with amplified warming in polar regions and cooling in the stratosphere. Such changes, however, are not now obvious in the climatic record... If these estimates [of atmospheric carbon dioxide rise] are correct then it follows directly from the results of numerical models that the global climate should by now have warmed by several tenths of a degree Celsius. While a number of individual researchers have suggested that this has occurred,

many others disagree for a wide variety of reasons... Quite clearly...the temperature is not increasing exponentially from some baseline value as would be initially expected if CO_2 emissions were increasing exponentially and the temperature change varied logarithmically with CO_2 concentrations. Does this mean that the model predictions of the CO_2 effect on climate are wrong...? ...In summary, the search for definitive evidence that the climate is responding to increasing CO_2 concentrations as most models predict is not yet successful.[615]

It may surprise readers that even within the 'carbon dioxide community' it was not hard to find the view that the modelling of the carbon dioxide warming was failing validation against historical data and, further upon this admission, the suggestion that their predicted warming effect is wrong. In fact, there was much scepticism of the modelling freely expressed in and around the Carbon Dioxide Program in these days before the climate treaty process began.

Those who persisted with the search for validation got stuck on the problem of better identifying background natural variability. There did at least seem to be agreement that any recent warming was well within the bounds of natural variability, and so the common speculation that detection of the carbon dioxide signal might arrive by early the following century was usually based on arguments that, if the warming continued, it could somehow be identified (using current or improved data and techniques) as beyond these bounds. Or not. As we have seen, when Wigley soon came to write his 'empirical science' section in the SCOPE 29 report, he would not even speculate on that, and the 1985 Villach Impact on Climate Change working group was even more sceptical. This shows just how extraordinary was Hansen's claim three years later that he had detected the emissions signal in the recent warming of the global climate to a confidence level of 99%. Hansen came to this figure in 1988 by finding the likelihood of such a natural warming anomaly at around 1% based on a calculation of a normal random distribution around a mean in the instrumental temperature record. This rather crude and old-fashioned methodology did not win him any admirers. In fact, his detection claim generated an enormous backlash in the US climate science community that belies the public perception of a breakthrough.

An article in *Science* the following spring gives some insight into the furore. In 'Hansen vs. the world on greenhouse threat', the science journalist Richard Kerr explained that while 'scientists like the attention the greenhouse effect is getting on Capitol Hill', nonetheless they 'shun the reputedly unscientific way their colleague James Hansen went about getting that attention'. Kerr found the objections to Hansen's detection claim most evident at a five-day workshop on the enhanced greenhouse effect:

None of the select greenhouse researchers at the meeting could agree with him. 'Taken together, his statements have given people the feeling the greenhouse effect has been detected with certitude', says Michael Schlesinger, himself a modeller at Oregon State University. 'Our current understanding does not support that. Confidence in detection is now down near zero.'[616]

Tom Wigley told Kerr that Hansen had only shown that there had been some recent warming, while Tim Barnett from Scripps Institution of Oceanography explained the problem of natural variability:

> The variability of climate from decade to decade is monstrous...To say that we've seen the greenhouse signal is ridiculous. It's going to be a difficult problem.[617]

Kerr described how Hansen had become a scientist under siege, how, soon after Hansen's congressional testimony, a climate workshop in Washington had descended into 'a get-Jim-Hansen session'. Since then the celebrity scientist tended to arrive at conferences only to give his presentation, leaving without joining the discussion.[618]

Clearly, the scientific opposition to any detection claims was strong in 1989 when IPCC assessment got underway. When Houghton asked Wigley to lead the writing of the detection chapter for the first IPCC report, Wigley's scepticism was not particularly outstanding, nor was it moderated when Barnett joined him.

Detection in the first IPCC assessment and the Rio Supplement

In the first assessment, some of the empirical science that Wigley had addressed in SCOPE 29 was covered by others in separate chapters. Just before Wigley's contribution on detection, the first assessment has another chapter entitled 'Observed climate variation and change'. While in SCOPE 29 Wigley had been reluctant to speculate about the recent global trend, this chapter addresses it with confidence and even illustrates the discussion with global temperature trend graphs. Not that it comes out in any conclusive way on the main point under discussion. Indeed, it notes that some of the evidence in the historical trend data seems to run contrary to the hypothesis that CO_2 has been a significant driver of change. It even suggests that the half-degree warming across the 20th century could well be a natural rebound from the Little Ice Age.[619]

In the following chapter, on detection, Wigley had found that very little had changed in the science since SCOPE 29. In writing the chapter during 1989,

Wigley was helped by Tim Barnett from Scripps, someone who was not exactly going to moderate his scepticism; as we saw above, that same year Barnett was quoted in *Science* calling any detection claim 'ridiculous'. Wigley and Barnett found that the problem of natural variability meant that detection was a long way off. *But how long?* During the IPCC review process, Wigley was asked to answer the question that he had avoided in the SCOPE 29: *When is detection likely to be achieved?* He responded with an addition to the IPCC chapter that explains that we would have to wait until the half-degree of warming that had occurred already during the 20th century is repeated. Only then are we likely to determine just how much of it is human-induced. If the carbon dioxide driven warming is at the high end of the predictions, then this would be early in the 21th century, but if the warming was slow then we may not know until 2050 (see Figure 15.1). In other words, scientific confirmation that carbon dioxide emissions is causing global warming is not likely for decades.

These findings of the IPCC Working Group 1 assessment presented a political problem. This was not so much that the working group was giving the wrong answers; it was that it had got stuck on the wrong questions, questions obsolete to the treaty process. The IPCC first assessment was supposed to confirm the scientific rationale for responding to the threat of climate change, the rationale previously provided by the consensus statement coming out of the 1985 Villach conference. After that, it would provide the science to support the process of implementing a coordinated response. But instead of confirming the Villach findings, it presented a gaping hole in the scientific rationale. When its Working Group 1 foundered on the question of empirical validation, this at first did not seem to matter in the march towards the climate convention. Nevertheless, the scientists at the IPCC would find it hard to avoid empirical science while retaining their involvement in the treaty process.

The one great hold that the IPCC could hope to maintain in the policy process was that its assessment of the science provided policy action with its ultimate authority. The treaty talks could not survive without the authority of science. The IPCC had been instituted because the 1985 Villach scientific consensus was not considered sufficient; it had been instituted because of concerns about the policy entrepreneurism that had followed. Scientist-advocates would continue their activism, but political leaders who pledged their support for climate action had invested all scientific authority for this action in the IPCC assessment. *What did the IPCC offer in return?* It had dished up dubiously validated model projections and the prospect of empirical confirmation perhaps not for decades to come. Far from legitimising a treaty, the scientific assessment of Working Group 1 provided governments with every reason to hesitate before committing to urgent and drastic action.

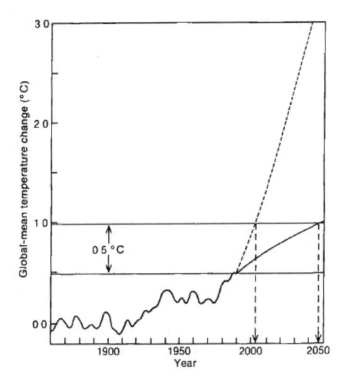

Figure 15.1: Detection not expected for decades.

The IPCC first assessment graphically explains its prediction of the date by which manmade global warming will be detectable using current detection techniques. The prediction that attribution to the human cause will be achievable sometime in the first half of the 21st century is itself dependent on the range of predicted rates of manmade global warming.[620]

And yet, the success of the Brazilian-led revolt that left the IPCC vulnerable at the end of its first assessment, also presented a way out. When the climate treaty negotiations began under an independent negotiating committee (the INC), the IPCC only remained in the game under the interim arrangement: it would continue to provide the treaty process with scientific advice until the first meeting of the parties to the convention, when the convention's own scientific advisory body, SBSTA, was supposed to take over. Under the interim arrangement, the INC made clear that it was not interested in such questions as empirical validation. Its interest in the science was not about *whether* danger-

264

ous manmade climate change is happening because it was already beyond that. As an independent policy organisation, the INC could play a role similar to that of the US EPA in the DDT controversy, where the uncertainties in the scientific justification for action need not interfere with the political expedience of acting decisively.[‡] One way the INC did this was to restrict the IPCC's role so that it was only to assess the science related to implementation of the climate convention. This included advice on assessing levels of emissions ('sources and sinks'), the setting of emissions-reduction targets, the methods and technologies of mitigation, and science related to the facilitation of technology transfer to poorer regions.[621] Under the interim arrangement this is all the intergovernmental negotiators ever asked of the IPCC. There were never any requests to reassess the basis science justifying all this activity. Thus the IPCC could have avoided such questions. After all, the problem was only with its Working Group 1, and only with a small part of that group's assessment. The modelling was still pointing to catastrophe and the other working groups had been operating from the start as though the scientific basis of action had already been established. Thus, it might have been prudent for the IPCC to leave alone the problems with empirical validation. However, it chose not to.

To support the discussions of the Framework Convention at the Rio Earth Summit, it was agreed that the IPCC would provide a supplementary assessment in response to six questions entirely within the confined scope of the interim arrangement. However, as part of its response to this request, Working Group 1 decided to also include an updated assessment of the basic science. This addition to the supplement remained just as *off-message* as the main (the first) assessment. This is how we have an assessment of the science behind manmade climate change published especially to support discussions of a treaty to mitigate manmade climate change that declares no improvement in the validation of the models predicting the change to be mitigated. Indeed, in its final chapter, we find that the attribution of any change to the manmade cause is still nowhere in sight. This 'Rio supplement' explains that 'progress is difficult' firstly due to various unresolved problems with the modelling, but then comes the problem of distinguishing various types of external climate forcing:

> ...the climate system can respond to many forcings and it remains to be proven that the greenhouse signal is sufficiently distinguishable from other signals to be detected except as a gross increase in tropospheric temperature that is so large that other explanations are not likely.[§]

[‡] See p. 23.

[§] The Supplementary Report is Working Group 1's response on the three tasks assigned to that group. Its Section B responds to Task 2, on 'predictions of the regional distribution of climate

In other words, we don't know how to distinguish greenhouse warming from other causes of warming unless the warming rate becomes unnaturally extreme: empirical validation is not here, and looks to be a long way off. This admission of failure to provide the scientific basis for climate action—provided by the IPCC *in addition* to what had been requested—could only serve as a reminder of the science/policy disjuncture when it was needed least. However, the momentum for action in the lead up to Rio remained sufficiently strong that these difficulties with the scientific justification could be ignored. The reception of this new IPCC report among treaty advocates at Rio was nothing if not positive.

The 1992 Rio 'Earth Summit' and the Framework Convention on Climate Change

The Rio Earth Summit was a monumental global event that attracted an enormous press contingent. From the outset, action on climate change was a major theme. In the opening address, the summit's Secretary General, Maurice Strong, was able to make a strong case for action based on the IPCC report:

> The IPCC has warned that if CO_2 emissions are not cut by 60 percent immediately, the changes in the next 60 years may be so rapid that nature will be unable to adapt and man incapable of controlling them.[623]

The policy action to meet this threat—the UN Framework Convention on Climate Change—went on to play a leading role as the headline outcome of the entire show. The convention drafted through the INC negotiation over the previous two years would not be legally binding, but it would provide for updates, called 'protocols', specifying mandatory emissions limits. Towards the end of the Earth Summit, 154 delegations put their names to the text. Signing for the USA, President Bush committed to early ratification of the convention and to the development of a national climate action plan by January 1993. In October 1992, the US Senate ratified the convention, becoming the fourth country to do so. The rapid pace of ratifications after that meant that the convention came into force much earlier than many expected, on 21 March 1994. The first session of the Conference of Parties (CoP1) was required to be convened within a year, and it was soon scheduled for Berlin at the end of March 1995. The date was therefore set for the expiry of the interim arrangement with the IPCC.

---

change', with low confidence in regional predictions and ends by discussing the difficulties with model validation even at the global scale. The following Section C on 'observed climate variability and change' is outside of scope of the tasks assigned to it by the INC. This section includes our quoted discussion of the attribution problem.[622]

The IPCC's troubled servitude to the INC

In the lead up to CoP1 in Berlin, the INC was frequently reminded that it should still consult the IPCC. The IPCC itself was badgering the negotiating committee to keep it involved in the political process, but tensions arose when it refused to compromise its own processes to meet the political need. It was tardy in responding to information requests, and it perpetually deferred, and then refused to answer, the most fundamental question of science underlying the climate convention.

The ultimate objective of the Framework Convention on Climate Change, outlined in Article 2, is the...

> ...stabilization of greenhouse gas concentrations in the atmosphere at a level that would prevent dangerous anthropogenic interference with the climate system.[624]

It seemed proper that the INC, as the body responsible for negotiating the convention, should ask its interim scientific advisory panel—the IPCC—to advise what it considered this 'dangerous' level to be. Bert Bolin's first responses were that the scientists could not yet provide an answer and he cooperated in efforts to address this issue in the hope of finding one. During the early 1990s, much expert effort was expended on this question, but no result was achieved. The INC kept pressing Bolin for an answer in the lead-up to CoP1, but he pushed back by arguing that it is not a question the IPCC could answer because it is a *political* question. While others on both the science and the policy sides disagreed, in his view the IPCC could only provide the science necessary to *inform* such a question. The best the IPCC could provide in this regard would be a series of probabilistic outcomes of preset scenarios. Even then, Bolin slipped into further doubt, wondering whether anyone in any sphere could provide a meaningful answer. With so many uncertainties in the relations between chaotic systems, both natural and social, it would be impossible to make a call on where the danger might lie.[||] This doubtful retreat from what was deemed to be the most fundamental scientific question behind the treaty negotiations was not seen as favourable to their progression.

Early friction over these matters did not amount to much and relations remained cordial until March 1993 when Raul Estrada-Oyuela, a career diplomat from Argentina, was elected to the chair of the INC. His chairmanship would

---

[||] Bolin recounts his concerns in his memoirs, where he suggests he held this view in the early 1990s;[625] see also a book chapter he wrote soon afterwards.[626] In the end, he offered the IPCC second assessment 'Synthesis Report' as a response.

be characterised by impatience with the IPCC. In preparing for CoP1, at that first session in Berlin (and on through to the third session in Kyoto) Estrada would make it all too clear that Bolin and his panel were irritations that the treaty process would be better off without. As soon as he attained the chair, he pushed back against Bolin's intransigence in the negotiations over how the IPCC might serve the convention process. One source of friction was the planning and preparations for a new full assessment. The IPCC had decided to undertake another complete assessment while under the interim arrangement to fully update the parties to the treaty, but it would be a long process and tension developed around Bolin's refusal to fast-track the process to have it ready in time for CoP1. Then Bolin deferred a long-planned joint meeting to discuss the 'dangerous levels' question. *What was it with these scientists?* Estrada was not happy, and he didn't mind saying so.

In a speech to the Royal Geographical Society on a visit to the UK in the spring of 1994, Estrada explained how he sensed that the scientists were suffering from a 'Dr Frankenstein syndrome.' The convention process, he said, is waiting...

> ...for (scientific) inputs from the IPCC but I wonder if they will come in time. Almost one year ago, explaining the needs of the convention to the IPCC Bureau, I had the feeling that the IPCC was suffering (some) kind of 'Dr Frankenstein syndrome'. After all, the idea of a convention was nourished by the IPCC, but now that convention starts to walk and begins to demand additional food, the IPCC answered that it had its own program of work and could not deliver products by client's request.[627]

The attack implicated not only Bolin but also the Working Group 1 with its base at the UK Met Office and so *New Scientist* asked John Houghton for a response. Houghton argued that the delay was necessary in order to undertake rigorous peer review (just as Bolin had repeatedly done in the past). Peer review is needed, Houghton said, 'to preserve the scientific unanimity that is essential to the IPCC's work.'¶ Anyway, he could not see how a new report on the science would change anything. According to *New Scientist* he said that...

¶It is doubtful that Bolin himself would have said that unanimity was essential to the IPCC's work as he was always encouraging chapter authors to identify points of controversy and any dissenting views. See, for example, the report of the fifth IPCC session;[628] see also 'Principles governing IPCC work' on p. 8 on the same document. This appears to be a point of difference between the chairman of the IPCC and the chairman of its scientific working group. Houghton was much more interested in presenting policymakers with a consensus. As for Bolin, he seemed to be concerned that sceptical and maverick views were being marginalised or excluded. In his memoirs, while discussing the new assessment procedures introduced after the shock of the revolt in Sundsvall, he makes a point of saying, 'It was also made clear in [the IPCC assessment] rules that different views that might arise in the course of the assessment should be explained and

> ...the report will contain little that is new. 'The past four years' work
> has underlined and confirmed most of what we said in 1990,' he says.
> The prediction that a doubling of carbon dioxide levels in the atmosphere
> within a century would raise global average temperatures by 2.5 degrees
> C is 'very likely to be the same' in the report.[630]

In the midst of the underlying tension and the sometimes-heated exchanges, Estrada and Bolin did agree on the preparation of a 'Special Report' responding to questions that the INC deemed pertinent to CoP1. Thus, during 1994 Houghton's Working Group 1 found itself in a strange position: for the Special Report it was responding to questions of a very advanced nature, but in preparing its second full assessment it was still preoccupied with questions most elementary. The critical questions remained around establishing the empirical ground for the entire global warming theory, including, in particular, the question of 'detection', upon which the first assessment had foundered. And on these questions the IPCC was giving no indication of any breakthroughs; indeed, early in 1994, a year before the new report was due to be released, Houghton was saying that little had changed.

The IPCC's dilemma was this: it could not yet say *scientifically* what was 'dangerous anthropogenic interference' because as yet it had no *scientific* confirmation of any interference at all. In the meantime, the nations could choose to act without waiting for the science to catch-up, *which was entirely up to them*. But this would be a *political* decision. And so therein we find one way to understand the position Bolin had come to with Estrada: without a shift in the science, *to answer this question of what is 'dangerous' would be to step into the political*.

As if the IPCC's relations with the treaty process were not bad enough leading into CoP1, at the meeting itself they only worsened, when there were further troubles in dealing with the poor-countries bloc.

The 'price of life' controversy at CoP1, Berlin, 1995

After the disastrous decision of the UN General Assembly to create the INC, and amid the resulting uncertainty about its future, the IPCC continued its attempts to win support from the poor countries. However, it was soon apparent that success in most of these initiatives could only serve to compromise the scientific integrity of the assessment process. The most significant change was to introduce additional working group co-chairs and vice-chairs, selected not for

---

recorded, i.e., unanimity was not required.'[629]

their level of expertise but to represent the viewpoint of industrially underdeveloped regions. Much more contentious was when later this policy was extended to chapter lead authors of the second (and subsequent) assessments.

Meanwhile, Bolin was lobbying hard for the long-term survival of the climate panel, mostly on the basis of its special ability to deliver an assessment of the latest science *independent of political influence*. But the political influence he made the most show of withstanding was from the wealthy countries that dominated its funding and personnel. To this end, he implemented, formalised and advertised new structures and processes. Technical support units for each of the working groups were established, each employing at least one poor-country nominee. So, by the beginning of 1993, when it was confirmed that the IPCC would proceed with a second assessment, and a two-year work plan was announced, the panel looked ready for an ongoing role, having fortified its processes to withstand political influence from its wealthy sponsors, while at the same time increasing the influence of poor country representatives.**

Alas, even if these efforts went any way to rebuilding the confidence of the developing countries, it all came crashing down in spectacular fashion in the wake of another controversy. For the Second Assessment, Working Group 3 was no longer dabbling dangerously with mitigation policy proposals; their brief had been completely transformed so as to open up the neglected economic dimension of the climate change problem. This time it was an element of the climate change damages assessment that proved politically explosive.

In order to give an assessment of the expected total damage bill from global warming, it is necessary to cost all damage arising, including injury and death. The estimates of the death toll among the next few generations were riding in the hundreds of thousands, and most of these would be in poorer regions. In the accounting generally used for such damage estimates, the price of life is estimated relative to economic means. On this basis, 10 to 15 deaths in a poor region were assessed to be worth one death among the rich. This devaluation of the poor was insulting enough, but, when the cost of damages was compared with the cost of mitigation given in another part of the report, it looked (to some eyes at least) that discounting poor lives got the wealthy countries off the hook: the total cost of the damage came in too low to justify all but minimal efforts at mitigation. Thus the new assessment could be used to justify sacrificing hundreds of thousands of poor folks in order to maintain the lifestyle of the

---

** For the new 'Principles for governing IPCC work' and Bolin's reasons for introducing them, see the report of the fifth session of the IPCC.[631] For the introduction of 'geographic representation' on the IPCC Bureau see the report of the sixth session.[632] This was not extended to lead authors until the ninth session of the IPCC in June 1993. The new structure was reviewed in 1992 and this review implemented the following year.

270

air-conditioned rich even as they continued to despoil the global climate.[††]

The final draft of the new Working Group 3 report was distributed to IPCC delegations early in 1995, just a few weeks before CoP1 in Berlin. A few days before delegates departed for the conference, the Working Group 3 cost–benefit analysis was highlighted for the poor-country delegations in a letter from the head of the Indian delegation. He described the analysis as not only 'absurd and discriminatory', but indicative of 'the bias which underpins [the IPCC] assessment intended to provide the basis for policy discussions at the CoP'. India called on other delegations to support them in their efforts to have the 'misdirection' of this 'faulty economics…purged from the process'. Thus was sparked a controversy that broke, not at a closed IPCC plenary (although it would continue there), but at the first big public event for all the nations already signed up to the climate convention.[634]

The IPCC had no formal role in CoP1; its leadership were only guests in Berlin, there at the invitation of the INC, so the damages assessment was not up for discussion. Anyway, the expert authors of the offending chapter saw these requests for changes to their 'final draft' as politically motivated. They had already refused to make any changes, pointing to the IPCC's newly formulated rules of assessment. Thus the demand during this independent political negotiation to have this part of the IPCC assessment 'purged from the process' was a direct hit on a defenceless IPCC. The effect of this controversy on the political negotiations was only to cause further polarisation on a question that had quickly become critical to obtaining an agreement in Berlin: whether poor countries should be exempt from any commitment to emissions reductions. Thus, at the first meeting of the parties to the climate convention, the IPCC found itself at the centre of a controversy that not only jeopardised agreement, but also seemed to vindicate the poor countries' push to have it marginalised.[635]

Worse for IPCC advocates at Berlin was that SBSTA was about to come into being as a potential advisory substitute for the IPCC. Details of the role and structure of the new body were duly decided. For its scientific and technical assessments, SBSTA was asked to compile and synthesise information provided by 'competent scientific and technical international bodies', including, *among others*, the IPCC. However, it was also instructed to establish for itself two 'intergovernmental technical advisory panels' to help it 'provide assessments of the state of scientific knowledge relating to climate change and its effects'.

The strongest statement supporting a role for the IPCC comes in an annex to the decisions of the parties. This outlines further interim arrangements, which should remain in place until the next CoP, in a year's time. Until SBSTA had

---

[††] For more on this controversy see 'Enter the economists'.[633]

established its own technical advisory panels, it was asked to make appropriate recommendations to the Convention parties after considering the IPCC's second report.[‡‡] In other words, now that the IPCC's interim advisory role had ended, and until SBSTA's technical advisory panels were up and running, the IPCC still had a role; this was to deliver its second assessment for consideration by SBSTA, which SBSTA would consider in its advice to the second session of the Convention of Parties in 1996. Thus the IPCC retained a tenuous link with the treaty process, but only through SBSTA, only until the next CoP, and only to deliver its (almost) completed second assessment report.

The aerosol cooling breakthrough for the IPCC's second assessment

When, in 1994, Houghton had told *New Scientist* that the (scientific part of) the IPCC second assessment report would contain little new, he meant that the policy-relevant headline finding would remain largely unchanged. In fact, in 1994, when the work got underway for Working Group 1, there were some very new developments in the modelling that the lead-authors were starting to incorporate into various chapters.

Through the early 1990s sceptics were continuing to hammer the modellers on the fact that their models were overpredicting warming. By 1994, this criticism was met by IPCC scientists, who underwent a remarkable volte-face. In the drafting of the IPCC second assessment, the fact that the models were overstating the rate of global warming would be explicitly acknowledged. Graphs were even produced to show this. However, this concession to their critics also came with an explanation. The reason the models overshot the observed 20th century warming was not that the sensitivity to greenhouse gases had been set too high but because another human influence had not been taken into account. When it was, all could be set right.

This is how finally in the early 1990s the 'warming' modellers came around to taking Reid Bryson's advice and incorporate the cooling effect of aerosol emissions. They introduced a damper on the warming to simulate the effect of, in particular, the sulphate aerosols released with carbon dioxide in the burning of fossil fuels. Although this is a local effect, and is seen mostly in the northern mid-latitudes, it was still assessed to have a significant global influence. Adjust the models they did... and... *voila!* They now matched the 20th century global temperature trend, including the pause in the warming during the 1960s and 1970s—before sulphate pollution controls kicked in.

[‡‡]See the decisions adopted by CoP1.[636] The establishment of SBSTA in Berlin and its relationship to the IPCC is covered in a *Nature* news story.[637]

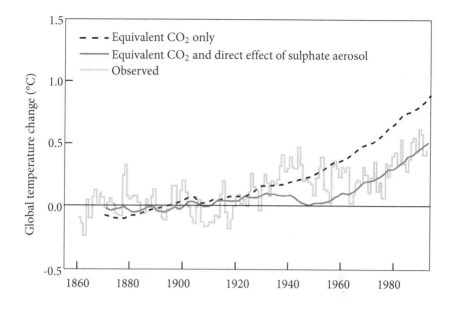

Figure 15.2: The aerosol fix.

Comparisons of the UK Met Office modelling results with the global temperature trend are presented in several places in the IPCC second assessment. This graph shows CO_2-only results shooting off above the observed temperature trend, while CO_2+sulphate runs provide a visibly impressive correlation.[638]

The combined effects of CO_2 and sulphate had been examined by several modelling groups, and their results were discussed in detail in the new IPCC report. In the Working Group 1 report's modelling chapter, special prominence was given to the recent, impressive-looking results from the team at the UK Met Office led by John Mitchell. But when the results were eventually published in *Nature*, it was revealed just how dubious they were. As Wigley explains in a highly critical review that he had been invited to write in the same issue of *Nature*, the Met Office models 'do not explicitly include aerosols at all'; rather 'they simulate their effect by changing the model's surface albedo or reflectivity'. Wigley's review is entitled 'A Successful Prediction?', and by the time one reads to the end of it, it is hard to see the answer could possibly be 'yes'.[639] Nevertheless, months before Mitchell's paper was published, its results were in the draft IPCC assessment that was sent around for peer review.

One of those suspicious that the new CO_2+sulphate results were all too neat and convenient was Pat Michaels. An atmospheric scientist at the University of Virginia, Michaels was already an outspoken sceptic and he was early in moving to use the new World Wide Web to promote his criticism. He came across the Met Office results as an assigned reviewer of the IPCC report's modelling chapter and was keen to see a latitudinal breakdown so that he could examine how they fared in the polar latitudes, where the aerosol cooling effect should be negligible but greenhouse warming pronounced. He wanted to see to what extent the Met Office results confirmed his own findings, which suggested that model runs set with CO_2-only low-climate sensitivity better matched the warming pattern than high-sensitivity CO_2+sulphates runs. Alas, his repeated requests for this breakdown were refused. These refusals were firstly on the grounds that the paper cited in the IPCC report was not yet published, and so under publisher embargo. Then, after it was published in August 1995, the request was still refused for various other reasons. A US House of Representatives hearing revealed that Mike MacCracken (by then director of the Office of the Global Change Research Program and head of the IPCC's US delegation) had expended considerable effort encouraging the Met Office to give up the data. Still Mitchell refused.[640] The story was covered in the science press, and an editorial in *Nature* expressed dismay (see Figure 15.3).[641] For a while, this little scandal threatened to blow up into a minor international incident. Suffice it to say that the zonal breakdown was never sent to Michaels and has never been made public. Nevertheless, the controversy soon blew over and the marvellous match of the CO_2+sulphate modelling results remained to deliver a triumph for the Met Office modelling group.

The UK Met Office had made a late start with the computer modelling of climate, but a move into this field was something that Houghton had been encouraging. With the financial support of the Department of Environment and the personal encouragement of Prime Minister Thatcher, by the 1990s their modelling of the climatic impact of carbon dioxide was seen as being on a par with the Americans and the Germans.[644] These other groups produced similar CO_2+sulphate results, and so things were looking much brighter for the modellers' contribution to the second assessment. Alas, the situation was not so rosy for those working on the 'detection' chapter, Chapter 8.

The second assessment finds no yardstick for detection

The writing of Chapter 8 got off to a delayed start due to the late assignment of its coordinating lead author. When the Working Group 1 preparations for a new full assessment got underway in 1994, Houghton circulated a table listing all the lead authors for the various sections of the new report. This presented a glaring omission: although Wigley and Barnett had been listed again as lead authors of the detection chapter, assignment of the coordinating role had been

Climate critics claim access blocked to unpublished data

Washington. Researchers who do not believe human activity is having a significant impact on global climate change have been denied access to the latest data from British climate models used by the Intergovernmental Panel on Climate Change (IPCC), a congressional hearing was told last week.

The allegation was made by Patrick Michaels of the University of Virginia, an outspoken sceptic about man-made climate change. But the group that carried out the research has justified not making the data available on the grounds that it had not yet been published 'n the scientific li' .ure.

Mich⌐ spe⌐¹ ⌐at ⌐g r⌐

Ho⌐ ⌐e⌐

Mitchell declined the request on the grounds that the data had not yet been published; the paper appeared in *Nature* in the issue of 10 August (**376**, 501; 1995). But Michaels, speaking after the hearing, argued that he should have been given access to the data earlier on the ground that the work had been accepted for publication and was, in any case, already being used by the IPCC, whose work he was seeking to review.

Tim Roemer (Democrat, Indiana) told Michaels that his "problems with the UK" w⌐re beyond the jurisdiction ⌐ the US ⌐ong⌐⌐ but Mic⌐ ⌐ls cou⌐ ⌐ that the ⌐ fun⌐ ⌐ni⌐ , t⌐xpayers.

Figure 15.3: *Nature* on the IPCC.

In November 1995, *Nature* reported the controversy over the Met Office's repeated refusal to provide an IPCC reviewer with detail of the modelling used in the report he was reviewing.[642] During 1995, *Nature*'s editorial line had become increasing critical of the IPCC, but an editorial in the same issue was by far the most scathing.[643] Raising the several 'rows' that it had reported through the year, *Nature* said that 'the global warming business is in danger of getting out of hand' and that it appeared scientists critical of its findings were being excluded from access to data by 'a charmed circle'. It recommended that IPCC Working Groups 2 and 3 be scrapped, while Working Group 1 should be persuaded 'to follow a more judicious course'. This recommendation appeared just four days before the opening of the Working Group 1 session in Madrid that is discussed below.

delayed after those who had first been invited had declined. It was not until April that someone agreed to take on the role. This was Ben Santer, a young climate modeller at Lawrence Livermore Laboratory who had gained his PhD at the CRU under Wigley.[645]

The chapter that Santer began to draft was greatly influenced by a paper principally written by Barnett, but it also listed Santer as an author. It was this paper that held, in a nutshell, all the troubles for the 'detection' quest. The Barnett paper was not published until 1996 and was one of the three key papers for this chapter that was written by the lead authors but unpublished at the time of drafting and review. It was a new attempt to get beyond the old stumbling block of 'first detection' research: to properly establish the 'yardstick' of natural climate variability. The paper describes how this project failed to do so, and fabulously so. In fact, it arrived at a higher level of scepticism than that reached by Wigley in the IPCC assessment the first time around.[646]

Barnett had set out to explore how to establish the variability on the longer centennial and millennial timescales necessary to give some perspective on the multi-decadal warming trend in question, and to establish the pattern of this warming as a background to the pattern studies now being employed to solve the detection question. Circulating under the title 'Estimates of low frequency natural variability in near-surface air temperature', the account of this investigation is an extraordinary tale of how every toehold for detection collapses into dust.

Barnett first considers the models. He finds that they tend to vary widely in their estimates of climate variability on longer time scales, and they generally tend to underestimate it. This is perhaps understandable because, as Barnett explains, they 'do not incorporate changes in solar output or changes in volcanically induced aerosol inputs to the atmosphere', a failing that is 'likely to inflate the statistical significance of typical detection metrics by under-representing the air-temperature variance that one should expect in nature.' The models also tell us very little about the natural spatial patterns of climate variability, and what they do tell varies widely from one model to another. 'This result is particularly worrying', Barnett explained, 'since most modern sophisticated detection methods try to find predicted spatial patterns of change'. In other words, his results show that there is no natural yardstick against which pattern analysis—or 'fingerprinting'—can be used to detect the human influence.

> All these facts make it difficult to say if *observed* spatial changes in climate are 'normal' or due to anthropogenic effects. One or both of these model flaws [in spatial distribution and understated variability] might bias the results of an objective detection study and lead us to believe confidently that an anthropogenic signal has been found when, in fact, that may not be the case.[647]

If the models are left aside, then there is no recourse to the instrumental record because it is too short for this purpose, and, anyway, it is likely contaminated by the human impact. So the only place to establish the long-range background climate variability is in the proxy climate records across the last millennium. And this is where Barnett's critique really bites, slamming available proxy reconstructions as inaccurate, contradictory and evidently tending to underestimate variability.

The conclusion of the paper heralds a warning:

> Our results should serve as a warning to those anxious rigorously to pursue the detection of anthropogenic effects in observed climate data.

And the warning is:

> The spectrum of natural variability against which detection claims, positive or negative, are made is not well-known and apparently not well represented in early CGCM [model] control runs. [648]

In other words, with the question of natural variability unresolved 'it is hard to say, with confidence that an anthropogenic climate signal has or has not been detected'.[649]

The detection chapter that Santer drafted for the IPCC makes many references to this study. More than anything else cited in Chapter 8, it is the spoiler of all attribution claims, whether from pattern studies, or from the analysis of the global mean. It is the principal basis for Santer's conclusion that...

> ...no study to date has both detected a significant climate change and positively attributed all or part of that change to anthropogenic causes.

The 'Concluding Summary' of the chapter is particularly sceptical and its final paragraph reads as though it is an elaboration of Barnett's chief warning:

> Furthermore, the large differences between the internally-generated noise estimates from different [models] translate into large uncertainties in estimates of detection time, even for a perfectly-known time-evolving anthropogenic signal. These noise estimates are the primary yardsticks that must be used to judge the significance of observed changes. They may be flawed on the century time scales of interest for detection of a slowly-evolving anthropogenic effect on climate. The burden of proof that this is not the case lies with climate modellers, experts in the analysis of paleoclimatic data, and with the scientists engaged in detection studies.[650]

So the conclusion of Santer's chapter—like Barnett's paper—is that we have no yardstick against which to measure the manmade effect. If long-range natural variability cannot be established, then we are back with the critique of Callendar in 1938, and we are no better off than Wigley in 1990.

Just before the Concluding Summary of Chapter 8, the perennial question is posed again: *When will unambiguous detection be achieved?*

> In the light of the very large signal and noise uncertainties discussed in this Chapter, it is not surprising that the best answer to this question is, 'We don't know'.[651]

Thus, we find the IPCC reporting for the *third* time that the science is failing to detect any positive sign of the catastrophe that the climate convention is supposed to be averting. Indeed, it seems we were even further from the answer than ever before.

In 1995, the IPCC was stuck between its science and its politics. The only way it could save itself from the real danger of political oblivion would be if its scientific diagnosis could shift in a positive direction and bring it into alignment with policy action. Without a positive shift in the science, it is hard to see how even the most masterful spin on another assessment could serve to support momentum towards real commitment in a binding protocol. With ozone protection, the Antarctic hole had done the trick and brought on agreement in the Montreal Protocol. But there was nothing like that in sight for the climate scare. Without a shift in the science, the IPCC would only cause further embarrassment and so precipitate its further marginalisation.

Equally bleak were the prospects for global warming science generally. With yet another assessment finding no sign of manmade warming, the credibility of this massively expanded field of research would surely come into question. Perhaps it was not human industry that had created a monster in the sky. Perhaps the message of sober science coming through the IPCC assessment was only reiterating the sceptics' warnings. Perhaps it was indeed overenthusiastic scientist-advocates in the late 1980s who had conjured up this treaty—Estrada's Frankenstein.

Nonetheless, as we know, the IPCC did survive. If its formal role in the treaty process remained marginal and open to challenge, it continued to respond to requests from the parties to the convention and it continued to periodically conduct full assessments. The challenge from SBSTA would subside when attempts to establish intergovernmental advisory panels were abandoned following failure to reach agreement on the fairest way to elect membership. While SBSTA did go on to undertake its own research and seek independent advice, it soon fell into its primary role which was as the medium of IPCC advice into the treaty process. In this way the IPCC survived. Indeed it would prosper. But that it would was by no means certain in 1995.

16 The catastrophe signal found

Santer's finding

For the second assessment, the final meeting of the 70-odd Working Group 1 lead authors was scheduled for July 1995 in Asheville, North Carolina. This meeting was set to finalise the drafting of the chapters in response to review comments. It was also (and mostly) to finalise the draft Summary for Policymakers, ready for intergovernmental review. The draft Houghton had prepared for the meeting was not so sceptical on the detection science as the main text of the detection chapter drafted by Santer; indeed it contained a weak detection claim. However, it matched the introduction to the detection chapter, where Santer had included the claim that 'the best evidence to date suggests'...

> ...a pattern of climate response to human activities is identifiable in observed climate records.[652,653]

This detection claim appeared incongruous with the scepticism throughout the main text of the chapter and was in direct contradiction with its Concluding Summary. It represented a change of view that Santer had only arrived at recently due to a breakthrough in his own 'fingerprinting' investigations. These findings were so new that they were not yet published or otherwise available, and, indeed, Santer's first opportunity to present them for broader scientific scrutiny was when Houghton asked him to give a special presentation to the Asheville meeting.[654]

In Santer's presentation to the Working Group 1 lead authors, the most impressive graph showed a pattern of warming in the vertical profile of the atmosphere that was similar to what the modelling predicted for the CO_2+sulphate effect (see Figure 16.1). It appeared that Santer had found a human 'fingerprint' in the sky. Curiously, he had used an old dataset that finished in 1987, even though more recent data were available. Later, when his work was finally published, his results would be challenged by showing how the tell-tale pattern disappeared when the more recent data were included.[658,659]

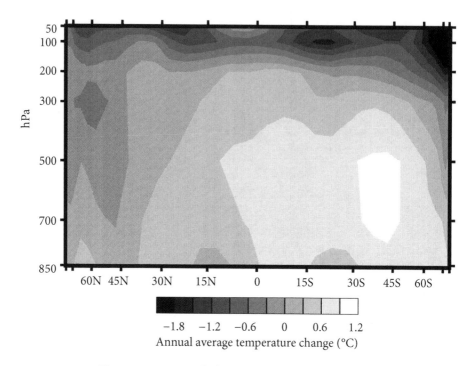

Annual average temperature change (°C)

Figure 16.1: Santer's 'human fingerprint' graph.

Using radiosonde data from 1963 to 1987, the graph presents a pattern of atmospheric warming (light) and cooling (dark) by latitude and height (measured in hPa air pressure). This pattern is similar to that expected due to industrial emissions of carbon dioxide and sulphate aerosols. While, with carbon dioxide alone, a warm patch is expected centred over the tropics, the addition of the local cooling effect of sulphate aerosols in northern industrial zones will give asymmetrical warming in the south. Santer considered the warm patch (white) over the mid-southern climes as a clear human fingerprint. When finally published just before CoP2 (Geneva, July 1996) the detection claim was very tentative.[655] Anyway, his evidence was challenged when the inclusion of more recent data caused the tell-tale southern hotspot to disappear.[656,657] Adapted from Santer's original.

However, the results were also challenged at Asheville: Santer's fingerprint finding and the new detection claim were vigorously opposed by several experts in the field. One of the critics, John Christy, recalls challenging Santer on his data selection.[661] Santer recalls disputing the quality of the datasets used by Christy.[662] Debates over the scientific basis of the detection claim dominated the meeting, sometimes continuing long after the formal discussions had finished and on into the evening.[663]

That the matter remained unresolved at the end of the meeting is reflected

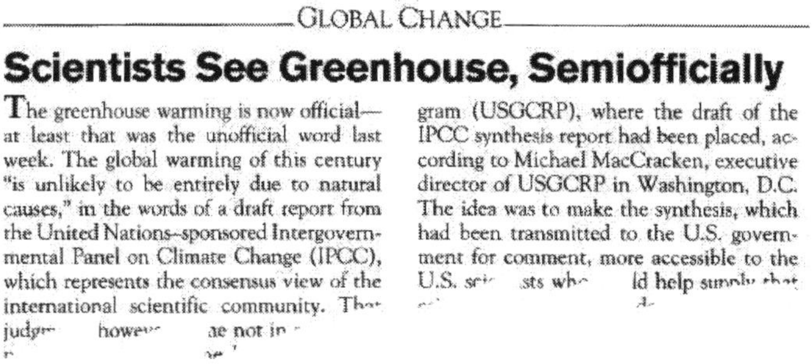

—————————GLOBAL CHANGE—————————

Scientists See Greenhouse, Semiofficially

The greenhouse warming is now official— at least that was the unofficial word last week. The global warming of this century "is unlikely to be entirely due to natural causes," in the words of a draft report from the United Nations-sponsored Intergovernmental Panel on Climate Change (IPCC), which represents the consensus view of the international scientific community. That judgm-   howev-   ae not in -   gram (USGCRP), where the draft of the IPCC synthesis report had been placed, according to Michael MacCracken, executive director of USGCRP in Washington, D.C. The idea was to make the synthesis, which had been transmitted to the U.S. government for comment, more accessible to the U.S. sci-   sts wh-   ld help suppl- th-t

Figure 16.2: Rising anticipation of a detection finding.

The leaking of the IPCC's draft detection finding by the *New York Times* led to widespread anticipation of a break through. This is Richard Kerr's report in *Science*.[660]

in the draft chapter Santer subsequently produced: this 'final' draft incorporated his new work, presented the detection claim in the introduction as a bottom line statement of all the evidence 'taken together', but it still failed to resolve the contradiction between this claim and the scepticism throughout the main text and in the Concluding Summary. The draft Summary for Policymakers also carried the detection claim as a bottom line statement, while the contradictory scepticism had been reduced to uncertainty caveats qualifying what was nevertheless a positive claim.

In September, a draft summary of the entire IPCC second assessment was leaked by the *New York Times*, the new detection claim revealed on its front page.[664] With legally binding emissions protocols set for consideration at the next session of the conference of parties, the business lobby was keen to have the detection claim moderated. Under the leadership of a Washington-based political lobbyist, Don Pearlman, a small team of experts prepared suggested changes for toning down the policymakers' summary based on the underlying report. These suggestions would be ready for use by country delegates at the next session of the working group, where the wording would be finalised.[665,666] This session was scheduled for November in Madrid. The stage was set for a showdown.[667]

Madrid: the bottom line

On the first day of the Madrid session of Working Group 1 in November 1995, Santer again gave an extended presentation of his new findings, this time to mostly non-expert delegates. When he finished, he explained that because of what he had found, the chapter was out of date and needed changing. After some debate John Houghton called for an *ad-hoc* side group to come to agreement on the detection issue in the light of these important new findings and to redraft the detection passage of the Summary for Policymakers so that it could be brought back to the full meeting for agreement. While this course of action met with general approval, it was vigorously opposed by a few delegations, especially when it became clear that Chapter 8 would require changing, and resistance to the changes went on to dominate the three-day meeting.* When the side group came back with a new version of the detection claim on the last day, it was presented as a 'bottom line' finding on all the evidence discussed in the chapter, including the Met Office simulation of the global temperature trend. However, Santer's fingerprint findings came last, representing the 'more convincing recent evidence' that tipped the 'balance of evidence' over to a positive claim.

In the debate the Saudis and Kuwaitis were insistent that these new unpublished findings should be presented as only 'preliminary'. As the debate continued without agreement, the Saudis argued, as they had earlier, that the best course of action at this late stage would be to revert to the findings as expressed in the underlying report; indeed, they suggested that the Summary for Policymakers should use the exact text of the chapter's Concluding Summary.[668] This suggestion was strongly opposed and procedural objections were to no avail. After further debate, at 10.30 pm—that is, hours after the session was supposed to close and after many delegations had departed—a final version of a 'bottom-line' detection claim was decided:

> The balance of evidence suggests a discernible human influence on global climate.[669]

With this weak declaration of human attribution in hand, but with so much more to decide, Houghton still managed to close the meeting just after midnight on the final night, and so pushing into Thursday, 30 November 1995.

Working group plenaries are closed to the press and the Working Group 1 report was still to be accepted by the full IPCC session in a few weeks' time. However, no time was lost in getting the news out to the world, as many science journalists knew the significance of this breakthrough and some had even published pieces anticipating it.[670,671] Later that Thursday morning, the freshly approved Working Group 1 summary was formally delivered to the world in a press release highlighting the human attribution statement. But some reporters

---

*See p. 2.

stalking the building the previous night managed to beat this announcement, getting the story out directly after the meeting, and so the first news of the human attribution finding appeared in morning papers only hours after the meeting closed. Thus began the fateful journey of this most important line of any IPCC report.

The UK *Independent* headlined 'Global Warming is here, experts agree' with the subheading:

> Climate of fear: Old caution dropped as UN panel of scientists concur on danger posed by greenhouse gases

The article explains the breakthough:

> The panel's declaration, after three days of torturous negotiation in Madrid, marks a decisive shift in the global-warming debate. Sceptics have claimed there is no sound evidence that climate has been changed by the billions of tonnes of carbon dioxide and other heat-trapping 'greenhouse gases' spewed into the atmosphere each year, mostly from the burning of fossil fuels and forests.
>
> But the great majority of governments and climate scientists now think otherwise and are now prepared to say so. 'The balance of evidence suggests a discernible human influence on global climate', the IPCC's summary of its 200-page report says. The last such in-depth IPCC report was published five years ago and was far more cautious.[672]

Houghton is quoted describing it as a 'big step', while Jeremy Leggett of Greenpeace declared the finding as 'a turning point' towards government action. How right he would be![673] The next day, the *Washington Post* referred to the 'breakthrough sentence' and explained its significance in the words of the head of the US delegation, Robert Watson, 'just off the plane from Madrid':

> …the tedium of moving representatives of 75 governments to a consensus is worthwhile because it means that 'these governments have basically bought into that statement. It will now be much harder for them to go and negotiate and say they don't agree with the science'.[674]

Watson knew well that intergovernmental involvement in the IPCC assessment had been designed to avert any retreat into doubt about the science during policy negotiations, such as had hampered the early ozone negotiations. In the syndicated story, Reuters cited the attribution finding as the 'lynchpin' of the report and quoted Bert Bolin saying 'The scientists have reached an agreement on the evidence, now the governments have to decide how to proceed'.[675]

Experts Agree Humans Have 'Discernible' Effect on Climate

Delegates From 75 Nations Reach Consensus in Madrid

By Kathy Sawyer
Washington Post Staff Writer

It's official. After years of alarms, an international panel of scientists and government experts have agreed in writing that human activities are affecting the global climate.

At the end of a contentious three-day session in Madrid, delegates at a meeting of Working Group I of the Intergovernmental Panel on Climate Change (IPCC) spent hours debating the wording of a single key passage. At last, they adopted b- .onsen·· the following l-- :

more emphasis on the technical uncertainties, according to scientists who attended. (The burning of fossil fuels such as coal and oil is a primary source of heat-trapping "greenhouse" gases such as carbon dioxide that are being added to the atmosphere.)

And after it was all over, there was still an open issue. A sentence in the same section refers to "more convincing recent evidence for the attribution of a human effect on climate . . ." It carries a footnote indicating that "two countries," identified by parti·· .nts ·- · idi Ar·· d K·· ait, "··

Figure 16.3: The rush to announce the detection finding.

The agreement of both experts and governments on the detection finding was announced in the press directly after the Madrid IPCC Working Group 1 meeting. (*Washington Post*, 1 December 1995)

Stories appearing in the major newspapers over the next few days followed a standard pattern. They told how the new findings had resolved the scientific uncertainty and that the politically motivated scepticism that this uncertainty had supported was now untenable. Not only was the recent success of the attribution finding new to this story; also new was the previous failure. Before this announcement of the detection breakthrough, attention had rarely been drawn to the lack of empirical confirmation of the model predictions, but now this earlier failure was used to give a stark backdrop to the recent success, maximising its impact and giving a scientific green light to policy action. Thus, the standard narrative became: *success after the previous failure points the way to policy action*. The syndicated Associated Press story of 30 November 1995 is typical of this narrative, with its headline 'Scientists tie global warming to human activity':

The Intergovernmental Panel on Climate Change, convened by the UN World Meteorological Organization, urged governments to take steps to lower the use of fossil fuels and other human activities believed to cause global warming. Climatologists from more than 100 countries confirmed that 'the balance of evidence suggests a discernible human influence on global climate.' The statement marked a change from a 1990 report by the same panel, which stated that 'detection of the enhanced greenhouse

284

effect from observations is not likely for a decade or more.'†

Only a few of the articles mentioned how recognition of the influence of sulphate aerosols had lowered the expected rate of warming by one third. The *New Scientist* story explained how factoring in sulphate emissions meant that it was the lack of observed global warming that had paradoxically 'delivered the guilty verdict':

> Like Sherlock Holmes's dog that didn't bark, it is the warming that hasn't happened that has finally convinced climatologists that human activity is probably to blame for global warming.[677]

The stories in the science press gave more coverage to the complexities and uncertainties behind the detection finding and the fight to get it through. This is especially evident in the *Nature* story, which nevertheless heralded a breakthrough in its headline, which claimed 'Climate panel confirms human role in warming', and then added 'fights off oil states'.[678] In *Science*, Richard Kerr's report ran under a similarly strong proclamation:

> It's official: First glimmer of greenhouse warming seen.[679]

In fact its claim was not official yet, and would not be until all three working group assessments were accepted at the eleventh session of the IPCC, scheduled for a meeting in Rome less than two weeks after Madrid.

Acceptance of the IPCC second assessment, Rome, December 1995

The IPCC re-convened in Rome on 11 December, 1995. Although the meeting was also closed to the press, this was a much more public affair. Delegates entering the meeting through throngs of chanting, placard-waving campaigners had a flyer thrust into their hand depicting Rome as a climate-change ravaged wasteland. Those unfortunate enough to be identified as politically suspect were surrounded by a group chanting 'climate criminal!'

Once inside, as the opening ceremonies began, a gloom hung over the celebration of a completed second assessment. The Price of Life scandal had not receded, and in fact it had raged unabated ever since Working Group 3's differential valuation of poor and rich lives had been publicised at the treaty negotiations in Berlin back in March. Protests against the low valuation of human life in poor countries had gone on to cause the collapse of the Working Group 3

---

†This was widely syndicated, for example by the *San Francisco Examiner*.[676]

session at which its report was supposed to have been accepted and the summary finalised. In the negotiations for a resolution, the authors held out against changes to their chapter that would have brought it into alignment with any summary that might win intergovernmental approval. Reconvening in October, intergovernmental agreement on a summary was finally achieved, but this consensus was not shared by the chapter authors, and the controversy continued in the science press. By November it had come to a stand-off between the chapter authors—who refused to accept the summary—and a campaign that had successfully recruited a small army of first-world supporters, including leading scientists and economists as well as activists.[‡] In late November, the coordinating lead author of the chapter had been forced to don a disguise in order to escape a picket of his offices at University College London. By this time, both sides were calling on the IPCC to fully excise from the report, not just the disputed price of life calculations, but the entire chapter that contained them. Only then might the report be brought into line with an acceptable summary and the charge of political influence deflected. However, this avenue of resolution had not been acceptable to the IPCC leadership and so the offending price of life calculations remained in the final Working Group 3 report tabled for acceptance by all delegations in Rome. Thus, when the Rome session convened, there was every reason for concern that the finalisation of the second assessment might face difficulties similar to those with the first assessment in Sundsvall five years earlier.[681]

If the spectre of the Sundsvall disaster hung in the air at the opening of the Rome meeting, then it was dismissed by a triumphal chorus during the opening speeches. These contained barely a hint of the new Working Group 3 controversy, and the gathered delegations were instead regaled with news of the breakthrough announced by Working Group 1.

Following the welcoming addresses by the Italian President and Environment Minister, there first came Patrick Obasi, Secretary General of the WMO. At the conclusion of a speech mostly making recommendations for the future direction of the IPCC, he noted that the most important result in the current assessment is the evidence for a 'discernible human influence on global climate'.[682]

[‡]Recruits to the campaign against the price of life calculations included the diplomat Sir Crispin Tickell and the economist Michael Grubb. Two other prominent supporters, the physicist Hans-Peter Dürr and the astronomer Sir Martin Rees, where among the 38 who signed a protest letter published in *Nature*. In calling for the chapter's removal from the report, this letter cited the IPCC procedures where intergovernmental approval of the summary implies that it is consistent with the underlying report.[680] The Working Group 3 debate over what to do about this inconsistency was in full swing just as a similar problem was emerging in Working Group 1. Indeed, *Nature* published the letter on 30 November 1995, which happened to be the same day the Madrid meeting closed and the detection breakthrough was announced.

Next came the new head of UNEP, Elizabeth Dowdeswell, who opened with the now familiar narrative of triumph:

> A decade ago, the scientific community alerted the world to the likelihood that we humans are causing the global climate to change. Five years ago, you said you were very confident that this is indeed the case, but that it would be ten years before we would experience any consequences. Now, just five years later, you are reporting that effects of global warming are upon us. As you put it in your report, 'The balance of evidence suggests a discernible human influence on global climate'.[683]

Later in her speech, this key component of the report's message is summarised, without qualification, as 'human activities are affecting the global climate' and so...

> For the first time, we have evidence that a signal of global warming is beginning to emerge from the 'noise' of natural variability. In other words, you [the IPCC] have given the world a reality check. You have pinched us and we have realised we are not dreaming. Climate change is with us. The question is: what do we do with this knowledge?[684]

After the Rome session was declared open there was a much more nuanced address by Michael Cutajar, the executive secretary of the 'Interim Secretariat' of the climate convention, the organisation that had replaced the INC after CoP1. Before there was an INC, in the early days of the IPCC, Cutajar was on the Maltese delegation, which had been a staunch supporter of the WMO/UNEP process towards climate action. After the INC was formed and Cutajar took up his role, he sometimes attended the IPCC meetings instead of Estrada. Cutajar was clearly more sympathetic to the IPCC than Estrada, but in his Rome speech he was more direct than Obasi and Dowdeswell in addressing the threat of its exclusion from the treaty process. In this regard, he did not shy away from alluding to the troubles over the Price of Life. 'Your work on the economics of climate change has sparked controversy', he said, and he offered the panel some advice for when their work steps into 'the zone between science and politics, where the rules of the game are not crystal clear and where ethical questions arise'. From the perspective of the convention, he said that 'the credibility of the IPCC must remain unimpaired', and so he advised them again—just as he had in a speech to the previous IPCC session[685]—that they consider 'setting limits' to their involvement in this sensitive zone. Of course, the IPCC had already set limits in this regard, strictly ruling against stepping into the domain of policy proposals. But, according to Cutajar, and as demonstrated by the new scandal, even these limits were insufficient to prevent political controversy.[686] Cutajar also advised the IPCC to orientate its work more towards the needs of SBSTA and he said that 'the mandates and work programs' of its planned technical advisory panels should to be taken into account 'in defining the topics for future technical work by the IPCC'.

If all this might have made the IPCC leadership uneasy, at least it did give all the more consequence to Cutajar's singular point of unqualified approval:

> Many who are outside this community will be impressed by the conclusion of scientists, after careful assessment of the evidence, that the climate is changing and that the change can be attributed to human influences. This finding, in particular, should give momentum to current negotiations under the convention by removing a layer of scientific uncertainty.[687]

It would turn out that Cutajar was right: the finding of a human attribution 'in particular' would give momentum to negotiations when the convention parties convened again the following summer. However, it still had to get over the line as the accepted view of the IPCC. With the formalities out of the way, much of the Rome plenary was spent in the line-by-line approval of the summary of the working group summaries: the so called 'Synthesis Report'. Predictably, this process was delayed by some of the poor countries, who were still fuming over the Price of Life calculations. But there was no repeat of the revolt at the finalisation of the first assessment's summary of summaries in Sundsvall. This time Bolin's draft got through, albeit with many changes. What did not change was Working Group 1's detection bottom line. It survived another hammering from the oil-rich countries, so that the IPCC could deliver the climate treaty negotiators a much more impressive offering this time around.

The aftermath of Rome

The passing of the detection bottom line at the close of the IPCC's Rome meeting on 15 December 1995 finally made it official. But this was no longer news. The story of the breakthrough had already gone out the week before and so the coverage of the Rome meeting by the news wire services was not picked up by major newspapers. Instead the discussion was moving on to its policy implications.[§] And this was not only in the press but also at international policy forums. Just three days after the IPCC adjourned in Rome, British environment minister, John Gummer[‖] was the first to bring the new detection finding into

---

[§]For discussion of the policy implications of the detection finding in the press see, for example, George Monbiot's article in the *Guardian*.[688]

[‖]Now Lord Deben.

the international policy discussion. This was at a meeting of EU environment ministers in Brussels. In a press release, Gummer called for countries to commit themselves to the target of reducing greenhouse gas emissions by 5–10% by 2010. At the meeting, he said:

> The Intergovernmental Panel on Climate Change, comprising the world's leading climate scientists, has for the first time accepted that there is now enough evidence to detect human interference with the earth's climate system. It is now highly probable that we are already seeing the beginning of a process of climate change, the effects of which will be felt by generations to come. This makes action by the international community all the more urgent.[689]

In February the following year, at an OECD meeting of environment ministers, Gummer again ran with the IPCC line and the other ministers came out in accord. As chair of the meeting, Gummer launched a communiqué echoing the IPCC statement that recent findings suggest 'a discernible human influence on global climate'. This was said to underline 'the necessity for urgent action at the widest possible level'. On the authority of IPCC science, the OECD environment ministers called for the setting of quantified limits for emissions beyond 2000. At the press conference, Gummer warned of food and water shortages, while saying that the OECD countries have a 'special responsibility...because our prosperity has been bought at an environmental price'.[690]

The second session of the conference of parties (CoP2) was still several months away, scheduled for Geneva in July 1996. However, already early in that year the IPCC's detection bottom line was headlining the banners of those pushing for binding emissions targets to be agreed at that meeting. With all the press and political interest, it was starting to look like Santer's discovery of the emissions 'fingerprint' would play much the same role in the climate convention process that Farman's discovery of the Antarctic ozone 'hole' had played with the ozone convention. However, there were important differences. Farman's ozone 'hole' finding had been in print for more than two years when the 1987 meeting in Montreal delivered binding emissions reductions commitments in its famous protocol to the framework convention for ozone protection. During those years, the Antarctic ozone depletion and its proposed causation had been debated, challenged and tested. But in early 1996, Santer's global warming 'fingerprint' finding remained unpublished. It would not appear until July, which was twelve months after his presentation to the scientists in Asheville and only days before governments would consider a legally binding response at CoP2.

Thus, in the lead up to CoP2, Santer's discovery of the human fingerprint had no scientific standing outside the IPCC assessment process. In that process, it had been the critical piece of evidence that had caused country delegates to agree to a late change that delivered the detection findings. But after Madrid, it was presented in the report as one piece of evidence among others. Even if it was the more convincing evidence that had tipped the 'balance of evidence' over to the detection bottom line, outside the IPCC Santer's warming 'fingerprint' was virtually unknown. It had nothing of the public profile of Farman's world famous ozone 'hole' and so it was rarely given as the scientific basis for policy action. That role was totally given over to the detection bottom line in the IPCC second assessment report.

As for that assessment report, in early 1996 it had not been formally released, and nor had it been formally accepted into the convention process. Indeed, the IPCC had no direct formal relationship with that process. The only channel of its advice to the convention parties was via an extension to an interim arrangement with a bodies subordinate to the convention organisation. In other words, the IPCC assessment findings could only be formally conveyed to CoP2 as recommendations taken up and agreed by the full spectrum of country delegations at SBSTA.

Rejection of the IPCC second assessment by SBSTA

We will remember that CoP1 had asked SBSTA to establish technical advisory panels to provide it with ongoing assessments of the state of the science. But these panels could not be expected to provide advice before CoP2, so in the meantime it was also asked to make 'appropriate recommendations' to the convention parties after considering the completed second assessment of the IPCC. Thus, for its second session, SBSTA had two main tasks: on the one hand it was to consider the new report by the IPCC and agree on technical and scientific recommendations to the convention parties, while on the other hand it was to establish the panels that would replace the IPCC.

On Tuesday 27 February 1996, SBSTA convened to consider the IPCC Second Assessment Report. Actually, it did not consider the full report, as its three parts were still being prepared for publication. However, the delegations had been issued with more than enough reading material. Their document package included the Summary for Policymakers of each of the three working groups, which had been published in a stylish coloured pamphlet that led with the final version of the Synthesis Report agreed in Rome. Bolin had organised the Synthesis Report to inform the convention's main objective as defined by its Article 2, namely to stabilise atmospheric concentrations of greenhouse gases at a level that would prevent dangerous interference with the climate.[691] The presentation of the IPCC findings in this concise and applicable form, and its sur-

290

vival through the approval process, was an achievement of which Bolin was not unduly proud.

At SBSTA, Bolin opened the discussion with a broad overview of the major findings, placing special emphasis on the human attribution bottom line just in case its repeated appearance in the IPCC paperwork might have been missed. However, it was when this finding appeared in a list of twelve key findings in a draft of the SBSTA 'Concluding Statement' that it was brought into contention. The final agreed version of this statement would be the vehicle through which the IPCC's findings were conveyed to the convention parties. Thus, the discussion was quickly drawn to this statement, especially to the twelve highlighted findings, and debate over the precise wording of SBSTA commendation of the IPCC assessment ended up dominating the entire meeting over the following days.

A number of delegations were quick to jump in and recommend strong support for all twelve highlighted findings, including Switzerland and the EU. The EU delegation reportedly 'highlighted that these findings underlined the necessity for urgent action to address adverse effects of climate change, including mitigation and adaptation.'¶ However, a very different view was expressed by a group led by Saudi Arabia and mostly consisting of Arab oil-producing nations. The official report says they considered that it is...

> ...very premature for the SBSTA to attempt to highlight specific findings of conclusions contained in the Second Assessment Report. They stated their belief that the list of items identified by certain delegations was highly selective and reflected a very limited, and, therefore, biased view of what were important findings in the Second Assessment Report. They also pointed out that some of the items listed by others were taken out of context and failed to set forth important qualifications that the IPCC specifically stated.[694]

These critics then presented the meeting with an impressive, detailed analysis of the bias they claimed was evident in the highlighted findings.

Starting with the detection bottom-line finding, they argued that the draft statement fails to disclose the related finding that our ability to quantify the human influence is limited 'because there are uncertainties in key factors'. Moreover, it fails to disclose that one of these uncertainties 'concerns the magnitude and patterns of long-term natural variability'. They also mentioned that the list of findings fails to disclose that the estimates of the temperature increase by the

---

¶ While the official report often provides much detail of the discussion, it withholds the names of the delegations involved. The *Earth Negotiations Bulletin* is more frank in this regard and so the two reports together can be used to reconstruct the debate.[692,693]

year 2100 are one third lower than in the first assessment, and a quarter lower for sea-level rise. Another finding missing from the list relates to something that gets much attention in the climate debate, namely 'the issues of so-called extreme events.' On this point they argued that surely this finding of the IPCC is significant:

> There are inadequate data to determine whether consistent global changes in climate variability or weather extremes have occurred over the 20th century.

The draft statement does mention the impacts on food supply, they said, but only the negative ones, leaving undisclosed the finding that 'baseline production' could be maintained if carbon dioxide levels were to double. And so the examples go on (as recorded in the minutes) with an impressive list of additions, contradictions or alternatives quoted *line-and-verse* from the IPCC's own working group summaries and/or the overall synthesis. Theirs was a very different overview of the IPCC assessment to the one just delivered by Bolin.[695] The Chinese also joined the protest, suggesting that the recommendation of the IPCC report as 'the most authoritative and comprehensive assessment available' should be replaced by a reference to it being only 'useful and comprehensive.'

And so on it went. The debate over the IPCC report, which wore on through the week (in between the wrangling over how to select membership of SBSTA's own technical advisory panels), only served to highlight deepening divisions. The oil producers held out against the 'biased' list of highlights, while other delegations continued to push the other way. The Canadians, for example, wanted SBSTA not only to 'commend' the assessment, but to fully 'endorse' it and take it to CoP2 'for urgent action.'[696] At the end of the week, the formal meeting had to pause, but negotiations with the naysayers continued over the weekend, in an attempt to find some compromise on the content of the Concluding Statement before the meeting finally closed.

In the end, agreement was achieved, but it was a pyrrhic victory for anyone who held out hopes for SBSTA as an effective medium of scientific assessment and advice. The final version of the Concluding Statement was nothing like what had first been envisaged. SBSTA still 'commended' the Second Assessment to the convention parties, but it was no longer accepted from the IPCC with 'warm' appreciation. It was no longer 'the most authoritative' assessment, only the 'most comprehensive'. Most significantly, all the highlighted major findings had been stripped out, leaving only five short paragraphs. The oil producers and others had succeeded in blocking consensus on the drafted 'highlights'. With nothing agreed to replace them, the statement delivered no specific conclusions or recommendations.[697,698]

This, then, is how the science-policy advice process concluded. After intergovernmental acceptance of each of the IPCC working group reports, after line-by-line consensus on each of the summaries, after full agreement again on a further synthesis of these summaries, SBSTA had been asked to compile and synthesise the resulting information. But a further week of intergovernmental debate brought it all to nought. Not in its own assessment nor through reviewing the assessment of the IPCC did SBSTA have anything to say to world leaders about to decide on action to head off a climate catastrophe. In Madrid the Saudis had failed to block the new positive findings of IPCC Working Group 1, and they failed again in Rome, but at SBSTA they were presented with another chance, and this time they won.

Luckily for the IPCC, what was decided (and more likely not decided) at SBSTA hardly mattered to anyone. The arguments presented there against the prevailing interpretation of the IPCC assessment were summarily dismissed by scientist-activists, policy activists and the press as little more than the carping and blocking strategies of vested interests. But anyway, with no specific recommendations coming through the formal path of scientific advice, SBSTA could be easily ignored, and so, mostly, it was. Moreover, with agreement on its advisory panels also proving elusive, SBTSA was starting to take on an air of dysfunction. This could only be good news for the IPCC as it readied itself for the publication of the Second Assessment Report in early June. The release had been scheduled to coincide with World Environment Day, and this might have been the occasion for further celebration of its breakthrough finding. But it was not to be. In the last weeks before CoP2, it was not another celebration of the IPCC's detection finding that received publicity. Instead, all the fuss was about a newspaper editorial that would bring the probity of that finding into question.

A 'major deception' behind the IPCC detection finding

Under the title 'Major deception on "global warming"', the offending opinion editorial appeared in the *Wall Street Journal* on 12 June 1996. It begins:

> Last week the Intergovernmental Panel on Climate Change, a United Nations organization regarded by many as the best source of scientific information about the human impact on the earth's climate, released 'The Science of Climate Change 1995', its first new report in five years. The report will surely be hailed as the latest and most authoritative statement on global warming. Policy makers and the press around the world will likely view the report as the basis for critical decisions on energy policy that would have an enormous impact on US oil and gas prices and on the

international economy. This IPCC report, like all others, is held in such high regard largely because it has been peer-reviewed. That is, it has been read, discussed, modified and approved by an international body of experts. These scientists have laid their reputations on the line. But this report is not what it appears to be—it is not the version that was approved by the contributing scientists listed on the title page. In my more than 60 years as a member of the American scientific community, including service as president of both the NAS and the American Physical Society, I have never witnessed a more disturbing corruption of the peer-review process than the events that led to this IPCC report.[699]

The author, Frederick Seitz, was never any friend of warming alarmists, having co-authored an influential sceptical report just prior to the completion of the first assessment.** However, his authority as a scientist was hard to challenge as he had enjoyed a stellar career in solid-state physics before being elected to a number of leadership positions in peak scientific bodies. The 'disturbing corruption' to which Seitz referred related specifically to the detection finding as given in Chapter 8 of the Working Group 1 report. When comparing the final draft of the chapter with the version just published, he found that key statements sceptical of any human attribution finding had been changed or deleted. His examples of the deleted passages include:

- 'None of the studies cited above has shown clear evidence that we can attribute the observed [climate] changes to the specific cause of increases in greenhouse gases.'

- 'No study to date has positively attributed all or part [of the climate change observed to date] to anthropogenic [manmade] causes.'

- 'Any claims of positive detection of significant climate change are likely to remain controversial until uncertainties in the total natural variability of the climate system are reduced.'[701]

In fact, we now know that after the meeting in Madrid, Santer had travelled directly to the UK Met Office in Bracknell. Following the vague and disputed direction of the Madrid meeting, and under Houghton's guidance, he had proceeded to modify the chapter so that it would not directly contradict the bottom line finding that the evidence points towards human attribution. He had paid special attention to the many statements arising from the sceptical Barnett paper, which had reported the lack of any 'yardstick' of natural variability against

---

** I have been unable to obtain a copy of this report. It was, however, expanded into a book[700] by Seitz and two other figures mentioned already in our story, William Nierenberg and Robert Jastrow. It also included an essay by Robert White (see p. 143).

which the human influence could be measured. The chapter's Concluding Summary had been entirely removed.

These changes had been made before the full IPCC session in Rome. However, for Rome the old 'final' version of the report had been circulated, as required, well in advance of the December meeting, which, on the IPCC's tight schedule, was also well before the November meeting of Working Group 1 in Madrid.[††] A rare hint of the trouble to come is found in a *Nature* news story that had appeared just before the IPCC session in Rome. The article explains how, in Madrid, the IPCC 'fights off oil states' to confirm 'the human role in warming', but then moves quickly to the issue of 'last-minute revisions' and the suggestion that these had been undertaken without the consent of the entire intergovernmental panel. A number of those interviewed for the article played down the significance of the changes, including Houghton, who also denied that IPCC rules had been violated:

> The rules state clearly that background material must be consistent with the policymakers summary... This is not a fudge or a cover-up... The compromise has been made on presentational points. If the science had been changed, I would have abandoned the process.[705,‡‡]

This hint aside, right up until the full report was released just before the CoP2 meeting, all the discussion of the IPCC assessment at Rome, at SBSTA, in the press and in policy debates had been based on the three working group summaries and the Synthesis Report. The underlying chapters had not been seen since copies of the 'final' draft had been distributed to delegates before Madrid and Rome. In Madrid, the Saudis had kicked up a fuss about the strengthening of the detection finding in the summary, but the detail of what was in the chapter itself was not widely known. At SBSTA, the Arab oil producers had objected to bias in the determination of the report's 'key findings' but, in the summaries from which they were taken, there was nothing explicitly contradictory of the detection finding.

[††] John Houghton later claimed that he 'reported to the whole meeting in the proper way that the available version of Chapter 8 was not and could not be the final version, and that further editing was needed to remove inconsistencies as instructed by the Madrid meeting...'[702] No mention of this is made in the brief official report of the Rome meeting nor in the Australian Delegation report.[703,704]

[‡‡] After the controversy broke, the 'presentation' arguments, as well as the 'consistency' argument, would be widely used. There was no disputing that the Concluding Summary was not consistent with the detection claim in the Introduction. On the presentation argument, its removal was defended on the grounds of conformity with the other chapters, which did not have such a summary. See for example the article by Edwards and Schneider.[706]

In fact, the 'Chapter 8 controversy' had its beginnings before Seitz's article, when business lobbyists had obtained copies of the published version of the report in the weeks before it was to be officially released. They had identified and itemised the changes and used them in a campaign to halt the report's distribution. When that failed, they publicised the changes, so as to discredit the scientific authority of the report. This campaign had begun on 17 May with an anonymous fax entitled 'The IPCC: Institutionalized "Scientific Cleansing"', which had been circulated to US politicians and journalists and used a few days later as the basis of an article in an energy industry newsletter. Its accusation that the documents had been 'doctored' was only vaguely referred to in the scientific press before Seitz detailed some of the changes in his *Wall Street Journal* editorial. Seitz used strong and direct language to repeat the claim that the celebrated detection finding is explicitly contradicted by the final draft of the report as agreed in peer review by the scientists. Whatever their intent, the effect of the changes was, according to Seitz, 'to deceive policy makers and the public into believing that the scientific evidence shows human activities are causing global warming'.*

What followed was the biggest public controversy yet to hit the IPCC. It first played out in heated exchanges of emails, copied (and later re-copied) to dozens of witnesses. These soon involved Tom Wigley, who was early in stepping up to defend Santer against mocking responses from Fred Singer. Reports in the science press were mostly in sympathy with Santer as the victim of the meddling of lobbyists, as for example in a *Science* article entitled 'Industrial group assails climate chapter'. The controversy developed through science and industry press reporters seeking opinion from both sides of the debate. But most embarrassing of all, the controversy was advertised through prominent stories in leading newspapers. This forced the leadership of the IPCC and some of its supporters to defend the assessment process in the press and eventually also in congressional hearings.

Remarkably, the bandwagon rolled on. On 4 July, *Nature* finally published Santer's human fingerprint paper. As expected, the final version claims that the discovered warming pattern is likely partially due to human activities. How-

---

*The published version of the report became available on 14 May 1996. The anonymous nine-page fax sent on 17 May highlighting the Chapter 8 changes includes footnotes citing IPCC procedures.[707] The contents were apparently the basis for the article 'Doctoring the documents?' by Deniss Wamsted, which appeared in *Energy Daily* on 22 May. The matter was also mentioned in a *Washington Times* editorial around this time. Along with Don Pearlman, John Shlaes was also active, and on 30 May 1996, the two lobbyists formally raised the matter with Bolin in a letter copied to 14 US senators/representatives. An early report in the science press came on 6 June, when a *Nature* news story repeated Houghton's claim that the changes were made for clarity and that the meaning had not been altered.[708]

ever, it also emphasises that 'many uncertainties remain, particularly relating to natural variability'. Indeed, its title heralds no breakthrough *finding*, but instead only describes a *search* 'a search for human influences'.[709] There was a flurry of interest in the science press, but this was also sometimes very tentative. In *Science*, Richard Kerr quoted Barnett saying that he is not entirely convinced that the greenhouse signal had been detected and that there remain 'a number of nagging questions'. Kerr also interviewed a meteorologist not involved in the writing of Chapter 8 whose own work showed that the warming pattern predicted by the models for the effect of greenhouse gas emissions could have natural causes. He told Kerr that he is 'a bit sceptical of the idea that every time we see something in the recent record that we haven't seen before, it must be due to greenhouse warming.'[710] Later in the year a critique striking at the heart of Santer's detection claim would be published in reply,[711,712] but such casual and vague comments as these were the extent of the published discussion in the four days between Santer's publication and the opening of CoP2 in Geneva.

CoP2, Geneva, July 1996

At CoP2, all eyes were on the US delegation, which was widely rumoured to be about to announce a policy shift. Until CoP2, the USA had led a group, including Japan, Canada, Australia and New Zealand ('JUSCANZ'), that objected to the proposal that industrialised countries alone should commit to legally binding emissions targets. The rumour was that the USA would now move to support such targets. If they did so, it could swing the treaty process beyond the current toothless agreement and towards a convention protocol, and it was already being suggested that such a protocol could be made ready for signing at CoP3 in Kyoto. For many of those who looked to this goal like a prize, concern still lingered over the possibility that the fossil fuel lobby would persuade the US government to stand firm.

One promising sign for the environmentalists was what Jeremy Leggett of Greenpeace describes as: 'cracks...in the Carbon Club'.[†] The Carbon Club—the name was coined by environmentalists—was a coalition of fossil-fuel and big-business lobbyists who attended the climate negotiations. Its central figure was Don Pearlman, who was despised by Leggett as the coordinating figure behind the Saudi-led challenge in Madrid and one of the instigators of the Chapter 8 controversy. However, around this time Pearlman was losing his hegemony over

---

[†] At this time and through to CoP3, the loosely coordinated business lobby went by several different names. For this reason, I will use the general term provided by their enemies, the 'Carbon Club'.

the big-business influence on proceedings. An alternative business voice was starting to issue forth from various 'sustainable business councils'. These bodies involved not only green energy groups, but also gas producers who wished to distinguish gas as a cleaner alternative to coal. In addition, major chemical companies such as Dow and DuPont were scrubbing up their green images, as were the oil producers BP and Shell, and these multinationals were now beginning to distance themselves from Pearlman and the Carbon Club.[713] However, it was an industry sector not directly related to energy production that presented the strongest new business message at CoP2.

For some time Leggett had been working hard to get insurance companies involved on his side of 'the carbon war' and he had already achieved some success at the first CoP in Berlin. By CoP2, a coalition of global re-insurance firms had formed as the 'UNEP Insurance Initiative'. The initiative's statement to CoP2 was a call for action 'based on the current status of climate research and on their experience as insurers and reinsurers'. Its conclusion begins:

> Human activity is already affecting climate on a global scale, e.g. through the enhanced greenhouse effect. According to IPCC the balance of evidence suggests a discernible human influence on global climate. ...[714]

With mixed messages now coming from big business, change was in the air, and there was a great sense of anticipation when CoP2 formally opened in Geneva. As with the IPCC session in Rome, the opening speeches often implicitly or explicitly referred to the IPCC's human attribution bottom-line. The WMO Secretary General Patrick Obasi opened his speech by saying that 'the most significant development since the first session of the Conference of Parties was the finding by the IPCC that there was a discernible human influence on global climate'. He then said the time for debate is over, with the onus now on the parties to take decisive action.[715] Elizabeth Dowdeswell of UNEP was next. She said that the IPCC, in the conclusions reached in its second assessment, had been 'forthright and clear in its message to the world'. Also clear was the implication that further emissions of greenhouse gases...

> ...need to be regarded as deliberate acts of pollution which governments are ethically bound to control within limits that would not allow dangerous interference with the climate system.[716]

After the leaders of the IPCC's parent organisations had so clearly and proudly proclaimed the policy implications of its findings, Bolin's speech also referred to the detection statement, emphasising how it had been agreed after careful consideration by the government delegations.[717] The implication Bolin was making here, as he often would, is that the IPCC report had already achieved intergovernmental agreement, and so there was no need for it to undergo a further acceptance procedure, as had been attempted at SBSTA earlier in the year. However, it was already clear that what did, or did not, emerge via the formal process of advice was of little consequence. What really mattered was the extent to which the IPCC's scientific achievements impacted, by whatever means, on the listening delegates, and this would be revealed in the ministerial segment of the conference. As Jeremy Leggett reported, many seemed to be completely won over:

> The speeches began on 17 July. A sizable international press corps and hundreds of NGOs waited anxiously to hear whether the Carbon Club had shaken the faith of any governments in the IPCC Second Assessment. But as the day wore on it became clear that they hadn't. The report was ringingly endorsed in almost every speech. Costa Rica, speaking on behalf of the full G77 group of developing countries and China, referred to the report as 'the most authoritative scientific assessment of climate change in existence.' Ireland, speaking for the EU, said the report left 'no room for doubt about the expected adverse effects of climate change'.

With no room for doubting the premise for action, the consequences of inaction would be monumental, according to a delegate for one of the members of the EU, John Gummer. In a hard-hitting speech, he invoked meetings of the League of Nations so many years earlier in that very city; in fact, in the very same building. The League, said Gummer, had failed to prevent war by failing to take common action, by failing to stand up to powerful interests and by failing to call the bluff of the purveyors of falsehood who put their selfish concerns before the interests of the world community. Gummer's was a call to arms against a new peril, with its new set of naysayers, including particularly the Australians, who were arguing for special treatment as a rich country due to the dependence of their export income on fossil fuels, especially coal.[718]

Finally, it was time for the all-important contribution from the US delegation, which would be delivered by Tim Wirth, the senator who had helped to orchestrate that sweaty performance by James Hansen on Capitol Hill eight years earlier.[‡] The Democrats lost that campaign but won the next time around when Bill Clinton ran with the climate alarmist Al Gore. In Geneva, at the end of his speech, Wirth did indeed make the hoped-for policy change announcement, citing the IPCC's 'careful, comprehensive, and uncompromised work', but not before first denigrating those who attempted to undermine its authority. He said that the IPCC's efforts 'serve as the foundation for international concern'

[‡] See p. 223.

and its 'clear warnings about current trends are the basis for the sense of urgency which my government holds in these matters.' Referring implicitly to the Chapter 8 controversy, he continued:

> We are not swayed by, and strongly object to, the recent allegations about the integrity of the IPCC's conclusions. These allegations were raised not by the scientists involved in the IPCC, not by participating governments, but rather by naysayers and special interests bent on belittling, attacking, and obfuscating climate change science. We want to take this false issue off the table and reinforce our belief that the IPCC's findings meet the highest standards of scientific integrity.

Next he referred to the debacle at SBSTA:

> We also note with regret that…SBSTA, blocked by a very small group of countries, did not agree on how to use the IPCC report. Let me make clear the US view: The science calls upon us to take urgent action. The IPCC report is the best science that we have and we should use it.

This was followed by a note of pride in the billion-dollar annual investment that the US was making in 'global change research' and how this shows 'the seriousness with which we view these matters'. Then came the rationale for the shift of US policy towards binding emissions targets, based primarily on the breakthrough announced in the new IPCC report.

> The United States of America takes very seriously the IPCC's recently issued Second Assessment Report…[which] reported that 'the balance of evidence suggests that there is a discernible human influence on global climate.' This seemingly innocuous comment is, in fact, a remarkable statement. For the first time ever, the world's scientists have reached the conclusion that the world's changing climatic conditions are more than the natural variability of weather. Human beings are altering the earth's natural climate system…In our opinion, the IPCC has clearly demonstrated that action must be taken to address this challenge and that, as agreed in Berlin, more needs to be done through the convention. This problem cannot be wished away. The science cannot be ignored and is increasingly compelling. The obligation of policymakers is to respond with the same thoughtfulness that has characterized the work of the world's scientific community.[719]

The CoP2 Ministerial Declaration released a few days later echoed Wirth's speech, declaring support for the IPCC and using its authority as the basis for policy action. Its affirmation of the new IPCC assessment is worded as an implicit rejection of the SBSTA Concluding Statement: it recognises the IPCC's assessment as the 'most...authoritative', therefore providing the basis for 'urgent strengthening action'. Its second article begins by saying that the government ministers...

> *Recognize* and *endorse* the Second Assessment Report of the Intergovernmental Panel on Climate Change (IPCC) as currently the most comprehensive and authoritative assessment of the science of climate change, its impacts and response options now available. Ministers believe that the Second Assessment Report should provide a scientific basis for urgently strengthening action at the global, regional and national levels, particularly action by Parties included in Annex 1 to the Convention [i.e. rich countries]...to limit and reduce emissions of greenhouse gases, and for all Parties to support the development of a Protocol or another legal instrument; and note the findings of the IPCC, in particular the following:
>
> • The balance of evidence suggests a discernible human influence on global climate....[720]

The article continues with a summary of many of the key findings that had been removed from the SBSTA Concluding Statement, going on to recognise 'the need for continuing work by the IPCC to further reduce the uncertainties....'.

With the change of US policy, the Australians seemed genuinely caught off-guard. Their Prime Minister, John Howard, immediately issued a public protest.[721] Japan and Canada had already been sufficiently won over, and so the small but powerful JUSCANZ alliance was in tatters. With all those in the poor-country bloc signed up to the suggestion that only the rich would be making the cuts, Australia and New Zealand were now cast into isolation as the only delegations, apart from Russia and the Arab oil producers, who were still refusing to sign the declaration.

How far the IPCC had come! History tells us that the Clinton Administration would soon experience a monumental backlash from the stand it took at CoP2, with the US Senate voting unanimously against the Union ever entering into such a commitment. Nonetheless, the chorus of support for binding targets at Geneva did project the treaty talks onward to a protocol that would be signed in Kyoto by an indexNierening majority of countries. The IPCC's human attribution claim had provided the science-to-policy linkage so that the geopolitical policy collaboration could be pulled up onto this next step. According to Houghton, without the detection finding achieved in Madrid, the Kyoto Protocol would not have been possible.[722,§] With so many political actors using the authority of the IPCC's detection finding to justify advancing in that direction, it is hard to disagree with his assessment. Another authority might well have been used to carry the treaty politics forward, but the fact that this particular authority was available, and was used, meant that the IPCC was hauled back into the political picture, where it remains the principal authority on the science to this day.

Thus, it might be said that by successfully guiding through this human attribution finding, the team of scientists active in Madrid had saved the IPCC, they had saved the treaty process and they had saved the role of science in one of the greatest global environmental policy initiatives of all time. *But what of the controversy over their actions in Madrid? How did Santer and Houghton handle the criticism? And how did the IPCC manage to survive it?*

A discernible political influence

As the Chapter 8 controversy developed, much would be made of a covering letter attached to the US comments on the policymakers summary that had been submitted just before Madrid. This letter, from the US Department of State to Houghton, mentioned that there were inconsistencies between the summary and the chapters and within the chapters. In order to resolve these issues, Houghton was told that it is 'essential' not to finalise the chapters prior to Madrid and that it is also essential…

> …that the chapter authors be prevailed upon to modify their text in an appropriate manner following the discussion in Madrid.[723]

This request was highly irregular. Earlier in the process, the draft chapters of the report had been circulated to the country delegations for comment, so that the lead authors could consult them in their redrafting. For the second assessment, the deadline for such comment was three weeks before the lead authors' meeting in Asheville, where US commentary on the chapters had been duly received and considered. After Asheville, the chapter drafts were finalised and circulated again to the delegations, but this time only so that they could be consulted for commentary on their fidelity with a draft summary. It was at this point that the Americans had asked for further changes to the chapters. As the Chapter 8 controversy developed, Fred Singer and others would take this US request as evidence confirming Seitz's case that the expert report had been altered under political influence. Yet this cover letter would not have had nearly

§See p. 3.

so much controversial force (it might not have been noticed at all) if it were not for Santer and Houghton referring to the US request as the motive for and justification of their actions.

We will recall Houghton's initial responses to charges of late and unauthorised changes were to downplay their significance.‖ This was before the details of the changes were widely known. After the details had been made public, Ben Santer was the first to use the US government cover letter to justify them. The same day that Seitz publicised the controversy in his op-ed, Santer sent a widely circulated email to all the Working Group 1 lead authors and all the contributors to Chapter 8. In it he quotes the US cover letter, where it asks 'that chapter authors be prevailed upon to modify their text'. He then comments:

> Clearly, the official view of the United States was that chapters should NOT be finalised prior to Madrid.[724]

A few days later, Santer was quoted in a *Nature* news article making just this deference to the US government request as though it were a directive in the line of command,[725] and for some time both Santer and Houghton continued to justify their action as though the request were a directive.

A better understanding of Houghton's rationale for taking the US request in this way is found later, at the height of the controversy, in a letter he had published in *Nature* under the heading 'Justification of Chapter 8'. This letter begins with an explanation of the IPCC procedure. The main task of the Madrid meeting, he says, was to accept the chapters and approve the summary. He then explains 'acceptance' thus:

> According to the rules of procedure, 'acceptance' by the plenary means that the meeting is satisfied that they have undergone a thorough process of peer review by experts and by governments and that they present a 'comprehensive, objective and balanced view' of the science.[726]

After the peer-review process, he explains, the report had been sent out to the delegates ahead of the conference with the request for comments on the summary. However, as Houghton implies, he continued to receive further comments on the chapters. He does not specify how many countries requested changes to *the chapters,* but he then proceeds with a single example:

> For instance, the US government, in submitting its points for review, commented on 'several inconsistencies' and stated that 'it is essential that the chapters not be finalized prior to the completion of the discussion at the

---

‖ See p. 295.

IPCC Plenary in Madrid, and that the chapter authors be prevailed upon to modify their text in an appropriate manner following discussion in Madrid.[726]

What he implies is that the US delegation could not 'accept' the report in Madrid without changes. However, with the changes agreed during the Madrid meeting, the Americans (and all other delegations) finally agreed to accept the report. Nowhere in the IPCC procedures is there any clear direction for what to do when delegations refuse to accept a report, a problem that had already caused great difficulties for Working Group 3. Anyway, the need to bring the US government towards acceptance of the report constitutes Houghton's main 'justification' for altering it. For Houghton, this did not constitute political interference in the assessment. The changes he says were 'not in any way motivated by political... considerations'. But whenever Houghton discusses political pressure brought to bear on the IPCC process, he is invariably referring to the pressure from the Arab oil-producers and their big business supporters such as when they tried to block the detection finding. It is their pressure that he seems to have in mind when his justification concludes that 'despite pressure from those with political agendas... the IPCC has stuck strictly to its brief' refusing 'to compromise its science for any political reason...'[727]

In Houghton's justification, the US cover letter remains key to the events that led to the controversy. However, it is a little vague, and only hints at the nature of the 'inconsistencies' requiring resolution. To find out exactly what he is referring to, we need to look at the commentary to which the cover letter is attached.

This US commentary on the Summary for Policymakers, submitted just before Madrid, was compiled by Robert Watson, the English chemist who had moved from a prominent role in ozone protection into the global warming controversy, becoming head of the US delegation in the IPCC 'scientific' working group.¶ After Clinton came to power in 1993, Watson left NASA to take up a position in the White House Office of Science and Technology Policy. In his commentary, Watson identified and repeatedly discussed two main inconsistencies in the report. The first was the continuing discussion of CO_2-only results, when the emphasis should have shifted entirely to the new findings on the combined human effect of CO_2+sulphates. We will recall that since the first assessment the modelling of the combined effect of CO_2 and sulphate aerosol emissions had been found to better match both the global temperature trend and the pattern of the recent warming, and that this match was the basis for the

---

¶See p. 104.

detection finding. However, several chapters in the report continued to discuss CO2-only scenarios.

The report's incomplete transition away from the CO2-only results was also a factor in the second inconsistency identified by Watson. This was the highlighting of uncertainties in CO_2+sulphate pattern matching, which undermined the impact of the detection claim. One of the justifications for continuing to work with CO_2-only modelling scenarios was (as Watson put it) because the sulphate 'aerosol forcing is highly uncertain and poorly understood'. However, Watson found this view 'inconsistent with the importance of aerosol forcing to the recent results on pattern matching'. He summarised the problem thus:

> We clearly cannot use aerosol forcing as the trigger of our smoking gun, and then make a generalized appeal to uncertainty to exclude these effects from the forward-looking modelling analysis.[728]

In the USA, the 'smoking gun' analogy is often used to suggest conclusive evidence of guilt. In this case, Watson used 'smoking gun' to imply human attribution of the warming, just as previously he (and others) had used it to declare human attribution of the ozone holes, south and north. Here we should also recall the important immediate context of the commentary, which was the widespread expectation of a human attribution finding, especially since the draft of the summary had been leaked to the US press two months before Madrid. Whether or not Houghton and Watson had been involved in any collaboration over using 'aerosol forcing as the trigger of our smoking gun', Watson's comments certainly suggest that he was keen to help smooth the way for the IPCC to meet this expectation.

Later, in a specific comment, Watson made his case more plainly:

> This text is not fully consistent with the rest of the [summary for policy-makers] and various parts of Chapter 8; because this is such a new and important aspect of the report, we believe particular care must be taken. We believe the text here, with some clarification, does represent current understanding as contained in the body of the chapter, but that the executive summary and concluding sections of the chapter may need to be revised.[729]

For Watson, the excessive levels of uncertainty about the new CO_2+sulphate modelling that remained expressed in various places served to weaken the report. Later, in a specific comment he said that the statements about the 'unreliability and uncertainty' of the sulphate effect in the modelling should be removed as inconsistent with the 'focus on the solidity of [sulphate] aerosol results for reaching significant new conclusions in the pattern matching'.[730]

Thus, we can see what the Saudis were up against when in Madrid they came up with their way to resolve the inconsistencies. The simplest resolution would have been to drop the detection claim and to revert to the doubt and uncertainty encapsulated in Chapter 8's Concluding Summary. The much more difficult way to resolve the inconsistencies was to reorientate the entire chapter to a significant new conclusion by removing the many statements of unreliability and uncertainty undermining it. Already before Madrid, in the US State Department cover letter and in Watson's specific comments, we can see a clear push to do just that. The US delegation was suggesting that the generalised appeal to uncertainty be swept aside to allow the new CO_2+sulphate 'pattern matching' to be used as solid evidence—indeed, as the *smoking gun*—of human culpability. And this is exactly the view that triumphed at Madrid. After Madrid, rather than mess with modifications to the Concluding Summary, it was simply removed.

In his defence, Houghton argued that he had been asked by the US government to make the chapters consistent with the summary. This is true. The US did use the occasion of providing commentary on the Summary for Policymakers to ask for the changes to be made to the chapters to make them more consistent. The existence of inconsistencies in Chapter 8 and between Chapter 8 and the summary is not the heart of the controversy. It is, rather, how the inconsistencies were dealt with and on whose advice. In Madrid, they could have been dealt with by changing the Summary for Policymakers, just as the Saudis had suggested. That would have preserved the scientific integrity of the chapter. However, the detection finding would have been lost. The tragedy is that it had been Houghton himself who had removed the intergovernmental panel from the writing of the assessments so that he could introduce an expert review process entirely independent of the potential political influence of country delegations. In the first assessment he followed this practice. For that assessment, acceptance by the delegations in his working group had been sufficient, and he sent the report off for printing so that commercially published copies could be distributed in Sundsvall, even before the report had been formally accepted by the IPCC plenary. However, this produced findings that did little to support the treaty process. For the second assessment, Working Group 3's attempts to conform to Houghton's independent expert review process caused its final approval plenary to collapse when chapter authors asserted their independence and refused to change their findings. The equivalent meeting of Working Group 1 in Madrid did not collapse, and the detection finding survived to support the treaty process, but only after the man who had instituted the assessment's scientific independence had successfully coordinated its violation.

The impact and aftermath of the Chapter 8 controversy

When the Chapter 8 controversy broke, many were exposed to its sting. In the six months since Madrid, many prominent public figures, groups and institutions had nailed their colours to the IPCC mast, directly under the plaque announcing the triumph of human attribution. There was not going to be any easy backing down. So much was at stake for the reputation of the IPCC, for the reputation of science in the public policy arena, as well as for the sympathetic politicians. What we witness, almost from when the controversy broke, is a closing of ranks in the scientific community. Many scientific institutions, including key journals, would no longer espouse or countenance scepticism in the easy liberal way they had previously.

One striking outcome was the termination (if a little hesitantly) of *Nature*'s editorial line critical of the IPCC. Its editorial, published the day after Seitz's

NATURE · VOL 381

nature

13 JUNE 1996

Climate debate must not overheat

Charges by parts of the US energy industry that a recent report on global climate change has been 'scientifically cleansed' should not be allowed to undermine efforts to win political support for abatement strategies.

Figure 16.4: *Nature* encourages support for the IPCC.

The editorial appeared on 13 June 1996,[731] the day after Seitz's *Wall Street Journal* op-ed delivered news of the Chapter 8 changes to those beyond the scientific community.

Wall Street Journal article, was not a response to it, but to the earlier efforts of the Carbon Club to get the controversy going.** No doubt recognising the explosive potential of these claims, it was essentially an appeal by Britain's leading science journal to keep voices down on this controversy in view of the potential impact on the forthcoming US elections, and, in turn, on the treaty negotiations. But this did not stop *Nature* continuing (if more gently) its critical line on the IPCC:

** See p. 297.

The complaints are not entirely groundless. IPCC officials claim that the sole reason for the revisions was to tidy up the text, and in particular to ensure that it conformed to a 'policymakers' summary' of the full report that was tortuously agreed by government delegates at the Madrid meeting. But there is some evidence that the revision process did result in a subtle shift in the relative weight given to different types of arguments...[732]

This gentle agreement with the Carbon Club's criticism, followed by an uncharacteristic deferral to the political world, was pounced upon by perhaps the most active sceptic at the time, Fred Singer. His response surely only served to illustrate *Nature*'s argument that it is best to keep quiet on the matter.[733] The previous November, Singer had organised a meeting of sceptics, who then attempted to influence the debate by issuing a statement opposing the human attribution finding: the so-called 'Leipzig Declaration'. Alas, this was a publicity failure. The Carbon Club was also failing. Pearlman and his team had been most effective in raising questions about the detection finding right from the beginning, but they won no kudos for triggering the Chapter 8 controversy. Through 1996–97, their subscription base collapsed as corporations raced to paint their image green.[734]

With very few speaking out against the IPCC detection finding, the controversy was allowed to subside and those behind the triumph at Madrid suffered no reputational damage. In the heat of the controversy, Santer fretted that he was being 'taken out' as a scientist, but it turned out that his role in delivering the IPCC detection finding gave a tremendous boost to his career. In 1998 he received a MacArthur Fellowship, or 'Genius Grant', specifically for the 'fingerprinting' breakthrough behind the detection claim.

Besides Santer, the other scientists instrumental in delivering the IPCC's first detection finding were Robert Watson, who suggested the chapter changes on behalf of the US government, and John Houghton, who ushered through those changes. These two British scientists went on to play even more prominent roles in the next assessment. Houghton would again oversee the scientific part of the assessment, but this time under Watson, who had taken up the chairmanship of the IPCC when Bolin retired.

Under Houghton and Watson the IPCC third assessment would champion the work of another young scientist who in 1998 produced a temperature trend graph that seemed to have solved Barnett's problem of a natural variability 'yardstick'. Using proxy data stretching back to the end of the Medieval Warm Period and instrumental data for the last 100 years, Michael Mann's results showed such a rapid general warming trend over the last 100 years that it towered over previous fluctuations, thus leaving no room for doubt that something extraordinary is now underway.[735] Mann soon extended his study back across an entire millennium and this so-called 'Hockey Stick' graph is what featured in the IPCC third assessment report. When the report was released in 2001, the graph was the most spectacular vehicle for its promotion; it was also later widely used by governments promoting emissions-reduction policies. These campaigns were not unduly affected by the concerns that were soon raised about the methodology of the graph's construction, nor by the ensuing Hockey Stick controversy, which would grow to be much larger and endure much longer than the Chapter 8 controversy.[736] Instead, the visual impact of the Hockey Stick continued to overwhelm any doubt that there was already a discernible human influence on the global climate.

If we consider the other lead authors of Chapter 8, we find that they would suffer little from the controversy, but they won none of the accolades afforded Santer, which is hardly surprising given that they were not always entirely in accord with the IPCC line. Tom Wigley's expressed scepticism of the science behind climate action extended beyond the determination of natural variability. We will remember that just after the lead author meeting in Asheville he had published a commentary on the Met Office's neat tracking of the recent global temperature trend, questioning the simulation of the sulphate effect and the apparent success of the modelling prediction.[††] But even before Asheville he also questioned the scientific-economic rationale behind the rush towards emissions reduction. Collaborating with energy economists on a study partly funded by the energy industry, he concluded that it is not advisable to start curbing emissions for another 30 years.[‡‡] Still, he remained fiercely loyal Santer during the Chapter 8 controversy and to all the scientists working under the funding generated by the scare. His continuing 'loyal opposition' is particularly evident in emails leaked in 2009, which show that during the Hockey Stick controversy he was at the same time working hard behind the scenes to fend off sceptics while privately agreeing with much of the criticism of Mann's work.[*,738]

As for Tim Barnett, after the Chapter 8 controversy had subsided, he brought together a loose collaboration of scientists called the 'International Detection and Attribution Group', most of whom had made a contribution to the chap-

---

[††] See p. 273.

[‡‡] The paper was submitted to *Nature* in May 1995 but did not appear until the following year because, according to *New Scientist*, it 'received a hostile reaction from some reviewers, resulting in counter-accusations of censorship'.[737]

[*] In November 2009 thousands of emails and computer files were leaked from the Climatic Research Unit at the University of East Anglia. The emails revealed the efforts of a small group of climate researchers to restrict the climate science debate and especially the debate over the Hockey Stick graph. The controversy that erupted came to be known as 'Climategate'.

ter. A status report released in 1999 extended the work on the natural variability 'yardstick' by showing how the problem remained when comparing Mann's graph with other graphs involving widely variant proxy results.[739] Barnett was never one for negotiation, let alone controversy, and he avoided involvement with the IPCC assessment the next time around.

What we can see from all this activity by scientists in the close vicinity of the second and third IPCC assessments is the existence of a significant body of opinion that is difficult to square with the IPCC's message that the detection of the catastrophe signal provides the scientific basis for policy action. Most of these scientists chose not to engage the IPCC in public controversy and so their views did not impact on the public image of the panel. But even where the scientific basis of the detection claims drew repeated and pointed criticism from those prepared to engage in the public controversy, these objections had very little impact on the IPCC's public image. The impact it did have only served to consolidate support behind the panel and its strengthening scientific case for policy action. Criticism would continue to be summarily dismissed as the politicisation of science by vested interests, while the panel's powerful political supporters would ensure that its role as the scientific authority in the on-going climate treaty talks was never again seriously threatened. During the Hockey Stick controversy, support from scientific institutions only strengthened, with most of the world's great scientific academies and associations coming out with statements explicitly supporting the panel and the scientific foundations of the warming scare.[740,741] Houghton and Watson were rewarded with knighthoods, Gummer a peerage, and the IPCC would share the 2007 Nobel Peace Prize with Al Gore. Today, after five full assessments and with another on the way, the IPCC remains the pre-eminent authority on the science behind every effort to head off a global climate catastrophe.

Acknowledgements

This book would not have been possible without the documents obtained via Mike MacCracken and John Zillman. Their abiding interest in a true and accurate presentation of the facts prevented my research from being led astray. Assistance with research came from librarians Rosa Serratore, Lily Gao, Galina Brejneva at BoM, Laetitia Rey at WMO, Emily Hewitt, Robert Perks at the British Library, Mark Beswick and others at the UK Met Office. Historians who offered assistance include James Fleming, David Hirst, Spencer Weart and Andrew Montford. Others who assisted with research include Alan Overton (UEA), Tim Osborn (CRU), Ken Haapala (SEPP), Jonathan Lynn (IPCC). Many of those who participated in the events here described gave generously of their time in responding to my enquiries, they include Ben Santer, Tim Barnett, Tom Wigley, John Houghton, Fred Singer, John Mitchell, Pat Michaels, Chip Knappenberger, John Christy, Eduardo Olaguer, Bettye Dixon, John Zillman, Aubrey Meyer, Richard Tol, Sam Fankhauser, Erik Haites, William Nordhaus, Vincent Grey, Bill Kininmonth, Kevin Trenberth, Garth Paltridge, Bob Carter, Tim Ball, David Unwin, Ray Bates, Steve McIntyre, Richard Lindzen, Brian O'Brien and many more. Support and encouragement came from Aynsley Kellow, John Zillman, Hilary Ostrov, Ray Evans, Alan Moran, Benny Peiser, David Henderson, Martin Cohen, Simon Scott, Eddie Schubert, Lucy Skywalker and especially Kelsey Hegarty. For feedback on the manuscript at various stages, thanks go to Richard Lindzen, John Zillman, Aynsley Kellow, Christopher Essex, Ross McKitrick, Antonino Zichichi, Erl Happ and Kelsey Hegarty. Finally, thank you to Benny Peiser for going with the idea of a full length book and to Andrew Montford for his close, careful and informed copyediting, which transformed my original manuscript into a publishable form.

Dramatis personae

Bolin, RJ 'Bert' (1925–2007) Swedish meteorologist, expert in atmospheric carbon dioxide. From the later 1960s he was active in coordinating international meteorological research and assessment. He helped establish GARP and chaired its Joint Organizing Committee from 1968 to 1971. He coordinated the first international assessment of the global warming threat ('SCOPE 29') and was founding chair of the IPCC from 1988 to 1997.

Bruce, James P 'Jim' (1928–) Canadian hydro-meteorologist. As the head of Canada's meteorological service, he served on the WMO Executive Committee and was elected third Vice President in 1983. During 1985 he chaired the Villach conference, became secretary of the AGGG and retired from the Canadian public service to take up employment in the WMO Secretariat in Geneva. In 1986 he became acting Deputy Secretary General of the WMO. From this position he helped establish the IPCC. During the IPCC Second Assessment he steered Working Group 3 through the 'Price of Life' controversy.

Dobson, Gordon (1889–1975) British meteorologist, who helped found the field of ozone layer studies. In the 1920s he developed the standardised measurement of ozone layer thickness with his ozone spectrophotometer. In 1956, he established the first ozone monitoring station in Antarctica and in the early 1960s he identified and explained the Antarctic winter depletion anomaly.

Estrada-Oyuela, Raul (1938–) Argentinian diplomat. When the INC was established in 1991 he was elected vice chair, and in 1993 ascended to the chair. He went on to lead the secretariat of the FCCC through successful negotiation of the 1997 Kyoto protocol.

Gummer, John (1939–) British conservative politician. In 1993 he became Secretary of State for the Environment and was soon advocating for a cli-

313

mate treaty at international forums. Within days of the IPCC announcing its detection finding in December 1995, he was using it to promote this cause. He won a peerage in 2010 and, as Lord Deben, he continues his advocacy of climate action in the House of Lords.

Hansen, James 'Jim' (1941–) US atmospheric scientist. In 1967 he took up a position at the Goddard Institute for Space Studies in New York under Robert Jastrow. When Jastrow resigned in 1981 the institute was threatened with closure, but continued under Hansen's leadership just as his research interest was shifting from the atmosphere of Venus to the atmosphere of the Earth. In 1988 Hansen became an outspoken advocate of global warming mitigation. That year he confidently claimed that he had detected the human influence on the climate in a Congressional testimony that launched the global warming mitigation movement in the USA.

Houghton, John T (1931–) British meteorologist. He began his career at Oxford, researching the ozone layer with Gordon Dobson, and went on to develop remote sensing devices for satellites. From 1976 to 1978 he was president of Royal Meteorological Society. When WCRP started in 1980, he was vice chair of its joint scientific committee and took up the chair when Joseph Smagorinsky departed. In 1983 he became Director General of the Met Office and a member of the WMO Executive Council. In 1988, as Chairman of IPCC Working Group 1, he led the transformation of the assessment process so that it was undertaken not by an intergovernmental panel but by expert lead-authors. In 1995 he helped steer through the IPCC Second Assessment's detection finding.

Johnston, Harold (1920–2012) US atmospheric chemist. In 1971 he projected that the NOx emissions of proposed fleets of supersonic aircraft would destroy the ozone layer. This speculation was influential in establishing funding for stratospheric research in the early 1970s. While Johnston's claims were never supported by the subsequent research, he remained influential with others involved in later ozone scares. Two chemists prominent in the ozone and global warming scares, Bob Watson and Susan Solomon, worked under him early in their research careers.

Lovelock, James 'Jim' (1919–) British atmospheric chemist. In 1957 he invented the electron capture detector that could detect trace amounts of chemical compounds in gas samples. This invention's links to the DDT scare and the CFC scare were due to its uses in showing the distribution of these compounds around the global atmosphere at concentrations otherwise undetectably low. His Gaia hypothesis proposed that the biosphere acts to control the atmospheric chemistry in a stable state favourable to biological life, and this challenged alarmist views of natural systems in 'a delicate balance'. Lovelock's suggestion early in the CFC controversy that there might be significant natural sources of stratospheric chlorine was confirmed by later research.

Mitchell, J Murray (1928–1990) US meteorologist. He was a highly respected researcher who took an active part in all the major climatic change controversies of the 1960s and 1970s, but his contribution usually tended to moderate alarm.

Singer, Siegfried 'Fred' (1924–) US atmospheric physicist. His research covered cosmic rays, the ozone layer and the development of satellite remote sensing technology. From the late 1980s he became publically sceptical of the science behind the CFC-ozone scare and the global warming scare. In 1990 he established the privately funded Science and Environment Policy Project (SEPP), which continues today.

Sullivan, Walter (1918–1996) Science editor of the *New York Times* 1964–1987. Sullivan began an interest in science reporting in 1957 with extensive coverage of the International Geophysical Year. During the late 1960s and 1970s he had a commanding influence on the currency of science stories in the US press, and this in turn influenced the direction of science funding. Articles by him launched several ozone scares.

Tolba, Mustafa (1922–2016) Egyptian microbiologist and UNEP Executive Director. He led the Egyptian delegation to the UN Human Environment Conference (Stockholm, 1972) and was elected Deputy Executive Director of UNEP. Two years later he ascended to the position of Executive Director after Maurice Strong retired. In this role he coordinated negotiations for international agreement on ozone protection from their beginnings in the late 1970s and all the way through to their success in the late 1980s. While pursuing ozone protection, he was also active in drawing attention to the need for climate protection. Following the first World Climate Conference in 1979 his efforts to have the WCP address the carbon dioxide question failed, but in 1983 he succeeded in initiating the first international assessment. At the Villach 1985 conference he launched the climate treaty movement. He then went on to coordinated the preparations for treaty negotiations, at least until 1991, when a poor-country revolt saw an independent negotiating committee (the INC) established. The following year he resigned.

Watson, Robert 'Bob' (1948–) British atmospheric chemist working mostly in the USA. After completing his doctorate in 1973, he worked under Harold Johnston on projects funded by CIAP. When international negotiations for ozone protection began in 1977 he chaired its expert advisory group, the Coordinating Committee on the Ozone Layer. While leading stratospheric research at NASA, he was in charge of the first US ozone hole expedition to Antarctica and he chaired the international Ozone Trends Panel. For the IPCC first assessment he led the US delegation to Working Group 1 and was one of its lead authors. In 1993 he took up an environmental policy role in the Clinton White House and was also the chair of the IPCC Working Group 2, which assessed the impacts of climate change. He wrote the US government official commentary on the draft summary of the IPCC Second Assessment. These comments, submitted in November 1995, recommended changes to the report to better accommodate a 'smoking gun' detection finding. In September 1996 he was elected to replace Bolin as chair of the IPCC for its third assessment.

Weinberg, Alvin (1915–2006) US nuclear physicist. He worked on the Manhattan Project and then after the War moved to Oak Ridge National Laboratory; he was appointed director in 1955 and remained in that position until he resigned in 1973. The following year he helped establish the Energy Research and Development Administration and then returned to Oak Ridge to direct the Institute for Energy Analysis. At this time and in response to a rising campaign against nuclear energy, he raised concerns about the continuing expanding reliance on fossil fuel burning and its potential effects on climate. After the ERDA decided to investigate the carbon dioxide climate connection, Weinberg chaired a study group and also helped prepare the first workshop dedicated to addressing the carbon dioxide question. In the subsequent DoE Carbon Dioxide Program, the IEA and Weinberg continued to play a coordinating role until his retirement in 1984.

Wigley, Thomas 'Tom' (1940–) Australian atmospheric physicist. In 1975 he joined the Climatic Research Unit at the University of East Anglia and he became director when CRU's founder, Hubert Lamb, retired in 1978. He immediately reorientated the unit's activities towards global warming research. The empirical science section of the 'SCOPE 29 report of 1985 was written by Wigley and he went on to write the detection chapter of the

IPCC First Assessment report. For the writing of the Second Assessment during 1995 he took a back seat role, while publishing an article critical of the new modelling that had been used to show how the impact of carbon dioxide and sulphate emissions matched the global temperature record. Another article that he co-wrote in early 1995 argued against emissions controls for the next 30 years, but it was held up in peer review until a modified version appeared the following year.

Notes

Madrid, 1995

1. Australian Delegation to the IPCC (1995a, p. 9).
2. Australian Delegation to the IPCC (1995a, p. 9).
3. Houghton (2008, p. 738).

Chapter 2 The postwar boom and its discontents

4. Bush (1945).
5. Bush (1960, pp. vii-xxvi).
6. Fleming (1998, pp. 21–4).
7. Kennedy (1961).
8. Fleming (2010, pp. 177–182).
9. Fleming (2010).
10. Fleming (2010, p. 191).
11. von Neumann (1955).
12. Fleming (2010, p. 191).
13. von Neumann (1955).
14. Fleming (2010, pp. 177–182).
15. Fischer (1948).
16. USCDCP (2010).
17. Carson (1962).
18. Sladen et al. (1966).
19. Carson (1962, pp. 219–243).
20. Edwards (1992).
21. Zubrin (2012).
22. Rubin (1994, pp. 38–44).
23. Jukes (1972).
24. Sweeney (1972).
25. Jukes (1972).
26. Wurster (1968).
27. Jukes (1972).
28. Committee on Research in the Life Sciences (1970, p. 432).
29. Ruckelshaus (1979).

30. Mellanby (1992, p. 90).
31. Lovelock (1979, p. xi).
32. Dotto and Schiff (1978, p. 24).
33. Kennedy (1963).
34. Conway (2005, pp. 120–128).
35. Conway (2005, pp. 139–140).
36. Conway (2005, p. 140).
37. Harwood (1970).
38. Conway (2005, p. 141–142).
39. Osmundsen (1963).
40. Sullivan (1965).
41. Committee on Atmospheric Sciences (1966, vol. 2, p. 88).
42. Committee on Atmospheric Sciences (1966, vol. 2, pp. 97–99).
43. Dobson (1968, p. 398).
44. Sullivan (1970).
45. Mastenbrook (1971).
46. Webster (1970).
47. Lydon (1971a).
48. Webster (1970).
49. SCEP (1970, p. 16).

Chapter 3 Ozone layer jitters

50. Dobson (1968).
51. Dobson (1968, p, 392).
52. Dobson (1963, p. 116).
53. Dobson (1968, p. 401).
54. Dobson (1963, p. 113).
55. Dobson (1963, p. 118).
56. Dobson (1968, p. 401).
57. Chapman (1934).
58. Fleming (2010, p. 219).
59. Sullivan (1975).
60. *Los Angeles Times* (1962).
61. Cowen (1962).
62. Sumner (1962).
63. Fleming (2010, p. 220–222).
64. *Los Angeles Times* (1962).
65. Harrison (2003, p. 10).
66. Harrison (2003, pp. 12–13).
67. Harrison (2003, p. 13–17).
68. Harrison (2003, p. 19–20).
69. Harrison (1970).
70. Dotto and Schiff (1978, pp. 40–42).

71. Abramson (1970).
72. Abramson (1970).
73. Dotto and Schiff (1978, p. 61).
74. US Congress House Appropriations (1971, pp. 324–325).
75. Lydon (1971c).
76. Lydon (1971c).
77. Lydon (1971c).
78. Central Unit on Environmental Pollution (1976, pp. 45–51).
79. US Congress House Appropriations (1971, p. 319).
80. Dotto and Schiff (1978, pp. 44–45).
81. Dotto and Schiff (1978, pp. 47, 54).
82. SCEP (1970, p. 69).
83. Dotto and Schiff (1978, pp. 50–54).
84. Dotto and Schiff (1978, pp. 54–56).
85. Dotto and Schiff (1978, pp. 55–57).
86. Dotto and Schiff (1978, p. 59).
87. Dotto and Schiff (1978, p. 60).
88. Lydon (1971c).
89. Dotto and Schiff (1978, p. 60).
90. Sullivan (1971).
91. Reuters (1971).
92. Lydon (1971b).
93. Sullivan (1971).
94. Johnston (1971).
95. Dotto and Schiff (1978, p. 61).
96. Conway (2005, p. 165).
97. Conway (2005, p. 171).
98. Dotto and Schiff (1978, p. 321).
99. Ridley et al. (1973).
100. Ackerman et al. (1973).
101. Howard and Evenson (1977).
102. Ellsaesser (1980).
103. Ellsaesser (1980, p. 384).
104. Hofmann and Solomon (1989).
105. Conway (2005, pp. 230–232).
106. Mormino et al. (1975, pp. v–vi).
107. Valery (1975).
108. Rowlands (1995, p. 46).
109. AWST (1971).
110. Conway (2005, p. 166).
111. Grobecker et al. (1974).
112. Dotto and Schiff (1978, p. 93).
113. Committee to Study the Long-Term Worldwide Effects of Multiple Nuclear-Weapons (1975).
114. Donahue (1975).

115. Dotto and Schiff (1978, p. 94).
116. Committee on Government Operations (1976).

Chapter 4 Scepticism of the SST scares

117. Goldwater (1970).
118. Swihart (1971).
119. Hotz (1971).
120. Conway (2005).
121. Oreskes and Conway (2010, pp. 107–111).
122. Dotto and Schiff (1978, pp. 66–7).
123. SMIC (1971, p. 265).
124. SMIC (1971, p. 272).
125. See Dotto and Schiff (1978, p. 121–3) and p. 80).
126. Harrison (2003, p. 27).
127. Priestly (1972, p. 14).
128. Priestly (1972, p. 10).
129. Priestly (1972, p. 14).
130. Scorer (1975).
131. Scorer (1975).
132. Schmidt (1979, pp. 165–6).
133. Hatch (1995, p. 420).
134. Schmitt (1980, p. 219).
135. Ellsaesser (1974).
136. Ellsaesser (1983).
137. Roan (1989, p. 88).
138. Roan (1989).
139. Ellsaesser (1983).
140. Scorer (1975).
141. Dotto and Schiff (1978, pp. 212–14).
142. Dotto and Schiff (1978, p. 214).
143. Dotto and Schiff (1978, pp. 214, 157).
144. Dotto and Schiff (1978, p. 312).
145. Gleick (1988, pp. 48, 292).
146. Lorenz (1971).
147. Mitchell (1971, p. 136).
148. Lamb (1972).

Chapter 5 Apocalypse in a spray can

149. Lovelock (1979, p. 7).
150. Lovelock (1979, pp. 84–93).
151. Lovelock (1979, p. 50).
152. Lovelock et al. (1973).

153. Lovelock et al. (1973).
154. Dotto and Schiff (1978, p. 10–15).
155. Roan (1989, pp. 1–8).
156. Royal Swedish Academy of Science (1995).
157. Lovelock (1975a).
158. Crutzen (1970).
159. Matthews et al. (1971, p. 163).
160. Lovelock (1975b).
161. Nature–Times News Service (1975).
162. Scorer (1974).
163. *New Scientist* (1975).
164. Dotto and Schiff (1978, pp. 156–8).
165. Dotto and Schiff (1978, pp. 123–33).
166. Sullivan (1974).
167. Roan (1989, pp. 31–32).
168. Roan (1989, pp. 44, 47).
169. US Congress House Committee on Interstate and Foreign Commerce, House of Representatives (1974, pp. 3–8).
170. IMOS Task Force (1975, p. iii).
171. Dotto and Schiff (1978, p. 248).
172. Dotto and Schiff (1978, p. 186).
173. SCEP (1970, p. 101).
174. Ramanathan (1975).
175. Dotto and Schiff (1978, pp. 256–7).
176. Committee on Impacts of Stratospheric Change (1976, p. 7).
177. Schmeck Jr (1976).
178. *New Scientist* (1976).
179. Roan (1989, pp. 105–109).
180. Tolba and Rummel-Bulska (1998, p. 64).
181. Benedick (1998, pp. 26–34).
182. Tinker (1976).
183. Benedick (1998, p. 39).
184. Roan (1989, p. 146).
185. Roan (1989).
186. Directorate of Air, Noise and Wastes (1976).
187. Directorate of Air, Noise and Wastes (1976, p. 71).
188. Directorate of Air, Noise and Wastes (1976, p. 11).
189. Directorate of Air, Noise and Wastes (1976, pp. 7, 37–38).
190. Directorate of Air, Noise and Wastes (1976, p. 12).
191. See for example Benedick (1998, p. 38).
192. Directorate of Air, Noise and Wastes (1976, p. 16).
193. Directorate of Air, Noise and Wastes (1979).
194. Kenward (1979).
195. NRC Committee on Chemistry and Physics of Ozone Depletion and NRC Committee on Biological Effects of Increased Solar Ultraviolet Radiation (1982).

196. NRC Committee on Causes and Effects of Changes in Stratospheric Ozone (1984).
197. NRC Committee on Causes and Effects of Changes in Stratospheric Ozone (1984, p. 101).
198. Dembart (1984).
199. *Wall Street Journal* (1984).
200. Lovelock (1984).
201. Lovelock (2000, p. 157).
202. Lovelock (2000, pp. 55–161).
203. Benedick (1998, p. ix).
204. Roan (1989, p. 117).
205. Benedick (1998, p. x).
206. Benedick (1998, pp. 19–20).
207. Tolba and Rummel-Bulska (1998, pp. 61, 84–5).
208. Benedick (1998, p. 56).

Chapter 6 The Antarctic ozone hole

209. Farman et al. (1985).
210. Farman et al. (1985).
211. Darwall (2013).
212. Gribbin (1985).
213. Walgate (1985).
214. *Economist* (1985).
215. Andersen et al. (2002, p. 299–300).
216. Sullivan (1985).
217. Sullivan (1985).
218. Solomon et al. (1986).
219. McElroy et al. (1986).
220. Roan (1989, pp. 151–2).
221. Kerr (1986).
222. Roan (1989, pp. 171–176).
223. Schoeberl and Krueger (1986).
224. Benedick (1998, p. 108).
225. Anderson et al. (1989).
226. Scotto et al. (1988).
227. Tolba and Rummel-Bulska (1998, p. 74).
228. International Ozone Trends Panel (1988).
229. Roan (1989, p. 260).
230. Maugh (1988).
231. Dumanoski (1988).
232. Shabecoff (1988b).
233. WMO (1988, p. 4).
234. International Ozone Trends Panel (1988).
235. Trenberth (1988).

236. Singer (1989).
237. Singer (1989).
238. Haas et al. (1993, p. 33, n. 14).
239. SMIC (1971, pp. 258–259).
240. London and Kelley (1974).
241. WMO (1988).
242. For example, see Watson (1988, p. 88).
243. Tully et al. (2015).
244. Atkinson et al. (1989).
245. *Ottawa Citizen* (1988).
246. Cordero (2000, p. 91).
247. Dobson (1963, p. 118–119).
248. Farman et al. (1985).
249. Watson et al. (1988, p. 96).
250. Keating (1987).
251. Solomon et al. (1988).
252. Dayton (1988).
253. *Le Monde* (1989).
254. Detjen (1989).
255. Reuters (1989).
256. *Toronto Star* (1989).
257. Hilts (1989).
258. Hotz (1989).
259. Evans (1989).
260. Hofmann et al. (1989).
261. Pearce and Anderson (1989).
262. Kerr (1989a).
263. Andersen et al. (2002, p. 100).

Chapter 7 The carbon dioxide question

264. US DOE (1980, Preface).
265. Schmitt (1980, p. ix).
266. US DOE (1980, pp. 74–76).
267. US DOE (1980, p. 5).
268. Schmitt (1980, pp. ix–x).
269. Weart (2003b).
270. Slade (1980).
271. Carbon Dioxide Research Division (1983, Appendix 4).
272. Keeling (1960).

Chapter 8 Energy and climate

273. Darwall (2013, pp. 80–89).
274. Ehrlich (1968).
275. Club of Rome (1972).
276. Carter (1977).
277. Nixon (1973).
278. AEC (1974).
279. Nixon (1973).
280. Weinberg (1972).
281. Weinberg (1974).
282. Weinberg (1974).
283. Weinberg (1974).
284. Whittle et al. (1976, p. 70).
285. Whittle et al. (1976, pp. 50–51).
286. Whittle et al. (1976, p. 49).
287. MacCracken, pers. comm.
288. Elliott and Machta (1979).
289. Elliott and Machta (1979).
290. Schmitt (1980, p. x).
291. Weart (2003a).
292. Weart (2003a, p. 140–141).
293. Schmitt (1980, p. 178–179).
294. Weart (2003b).
295. CRU (1979, p. 32).
296. CRU (1982, p. 31).
297. CRU (1984, p. 38).
298. Lewin (2015a, p. 23).
299. Weart (2003a, pp. 28–30).
300. Revelle and Suess (1957).
301. Keeling (1960).
302. Keeling (1998, pp. 45–46).
303. SMIC (1971, p. 234).
304. US Committee for GARP (1975).
305. Schmitt (1980, p. xvii).
306. Weart (2003a, p. 97).
307. Fleming (2010, pp. 183–185).
308. Bolin (2008, p. 33).
309. Geophysics Study Committee (1977, p. viii).
310. Geophysics Study Committee (1977, p. vii).
311. Sullivan (1977).
312. Geophysics Study Committee (1977, pp. 5–8).
313. Geophysics Study Committee (1977, p. viii).
314. Lewin (2015a, p. 33–35).
315. White (1989, p. 1125).

316. Jastrow et al. (1989).
317. Carbon Dioxide Effects Research and Assessment Program (1980).
318. Carbon Dioxide Research Division (1985a).
319. Carbon Dioxide Research Division (1985b).
320. Carbon Dioxide Research Division (1985c).
321. Carbon Dioxide Research Division (1985d).

Chapter 9 Meteorological interest in climatic change

322. Commission for Climatology (2011, p. 8).
323. Fleming (1998, pp. 21–32).
324. Grove (1995).
325. Stehr and von Storch (2000).
326. Brooks (1926).
327. Lewin (2015a, pp. 4–6).
328. UNESCO (1963).
329. Mitchell (1966).
330. Fleming (2010, p. 236).
331. Revelle et al. (1965).
332. Fleming (1998, pp. 74–82).
333. *Time* Magazine (1968).
334. Revelle et al. (1965, p. 123).
335. *New York Daily News* (2010).
336. *New York Daily News* (2010).
337. Kellogg (1987, p. 122).
338. SMIC (1971, p. 10–11).

Chapter 10 The global cooling scare 1972–76

339. Matthews (1976, p. 615).
340. Lewin (2013b).
341. Dickson et al. (1975).
342. Sanderson (2002, p. 285).
343. Walker (2012, p. 404).
344. Tendler (1977).
345. Dansgaard et al. (1972, p. 397).
346. Ewing and Donn (1956).
347. Weart (2003a, pp. 118–141).
348. Emiliani (1972, Abstract).
349. Bryson (1968).
350. Peterson and Bryson (1968).
351. Broecker et al. (1960).
352. Broecker (1975).
353. Broecker (1999, p. 6).

354. Rasool and Schneider (1971).
355. Schneider (1989).
356. Kukla and Matthews (1972a).
357. Lamb (1982, pp. 296–297).
358. Kukla (1972).
359. Kukla and Matthews (1972b).
360. Kukla and Matthews (1972b).
361. Reeves and Gemmill (2004, p. 4).
362. Letter from Rogers C.B. Morton, the chairman of the White House Environmental Resources Committee, to Frederick Dent, the Secretary of Commerce, 1 August 1974. Quoted in Reeves and Gemmill (2004, p. 3).
363. Center for Environmental Assessment Services (1982).
364. Kukla et al. (1977).
365. Sullivan (1978).
366. Mitchell (1972).
367. Reeves and Gemmill (2004, pp. 1–5).
368. Hecht and Tirpak (1995, p. 377).
369. UN (1978).
370. IFIAS (1974).
371. Hare et al. (1976).
372. Hare et al. (1976).
373. Walker (2012, p. 404).
374. *New Scientist* (1978).
375. Calder (1974b).
376. Calder (1974a).
377. Calder (1974c, pp. 134–135).
378. Calder (1974a).
379. Jones (1974).
380. Walker (2012, p. 403).
381. CRU (1973, p. 21).
382. Calder (1974c, p. 98).
383. Lewin (2015a, p. 21).
384. Mason (1974).
385. Lamb (1975).
386. Hawkes (1975).
387. Petit et al. (2001).
388. US Committee for GARP (1975, p. 188).
389. AAS Committee on Climatic Change (1976, p. 13).

Chapter 11 The WMO response to cooling alarm

390. Kissinger (1974).
391. Kissinger (1974).
392. Kissinger (1974).

393. US Bureau of International Organization Affairs (1974).
394. Davies (1974, p. 2).
395. UN (1973).
396. WMO Executive Committee (1970, Annex XVIII, IV. (a), p. 216).
397. WMO (1975, Res 25).
398. WMO Executive Committee (1975, Res. 10).
399. WMO (1976).
400. WMO (1976).
401. WMO (1976).
402. Zillman (2009, p. 143).
403. *Times* (1976).
404. Sullivan (1976).
405. Sullivan (1976).
406. UN (1978).
407. WMO (1977, Res 14, Res 16).
408. WMO (1977, Res. 15).

Chapter 12 The greenhouse warming scare begins

409. WMO Conference (1979).
410. WMO Conference (1979, pp. 3–6).
411. WMO Conference (1979, p. viii).
412. WMO Conference (1979, p. 714).
413. WMO Conference (1979, p. 713).
414. WMO Conference (1979, p. 739).
415. WMO Conference (1979, p. 743).
416. Alexander (1979).
417. Congress (1979, p. 56).
418. WMO Committee (1979, Secs 4.1.18, 7.15, Annex I).
419. Wiin-Nielsen (1999).
420. Bolin (2008, p. 136).
421. Joint Scientific Committee (1980, Annex F).
422. Joint Scientific Committee (1980, p. 21).
423. Joint Scientific Committee (1980, p. 21).
424. Strong (1973).
425. UNEP Governing Council (1974, p. 67).
426. UNEP Governing Council (1979, p. 102).
427. Howard (1980, p. 21).
428. Joint Scientific Committee (1981, p. 16).
429. WMO Executive Council (1984, p. 76).
430. WMO Executive Committee (1980, p. 58).
431. UNEP Governing Council (1980, p. 52).
432. WMO (1981, p. 28).
433. WMO Executive Committee (1981, p. 53).

434. WMO Executive Committee (1982, p. 60).
435. WMO Executive Council (1985, p. 37).
436. Wright (1976).
437. Hansen et al. (1981, p. 957).
438. Hansen et al. (1981, p. 961).
439. Hansen et al. (1981, p. 965).
440. Hansen et al. (1981, p. 965).
441. Sullivan (1981).

Chapter 13 UNEP and the push for a climate treaty

442. Bolin (2008, p. 36).
443. Bolin (2008, pp. 35–36).
444. Bolin (2008, p. 36).
445. Bolin (1979).
446. Ivanov and Freney (1983, p. xiv).
447. ICSU (1986, pp. 394–408).
448. Clark, William C. (1982, pp. 23–25).
449. ICSU (1986, pp. xxvii–xxviii).
450. ICSU (1986, p. 277).
451. IPCC (1990a, p. 202).
452. Lewin (2013c).
453. Lewin (2013b).
454. Wigley (2014).
455. Hecht and Tirpak (1995, p. 380).
456. Tolba and Rummel-Bulska (1998, p. 85).
457. WMO (1986, p. 11).
458. Tolba and Rummel-Bulska (1998, pp. 57–60).
459. ICSU (1986, p. xxviii).
460. WMO (1986, pp. 2, 56).
461. Hulme (2009).
462. Lewin (2010).
463. Nordhaus (1977).
464. Geophysics Study Committee (1977, p. ix).
465. Ekholm (1901, p. 61).
466. WMO (1986, p. 12).
467. Bolin (2008, p. 38).
468. WMO (1986, p. 15).
469. Zillman (2015, p. 111).
470. WMO (1986, p. 8).
471. WMO Executive Council (1984, p. 36).
472. Hirst (2014, p. 71).
473. Franz (1997, p. 13–14).
474. WMO (1986, p. 7).

475. Ramanathan et al. (1985).
476. Hecht and Tirpak (1995, p. 380).
477. Hirst (2014, p. 70).
478. WMO (1986, p. 3).
479. WMO (1986, p. 60).
480. WMO (1986, p. 56).
481. WMO (1986).
482. ICSU (1986, p. xxii).
483. ICSU (1986, p. xxiii).
484. ICSU (1986, p. xxiv).
485. WMO (1986, p. 12).
486. Bolin (2008, pp. 46–47).
487. Hirst (2014, pp. 81–82).
488. Darwall (2013, pp. 90–99).
489. Darwall (2013, pp. 73–79).
490. World Commission on Environment and Development (1987).
491. Darwall (2013, p. 98).
492. Hirst (2014, p. 98).
493. Darwall (2013, pp. 98–99).
494. Jäger (1988).
495. Jäger (1988, p. i).
496. Hirst (2014, pp. 83–85).
497. WMO (1986, p. 3).
498. Jäger (1988, p. 1).
499. Jäger (1988, p. 2).
500. Mead and Kellogg (1977, p. xxi).
501. World Commission on Environment and Development (1987, Ch. 7, Sec. 29).
502. Jäger (1988, p. 41).
503. Nordhaus (1977).
504. Jäger (1988, pp. 21–22).
505. Jäger (1988, p. v).
506. Hirst (2014, p. 85).
507. Kellow (1996, pp. 104–129).
508. Revkin (1988, p. 60).
509. WMO (1989, pp. 3–8).
510. WMO (1989, p. 20).
511. WMO (1989, p. 292).
512. WMO (1989, p. 292).
513. Zillman (2016).
514. Pearman et al. (1989, p. 62).
515. Pearman et al. (1989, p. 61).
516. *Frontline* programme (2007).
517. Hansen (1988).
518. Hansen (1988, p. 40).
519. Shabecoff (1988a).

520. Shabecoff (1988a).
521. Hansen (1988, p. 47).
522. Hecht and Tirpak (1995, p. 383).
523. World Commission on Environment and Development and Australia Commission for the Future (1990).
524. Tickell (1986, p. 65).
525. Bolin (2008, p. 57).
526. Bolin (2008, p. 55).
527. Hecht and Tirpak (1995, p. 394).
528. Hirst (2014, pp. 175–181).
529. Hague Declaration (1989).
530. G7 (1989).
531. IPCC (1988, App. III, p. 7).
532. IPCC (1990c, App. D).

Chapter 14 Origins of the IPCC

533. Hecht and Tirpak (1995, p. 380).
534. Ad Hoc Study Group on Carbon Dioxide and Climate (1979).
535. Carbon Dioxide Assessment Committee (1983, p. 4).
536. Seidel (1983, p. v).
537. Hecht and Tirpak (1995, pp. 380–381).
538. Hirst (2014, pp. 90–91).
539. Hallgren (1993, p. 8).
540. Hirst (2014, p. 111).
541. Zillman (2007, p. 871).
542. Zillman (2007, p. 871).
543. Zillman (2007, p. 872).
544. IPCC (1988, p. 7).
545. IPCC (1988, p. 4).
546. IPCC (1988, p. 5).
547. IPCC (1988, Annex V).
548. Houghton (2013, p. 118).
549. Houghton (2013, pp. 146–147).
550. IPCC (1990a, pp. 183–187).
551. Houghton (2013, p. 147).
552. Edwards and Schneider (1997, pp. 7–9).
553. Michaels (1995).
554. Laframboise (2011, pp. 43–47).
555. Joint Scientific Committee (1980, Annex F).
556. UN General Assembly (1988, Items 10d,e).
557. IPCC (1989, p. 1).
558. IPCC (1989, p. 2).
559. IPCC (1989, p. 16).

560. IPCC (1989, p. 20).
561. Agrawala (1998, p. 634).
562. From the summary in IPCC (1990c, p. 22).
563. IPCC (1990c, pp. 14–15).
564. Agrawala (1998, p. 634).
565. UN General Assembly (1989, p. 207).
566. IPCC (1990c, p. 3).
567. Darwall (2013, pp. 138).
568. IPCC (1990c, p. 3).
569. Tolba and Rummel-Bulska (1998, pp. 81–82).
570. IPCC (1988, Sec 6.3).
571. IPCC (1989, Sec 3).
572. IPCC (1990c, App F).
573. Brenton (1994, p. 182).
574. Zillman (2016).
575. Bolin (2008, pp. 67–68).
576. Australian Delegation to the IPCC (1990, Foreword).
577. Australian Delegation to the IPCC (1990, Summary).
578. Australian Delegation to the IPCC (1990, p. 3).
579. IPCC (1990b, Appendix).
580. UNEP Governing Council (1990, p. 25).
581. WMO Executive Council (1990, Res 8).
582. Bodansky (1994, p. 59).
583. Thatcher (1990).
584. UN General Assembly (1990).
585. Bolin (2008, p. 69).
586. Tolba and Rummel-Bulska (1998, pp. 92–96).
587. Hecht and Tirpak (1995, p. 391).
588. UN (1992, Article 9).
589. UN (1992, Article 21).
590. Zillman (1991, p. 4).
591. Zillman (1991, p. 7–9).
592. Zillman (1991, p. 10).
593. Lawson (1990).
594. Clark, William C. (1982, p. 213).
595. US DOE (1980, p. 63).

Chapter 15 Searching for the catastrophe signal

596. Callendar (1938, p. 223).
597. Ångström (1900).
598. Callendar (1961, pp. 237–240).
599. Brooks (1926, Chs. 17,22).
600. Quoted in Baxter (1953, p. 72).

601. Plass (1956).
602. Fleming (2007, p. 81).
603. Kaplan (1961).
604. Crowe (1971, p. 486).
605. Möller (1963).
606. Callendar (1961).
607. Fleming (2007, p. 79–81).
608. Fleming (2007, p. 32).
609. Schmitt (1980, p. xiv).
610. Schmitt (1980, p. 157).
611. Schmitt (1980, p. 183).
612. Schmitt (1980, p. 142).
613. Schmitt (1980, p. 188).
614. Schmitt (1980, p. 200).
615. Beatty (1982, p. 3–19).
616. Kerr (1989b, p. 1041).
617. Kerr (1989b, p. 1043).
618. Kerr (1989b, p. 1043).
619. IPCC (1990a, p. 203).
620. IPCC (1990a, p. 253).
621. UN (1992, Article 9).
622. IPCC Working Group I (1992, p. 162).
623. Strong (1992).
624. UN (1992).
625. Bolin (2008, pp. 87–90).
626. Bolin (1997, pp. 90–91, 98–99).
627. Quoted in Agrawala (1998, p. 636).
628. IPCC (1991a, 1.5.6 on p. 7).
629. Bolin (2008, p. 70).
630. Pearce (1994).
631. IPCC (1991a, pp. 6–9).
632. IPCC (1991b, pp. 5–6).
633. Lewin (2013a).
634. Lewin (2013a).
635. Lewin (2013a).
636. UNFCCC (1995, Decision 1 III, 4, 6 and Annexes I, IIA).
637. Abbott (1995).
638. Houghton (1996a, p. 297).
639. Wigley (1995).
640. Subcommittee on Energy and Environment (1996, pp. 1173–1177, 1188–1190).
641. MacIlwain (1995).
642. MacIlwain (1995).
643. *Nature* (1995).
644. Houghton (2013, p. 118).
645. IPCC Bureau (1994, Annex 4).

646. Barnett et al. (1996).
647. Barnett et al. (1996, p. 262).
648. Barnett et al. (1996, p. 262).
649. Barnett et al. (1996, p. 255).
650. Santer et al. (1995, p. 36).
651. Santer et al. (1995, p. 36).

Chapter 16 The catastrophe signal found

652. Santer et al. (1995, p. 3).
653. Zillman (1995, Appendix D, pp. 3, 23).
654. Zillman (1995, p. 5).
655. Santer et al. (1996).
656. Michaels (1996).
657. Michaels and Knappenberger (1996).
658. Santer et al. (1996).
659. Michaels and Knappenberger (1996).
660. Kerr (1995b).
661. Christy (2015).
662. Santer (2016).
663. Edwards and Schneider (1997).
664. Stevens (1995).
665. IPCC (1995a).
666. Olaguer (2012).
667. Reuters (1995).
668. Australian Delegation to the IPCC (1995a, p. 9).
669. Australian Delegation to the IPCC (1995a, p. 10).
670. Kerr (1995b).
671. Hawkes (1995).
672. Schoon (1995).
673. Schoon (1995).
674. Sawyer (1995).
675. Reuters (1995).
676. *San Francisco Examiner* (1995).
677. Pearce (1995).
678. Masood (1995).
679. Kerr (1995a).
680. Meyer (1995).
681. Lewin (2013a).
682. Australian Delegation to the IPCC (1995b, p. 7).
683. Australian Delegation to the IPCC (1995b, p. 72).
684. Australian Delegation to the IPCC (1995b, p. 73).
685. IPCC (1994, p. 3).
686. Australian Delegation to the IPCC (1995b, p. 80).

687. Australian Delegation to the IPCC (1995b, p. 80).
688. Monbiot (1995).
689. Department of Environment, UK (1995).
690. Kristiansen (1996).
691. IPCC (1996).
692. SBSTA (1996, p. 5, #24).
693. ENB (1996, p. 3).
694. SBSTA (1996, p. 6).
695. SBSTA (1996, #26).
696. ENB (1996, p. 4).
697. SBSTA (1996, #28–32).
698. ENB (1996, p. 13).
699. Seitz (1996).
700. Jastrow et al. (1989).
701. Seitz (1996).
702. Houghton (2013, pp. 182–183).
703. Australian Delegation to the IPCC (1995b).
704. IPCC (1995b, p. 2).
705. Masood (1995).
706. Edwards and Schneider (1997).
707. Global Climate Coalition (1996).
708. Masood (1996a).
709. Santer et al. (1996).
710. Kerr (1996).
711. Michaels (1996).
712. Michaels and Knappenberger (1996).
713. Leggett (1999, pp. 245–246).
714. UNEP Insurance Industry Initiative (1996).
715. UNFCCC (1996b, pp. 29–30).
716. UNFCCC (1996b, p. 30).
717. UNFCCC (1996b, p. 31).
718. Leggett (1999, pp. 246–247).
719. Wirth (1996).
720. UNFCCC (1996a).
721. Reuters (1996).
722. Houghton (2008).
723. US Delegation to the IPCC (1995).
724. Santer (1996).
725. Masood (1996b).
726. Houghton (1996b).
727. Houghton (1996b).
728. US Delegation to the IPCC (1995, p. 1).
729. US Delegation to the IPCC (1995, p. 4).
730. US Delegation to the IPCC (1995, p. 12).
731. *Nature* (1996).

732. *Nature* (1996).
733. Singer (1996).
734. Leggett (1999, pp. 245–246).
735. Mann et al. (1998).
736. Montford (2010).
737. Pearce (1996).
738. Lewin (2015b).
739. Barnett et al. (1999).
740. Montford (2012).
741. NASA (2013).

Bibliography

AAS Committee on Climatic Change (1976). Report of a committee on climatic change. Report 21, Australian Academy of Science.

Abbott A (1995). Climate change panel to remain main source of advice. *Nature*; **374**(6523): 584–585.

Abramson R (1970). Huge research project launched for SST. *Los Angeles Times*, 21 July 1970.

Ackerman M, D Frimout, C Muller, D Nevejans, JC Fontanella, A Girard, and N Louisnard (1973). Stratospheric nitric oxide from infrared spectra. *Nature*; **245**(5422): 205–206.

Ad Hoc Study Group on Carbon Dioxide and Climate (1979). Carbon dioxide and climate: A scientific assessment. Report, National Academy of Sciences.

AEC (1974). Nuclear power growth, 1974–2000. Report WASH-1139-74, USA Atomic Energy Commission.

Agrawala S (1998). Structural and process history of the Intergovernmental Panel on Climate Change. *Climatic Change*; **39**(4): 621–642.

Alexander G (1979). Somber scientists study climate changes: 450 experts focus on the possibility of a global warming trend. *Los Angeles Times*, 19 February 1979.

Andersen SO, KM Sarma, and L Sinclair (2002). *Protecting the Ozone Layer: the United Nations history*. Earthscan.

Anderson JG, WH Brune, and MH Proffitt (1989). Ozone destruction by chlorine radicals within the Antarctic vortex: The spatial and temporal evolution of ClO-O_3 anticorrelation based on in situ ER-2 data. *Journal of Geophysical Research: Atmospheres*; **94**(D9): 11465–11479.

Ångström K (1900). Ueber die Bedeutung des Wasserdampfes und der Kohlensäure bei der Absorption der Erdatmosphäre. *Annalen der Physik*; **308**(12): 720–732.

339

Atkinson RJ, WA Matthews, PA Newman, and RA Plumb (1989). Evidence of the mid-latitude impact of Antarctic ozone depletion. *Nature*; **340**(6231): 290–294.

Australian Delegation to the IPCC (1990). WMO/UNEP Intergovernmental Panel on Climate Change (IPCC), fourth session, Sundsvall, Sweden: Australian Delegation report. Report, Australian Delegation to the IPCC.

Australian Delegation to the IPCC (1995a). WMO-UNEP Intergovernmental Panel on Climate Change (IPCC), Working Group 1 (Science): Fifth session, Madrid, 27–29 November 1995. Report, Australian Delegation to the IPCC.

Australian Delegation to the IPCC (1995b). WMO/UNEP Intergovernmental Panel on Climate Change (IPCC), eleventh session of the IPCC, Rome, 11–15 December 1995. Report, Australian Delegation to the IPCC.

AWST (1971). Washington roundup. *Aviation Week and Space Technology*, 15 March 1971.

Barnett T, K Hasselmann, M Chelliah, T Delworth, G Hegerl, P Jones, E Rasmusson, E Roeckner, C Ropelewski, B Santer, and S Tett (1999). Detection and attribution of recent climate change: A status report. *Bulletin of the American Meteorological Society*; **80**(12): 2631–2659.

Barnett T, B Santer, P Jones, R Bradley, and K Briffa (1996). Estimates of low frequency natural variability in near-surface air temperature. *Holocene*; **6**(3): 255–263.

Baxter WJ (1953). *Today's Revolution in Weather*. International Economic Research Bureau.

Beatty NB (1982). *Proceedings of the Workshop on First Detection of Carbon Dioxide Effects, Harpers Ferry, West Virginia, June 8-10, 1981*. US Department of Energy.

Benedick RE (1998). *Ozone Diplomacy: New directions in safeguarding the planet*. Harvard University Press.

Bodansky D (1994). Prologue to the climate change convention. In: *Negotiating Climate Change. The inside story of the Rio Convention*. Cambridge University Press.

Bolin B (1979). *The Global Carbon Cycle*. Wiley.

Bolin B (1997). Scientific assessment of climate change. In: G. Fermann (ed.), *International Politics of Climate Change: Key issues and critical actors*. Scandinavian University Press.

Bolin B (2008). *A History of the Science and Politics of Climate Change: The role of the Intergovernmental Panel on Climate Change*. Cambridge University Press.

Brenton T (1994). *The Greening of Machiavelli: The evolution of international environmental politics*. Royal Institute of International Affairs.

Bibliography

Broecker WS (1975). Climatic change: Are we on the brink of a pronounced global warming? *Science*; **189**(4201): 460–463.

Broecker WS (1999). What if the conveyor were to shut down? Reflections on a possible outcome of the great global experiment. *GSA Today*; **9**(1): 2–5.

Broecker WS, WM Ewing, and BC Heezen (1960). Evidence for an abrupt change in climate close to 11,000 years ago. *American Journal of Science*; **258**(6): 429–448.

Brooks CEP (1926). *Climate through the Ages*. E. Benn.

Bryson RA (1968). 'All other factors being constant': A reconciliation of several theories of climatic change. *Weatherwise*; **21**(2): 56–94.

Bush V (1945). Science, the endless frontier: a report to the President. Report, United States Office of Scientific Research and Development.

Bush V (1960). Science, the endless frontier; a report to the President on a program for postwar scientific research. Report, United States Office of Scientific Research and Development.

Calder N (1974a). The snow blitz. *Guardian*, 20 November 1974.

Calder N (1974b). *The Weather Machine*. TV show, first aired BBC2, 20 November 1974.

Calder N (1974c). *The Weather Machine*. BBC Books.

Callendar G (1938). The artificial production of carbon dioxide and its influence on temperature. *Quarterly Journal of the Royal Meteorological Society*; **64**(275): 223–240.

Callendar GS (1961). Temperature fluctuations and trends over the earth. *Quarterly Journal of the Royal Meteorological Society*; **87**(371): 1–12.

Carbon Dioxide Assessment Committee (1983). Changing climate: report of the Carbon Dioxide Assessment Committee. Report, National Academy of Sciences.

Carbon Dioxide Effects Research and Assessment Program (1980). *Workshop on Environmental and Societal Consequences of a Possible CO_2-induced Climate Change, Annapolis, Maryland, April 2–6, 1979*. United States Department of Energy.

Carbon Dioxide Research Division (1983). The carbon dioxide research plan: a summary. Report, United States Department of Energy.

Carbon Dioxide Research Division (1985a). Atmospheric carbon dioxide and the global carbon cycle. Report DOE/ER-0239, United States Department of Energy.

341

Carbon Dioxide Research Division (1985c). Detecting the climatic effects of increasing carbon dioxide. Report DOE/ER-0235, United States Department of Energy.

Carbon Dioxide Research Division (1985b). Direct effects of increasing carbon dioxide on vegetation. Report DOE/ER-0238, United States Department of Energy.

Carbon Dioxide Research Division (1985d). Projecting the climatic effects of increasing carbon dioxide. Report DOE/ER-0237, United States Department of Energy.

Carson R (1962). *Silent Spring*. Houghton Mifflin.

Carter J (1977). Address to the United States. Broadcast live on television and radio from the White House Oval Office, 18 April 1977.

Center for Environmental Assessment Services (1982). Climate impact assessment: US economic and social impacts of the record 1976–77 winter freeze and drought. Report, United States Department of Commerce.

Central Unit on Environmental Pollution (1976). Chlorofluorocarbons and their effect on stratospheric ozone. Report, UK Department of the Environment.

Chapman S (1934). 'The gases of the atmosphere', Presidential address delivered before the Royal Meteorological Society on January 17, 1934. *Quarterly Journal of the Royal Meteorological Society*; **60**: 127–42.

Christy J (2015). Response regarding memories of the Asheville lead author meeting 1995. Email, 22 March 2015.

Clark, William C. (ed.) (1982). *Carbon Dioxide Review: 1982*. Oxford University Press.

Club of Rome (1972). *The Limits to Growth*. Earth Island.

Commission for Climatology (2011). *Commission for Climatology: Over eighty years of service*. World Meteorological Organization.

Committee on Atmospheric Sciences (1966). Weather and climate modification problems and prospects: Final report of the Panel on Weather and Climate Modification to the Committee on Atmospheric Sciences, National Academy of Sciences, National Research Council. Report, National Academy of Sciences, National Research Council.

Committee on Government Operations (1976). FAA certification of the SST Concorde. Hearings before a Subcommittee of the Committee on Government Operations, House of Representatives, 94th Congress, first and second sessions.

Committee on Impacts of Stratospheric Change (1976). Halocarbons: environmental effects of chlorofluoromethane release. Report, Assembly of Mathematical and Physical Sciences (US), Committee on Impacts of Stratospheric Change, National Academy of Sciences.

Committee on Research in the Life Sciences (1970). The life sciences: Recent progress and application to human affairs, the world of biological research, requirements for the future. Report, United States National Academy of Sciences.

Committee to Study the Long-Term Worldwide Effects of Multiple Nuclear-Weapons (1975). Long-term worldwide effects of multiple nuclear-weapons detonations. Report, National Academy of Sciences.

Congress W (1979). Eighth World Meteorological Congress: Abridged final report with resolutions. Report WMO 533, World Meteorological Organization.

Conway EM (2005). *High-speed Dreams: NASA and the technopolitics of supersonic transportation, 1945–1999*. Johns Hopkins University Press.

Cordero E (2000). Misconceptions in Australian students' understanding of ozone depletion. *Melbourne Studies in Education*; **41**(2): 85–97.

Cowen RC (1962). Space fuel: Weather maker? *Christian Science Monitor*, 17 January 1962.

Crowe P (1971). *Concepts in Climatology*. Longman.

CRU (1973). Second Annual Report, Climatic Research Unit in the School of Environmental Sciences, University of East Anglia, covering the academic year October 1972 to September 1973. Report, University of East Anglia.

CRU (1979). Eighth Annual Report, Climatic Research Unit in the School of Environmental Sciences, University of East Anglia, covering the academic year October 1978 to September 1979. Report, University of East Anglia.

CRU (1982). Eleventh Annual Report, Climatic Research Unit in the School of Environmental Sciences, University of East Anglia, covering the academic year August 1981 to July 1982. Report, University of East Anglia.

CRU (1984). Biennial report 1982–84, Climatic Research Unit in the School of Environmental Sciences, University of East Anglia. Report, University of East Anglia.

Crutzen PJ (1970). The influence of nitrogen oxides on the atmospheric ozone content. *Quarterly Journal of the Royal Meteorological Society*; **96**(408): 320–325.

Dansgaard W, SJ Johnsen, HB Clausen, and CC Langway (1972). Speculations about the next glaciation. *Quaternary Research*; **2**(3): 396–398.

Darwall R (2013). *The Age of Global Warming: A History*. Quartet.

Davies D (1974). Environmental pollution and other environmental questions: Implications of possible climatic changes. Report EC-XXVI/Doc.70; Twenty-sixth session of the Executive Committee, World Meteorological Organization.

Dayton S (1988). Canadians confirm ozone hole in Arctic. *New Scientist*; **1616**: 47.

Dembart L (1984). 'Very small' threat to public health: New study downplays ozone depletion peril. *Los Angeles Times*, 25 February 1984.

Department of Environment, UK (1995). Climate change – John Gummer calls for commitment to targets. Press Release, 18 December 1995.

Detjen J (1989). Arctic clouds linked to ozone decay. *Austin American-Statesman*, 14 January 1989.

Dickson RR, HH Lamb, SA Malmberg, and JM Colebrook (1975). Climatic reversal in northern North Atlantic. *Nature*; **256**(5517): 479–482.

Directorate of Air, Noise and Wastes (1976). Chlorofluorocarbons and their effect on stratospheric ozone. Report, Department of the Environment, Central Directorate on Environmental Pollution.

Directorate of Air, Noise and Wastes (1979). Chlorofluorocarbons and their effect on stratospheric ozone (second report). Report, Department of the Environment, Central Directorate on Environmental Pollution.

Dobson GMB (1963). *Exploring the Atmosphere*. Clarendon Press.

Dobson GMB (1968). Forty years' research on atmospheric ozone at Oxford: A history. *Applied Optics*; **7**(3): 387–405.

Donahue TM (1975). The SST and ozone depletion. *Science*; **187**(4182): 1144.

Dotto L and H Schiff (1978). *The Ozone War*. Doubleday.

Dumanoski D (1988). New study shows global ozone depletion. *Boston Globe*, 16 March 1988.

Economist (1985). Cold comfort. *The Economist*; **7402**: 76.

Edwards JG (1992). The lies of Rachel Carson. *21st Century Science and Technology Magazine*, Summer edition, pp. 41–52.

Edwards PN and SH Schneider (1997). The 1995 IPCC report: Broad consensus or 'scientific cleansing'. *Ecofable/Ecoscience*; **1**(1): 3–9.

Ehrlich PR (1968). *The Population Bomb*. Ballantine Books.

Ekholm N (1901). On the variations of the climate of the geological and historical past and their causes. *Quarterly Journal of the Meteorological Society*; **27**(117): 1–62.

Elliott WP and L Machta (eds) (1979). *Workshop on the Global Effects of Carbon Dioxide from Fossil Fuels, Miami Beach, Fla., March 7–11, 1977*. Carbon Dioxide Effects Research and Assessment Program. United States Department of Energy.

344

Ellsaesser HW (1974). The dangers of one-way filters. *Bulletin of the American Meteorological Society*; **55**: 1362–1363.

Ellsaesser HW (1980). Man's effect on stratospheric ozone. In: PA Trudinger et al. (ed.), *Biogeochemistry of Ancient and Modern Environments*. Springer.

Ellsaesser HW (1983). Response to WW Kellogg 'Carbon dioxide and climatic changes: Implications for mankind's future'. Published as a preprint by Laurence Livermore National Laboratory, October 1983.

Emiliani C (1972). Quaternary hypsithermals. *Quaternary Research*; **2**(3): 270–273.

ENB (1996). Report of the second meeting of the subsidiary bodies of the UN Framework Convention on Climate Change 27 February–4 March 1996. *Earth Negotiations Bulletin*; **12**(26): 1–14.

Evans WFJ (1989). A hole in the Arctic polar ozone layer during March 1986. *Canadian Journal of Physics*; **67**(2–3): 161–165.

Ewing M and WL Donn (1956). A theory of ice ages II. *Science*; **123**(3207): 1061–1066.

Farman JC, BG Gardiner, and JD Shanklin (1985). Large losses of total ozone in Antarctica reveal seasonal ClO_x/NO_x interaction. *Nature*; **315**(6016): 207–210.

Fischer G (1948). Award Ceremony Speech. Nobel Prize in Physiology or Medicine 1948, awarded to Paul Müller. http://www.nobelprize.org/nobel_prizes/medicine/laureates/1948/press.html.

Fleming JR (1998). *Historical Perspectives on Climate Change*. Oxford University Press.

Fleming JR (2007). *The Callendar Effect: The life and times of Guy Stewart Callendar (1898–1964)*. American Meteorological Society.

Fleming JR (2010). *Fixing the Sky: The checkered history of weather and climate control*. Columbia University Press.

Franz WE (1997). The development of an international agenda for climate change: Connecting science to policy. Working Paper IR-97-034, International Institute for Applied Systems Analysis.

Frontline programme (2007). Interviews: Timothy Wirth. PBS. 17 January 2007.

G7 (1989). Economic Declaration from Summit of the Arch. Paris, France. http://www.g8.utoronto.ca/summit/1989paris/communique/index.html.

Geophysics Study Committee (1977). Energy and climate. Report, United States Geophysics Research Board.

Gleick J (1988). *Chaos: Making a new science*. Cardinal.

Global Climate Coalition (1996). The IPCC: Institutionalized 'scientific cleansing'. Circulated to politicians in Washington by fax machine.

Goldwater BM (1970). The Big Lie and the SST. *New York Times*, 16 December 1970.

Gribbin J (1985). Antarctic stratosphere is losing ozone. *New Scientist*; **1457**: 7.

Grobecker AJ, SC Coroniti, and RH Cannon (1974). Report of findings: the effects of stratospheric pollution by aircraft (final report). Report, US Department of Transportation.

Grove R (1995). *Green Imperialism: Colonial expansion, tropical island Edens and the origins of environmentalism, 1600–1860*. Cambridge University Press.

Haas PM, RO Keohane, and MA Levy (eds) (1993). *Institutions for the Earth: Sources of effective international environmental protection*. MIT Press.

Hague Declaration (1989). The Hague Declaration. *Netherlands International Law Review*; **36**(01): 69–72.

Hallgren RE (1993). Natural hazards, global change and meteorology: World Meteorological Day address. Australian Bureau of Meteorology.

Hansen J, D Johnson, A Lacis, S Lebedeff, P Lee, D Rind, and G Russell (1981). Climate impact of increasing atmospheric carbon dioxide. *Science*; **213**(4511): 957–966.

Hansen JE (1988). The greenhouse effect: impacts on current global temperature and regional heat waves. Testimony before the US Senate Committee on Energy and Natural Resources, June 23, 1988.

Hare F, H Lansford, H Lamb, and H Johnson (1976). *Climate change, food production, and interstate conflict; (summary of discussion and conclusions and recommendations of) a conference held at the Bellagio Study and Conference Center, Italy Jun 4–8, 1975*. Rockefeller Foundation.

Harrison H (1970). Stratospheric ozone with added water vapor: Influence of high-altitude aircraft. *Science*; **170**(3959): 734–736.

Harrison H (2003). Boeing adventures, with digressions. http://www.atmos.washington.edu/~harrison/reports/b2707.pdf.

Harwood R (1970). Earth Day stirs nation. *Washington Post*, 23 April 1970.

Hatch MT (1995). The politics of global warming in Germany. *Environmental Politics*; **4**(3): 415–440.

Hawkes N (1975). When summer snow doth make sage weathermen dispute. *Observer*, 8 June 1975.

346

Hawkes N (1995). Mankind blamed for global warming. *The Times*, 27 November 1995.

Hecht AD and D Tirpak (1995). Framework agreement on climate change: a scientific and policy history. *Climatic Change*; **29**(4): 371–402.

Hilts PJ (1989). Arctic ozone also imperiled; Depletion not limited to South Pole. *Washington Post*, 18 February 1989.

Hirst D (2014). *Negotiating climates: the politics of climate change and the formation of the Intergovernmental Panel on Climate Change (IPCC), 1979–1992*. PhD thesis, University of Manchester.

Hofmann DJ, TL Deshler, P Aimedieu, WA Matthews, PV Johnston, Y Kondo, WR Sheldon, GJ Byrne, and JR Benbrook (1989). Stratospheric clouds and ozone depletion in the Arctic during January 1989. *Nature*; **340**(6229): 117–121.

Hofmann DJ and S Solomon (1989). Ozone destruction through heterogeneous chemistry following the eruption of El Chichón. *Journal of Geophysical Research: Atmospheres*; **94**(D4): 5029–5041.

Hotz R (1971). The ecological problem. *Aviation Week and Space Technology*, 12 April 1971.

Hotz RL (1989). Arctic atmosphere is 'primed' to destroy ozone, study shows. *Atlanta Journal and Constitution*, 18 February 1989.

Houghton J (ed.) (1996a). *Climate Change 1995: The science of climate change*. Cambridge University Press.

Houghton J (1996b). Justification of Chapter 8. *Nature*; **382**(6593): 665.

Houghton J (2008). Meetings that changed the world: Madrid 1995: Diagnosing climate change. *Nature*; **455**(7214): 737–738.

Houghton J (2013). *In the Eye of the Storm*. Lion Books.

Howard CJ and KM Evenson (1977). Kinetics of the reaction of HO_2 with NO. *Geophysical Research Letters*; **4**(10): 437–440.

Howard D (1980). Man and climatic variability. Report 543, World Meteorological Organization.

Hulme M (2009). *Why we Disagree about Climate Change: Understanding controversy, inaction and opportunity*. Cambridge University Press.

ICSU (1986). *SCOPE 29: The Greenhouse Effect, Climatic Change, and Ecosystems*. Wiley.

IFIAS (1974). Statement of the IFIAS workshop on the impact of climate change on the quality and character of human life adopted at Bonn, FRG, 10 May 1974.

IMOS Task Force (1975). Fluorocarbons and the environment: report of Federal Task Force on Inadvertent Modification of the Stratosphere (IMOS). Report NSF 75-403, Council on Environmental Quality, Washington.

International Ozone Trends Panel (1988). Executive summary of the Ozone Trends Panel Report. Report, NASA.

IPCC (1988). Report of the first session of the WMO/UNEP Intergovernmental Panel on Climate Change (IPCC). Geneva, 9–11 November 1988. Report, World Meteorological Organization.

IPCC (1989). Report of the second session of the WMO/UNEP Intergovernmental Panel on Climate Change (IPCC): Nairobi, 28–30 June 1989. Report, World Meteorological Organization.

IPCC (1990a). *Climate Change: The IPCC Scientific Assessment.* Cambridge University Press.

IPCC (1990b). Report of the fourth session of the WMO/UNEP Intergovernmental Panel on Climate Change (IPCC), Sundsvall Sweden 27–30 August 1990. Report, Intergovernmental Panel on Climate Change.

IPCC (1990c). Report of the third session of the WMO/UNEP Intergovernmental Panel on Climate Change (IPCC). Washington DC, 5–7 February 1990. Report, World Meteorological Organization.

IPCC (1991a). Report of the fifth session of the WMO/UNEP Intergovernmental Panel on Climate Change (IPCC). Geneva, 13–15 March 1991. Report, World Meteorological Organization.

IPCC (1991b). Report of the sixth session of the WMO/UNEP Intergovernmental Panel on Climate Change (IPCC). Geneva, 29–30 October 1991. Report, World Meteorological Organization.

IPCC (1994). Report of the tenth session of the WMO/UNEP Intergovernmental Panel on Climate Change (IPCC), Nairobi, 10-12 November 1994. Report, World Meteorological Organization.

IPCC (1995a). Collated comments on the draft Summary for Policy Makers (SPM) of the draft contribution of Working Group I to the IPCC Second Assessment Report submitted by non-government organisations before the meeting. WGI/5th/Doc5/26.XI.95. Report, Intergovernmental Panel on Climate Change.

IPCC (1995b). Report of the eleventh session of the WMO/UNEP Intergovernmental Panel on Climate Change (IPCC), Rome, 11–15 December 1995. Report, World Meteorological Organization.

Bibliography

IPCC (1996). IPCC Second Assessment: Climate Change 1995: A report of the inter-governmental panel on climate change. Report, Intergovernmental Panel on Climate Change.

IPCC Bureau (1994). Scientific Assessment Working Group of IPCC (WGI): Progress in implementation of the workplan. Submitted to the IPCC Bureau seventh session, Geneva, 4–5 February 1994. Report BUR/VII/Doc. 6 (4.II.1994), Intergovernmental Panel on Climate Change.

IPCC Working Group I (1992). *Climate Change 1992: The supplementary report to the IPCC scientific assessment.* Cambridge University Press.

Ivanov MV and JR Freney (1983). *The Global Biogeochemical Sulphur Cycle.* Wiley.

Jäger J (1988). Developing policies for responding to climatic change: A summary of the discussions and recommendations of the workshops held in Villach (28 September–2 October, 1987) and Bellagio (9–13 November, 1987), under the auspices of the Beijer Institute, Stockholm. Report WMO/TD-No. 225, World Meteorological Organization: United Nations Environment Programme.

Jastrow R, W Nierenberg, and F Seitz (1989). *Scientific Perspectives on the Greenhouse Problem.* George C Marshall Institute.

Johnston H (1971). Reduction of stratospheric ozone by nitrogen oxide catalysts from supersonic transport exhaust. *Science*; **173**(3996): 517–522.

Joint Scientific Committee (1980). Report of the first session of the Joint Scientific Committee, Amsterdam, 26 March–3 April 1980. Report, Joint Scientific Committee for the World Climate Research Programme and the Global Atmospheric Research Programme, World Meteorological Organization.

Joint Scientific Committee (1981). Report of the second session of the Joint Scientific Committee, Vienna, 17–26 March 1981. Report, Joint Scientific Committee for the World Climate Research Programme and the Global Atmospheric Research Programme; WMO/ICSU.

Jones T (1974). Voice of RAF confounds Po valley snails and Severn's Bean geese. *The Times*, 5 December 1974.

Jukes TH (1972). DDT stands trial again. *BioScience*; **22**(11): 670–672.

Kaplan L (1961). Reply (to Plass). *Tellus*; **13**(2): 301–302.

Keating M (1987). Lower levels of ozone found in experiments over Canada. *The Globe and Mail.* 24 March 1987.

Keeling C (1998). Rewards and penalties of monitoring the earth. *Annual Review of Energy and the Environment*; **23**(1): 25–82.

Keeling CD (1960). The concentration and isotopic abundances of carbon dioxide in the atmosphere. *Tellus*; **12**(2): 200–203.

Kellogg WW (1987). Mankind's impact on climate: The evolution of an awareness. *Climatic Change*; **10**(2): 113–136.

Kellow AJ (1996). *Transforming Power: The politics of electricity planning*. Cambridge University Press.

Kennedy JF (1961). Address by President John F Kennedy to the UN General Assembly. 25 September 1961.

Kennedy JF (1963). Remarks at Colorado Springs to the Graduating Class of the US Air Force Academy. Speech, 5 June 1963. http://www.presidency.ucsb.edu/ws/?pid=9254.

Kenward M (1979). Ozone: Cautious inaction needed. *New Scientist*; **84**(1178): 252.

Kerr RA (1986). Taking shots at ozone hole theories. *Science*; **234**(4778): 817–818.

Kerr RA (1989a). Arctic ozone is poised for a fall. *Science*; **242**(4894): 1007–1008.

Kerr RA (1989b). Hansen vs the world on the greenhouse threat. *Science*; **244**(4908): 1041–1043.

Kerr RA (1995a). It's official: First glimmer of greenhouse warming seen. *Science*; **270**(5242): 1565–1567.

Kerr RA (1995b). Scientists see greenhouse, semiofficially. *Science*; **269**(5231): 1667–1668.

Kerr RA (1996). Sky-high findings drop new hints of greenhouse warming. *Science*; **273**(5271): 34–34.

Kissinger HA (1974). Address to the sixth special session of the United Nations General Assembly, 15 April 1974.

Kristiansen J (1996). OECD countries urged stepped-up efforts on global environment. Agence France-Presse, 20 February 1996.

Kukla G (1972). Guest editorial: The end of the present interglacial. *Quaternary Research*; **2**(3): 261–269.

Kukla GJ, JK Angell, J Korshover, H Dronia, M Hoshiai, J Namias, M Rodewald, R Yamamoto, and T Iwashima (1977). New data on climatic trends. *Nature*; **270**(5638): 573–580.

Kukla GJ and RK Matthews (1972a). When will the present interglacial end? *Science*; **178**(4057): 190–202.

Kukla GK and RK Matthews (1972b). Letter to President Nixon, 3 December 1972. A facsimile can be found in a slide show entitled *The Origins of a 'Diagnostics Climate Center'*, by Robert Reeves and Daphne Gemmill, Available online at http://www.cpc.ncep.noaa.gov/products/outreach/proceedings/cdw29_proceedings/reeves.ppt.

Laframboise D (2011). *The Delinquent Teenager Who Was Mistaken for the World's Top Climate Expert*. CreateSpace.

Lamb HH (1972). A critical problem. *Nature*; **237**(5349): 53–54.

Lamb HH (1975). The weather: keep cool but don't freeze (Letter). *Guardian*, 6 January 1975.

Lamb HH (1982). *Climate, History and the Modern World*. Methuen.

Lawson H (1990). *The Greenhouse Conspiracy*. Cutting Edge series, Channel 4.

Leggett JK (1999). *The Carbon War*. Allen Lane.

Le Monde (1989). Environnement: La conférence de Londres Chinois et Soviétiques ne voient pas d'urgence à prendre des mesures énergiques pour protéger la couche d'ozone. *Le Monde*, 8 March 1989.

Lewin B (2010). The anatomy of virtuous corruption. *Enthusiasm, Scepticism and Science* blog.

Lewin B (2013a). Enter the economists: The price of life and how the IPCC only just survived the other chapter controversy. *Enthusiasm, Scepticism and Science* blog.

Lewin B (2013b). Hubert Lamb and the assimilation of legendary ancient Russian winters. *Enthusiasm, Scepticism and Science* blog, 22 September 2013.

Lewin B (2013c). Millennium idols: Smash the Hockey Stick but smash the others too! *Enthusiasm, Scepticism and Science* blog.

Lewin B (2015a). Hubert Lamb and the transformation of climate science. Report 17, The Global Warming Policy Foundation.

Lewin B (2015b). Tom Wigley: The skepticism and loyalty of CRU's second director. *Enthusiasm, Scepticism and Science* blog, 9 March 2015.

London J and J Kelley (1974). Global trends in total atmospheric ozone. *Science*; **184**(4140): 987–989.

Lorenz EN (1971). Climatic change as a mathematical problem. In: W. H. Matthews, W. H. Kellogg, and G. D. Robinson (eds), *Man's Impact on the Climate*. MIT Press.

Los Angeles Times (1962). Rocket exhaust threat told by meteorologist. *Los Angeles Times*, 29 March 1962.

Lovelock JE (1975a). Atmospheric halocarbons and stratospheric ozone (reply). *Nature*; **254**(5497): 275.

Lovelock JE (1975b). Natural halocarbons in the air and in the sea. *Nature*; **256**(5514): 193–194.

Lovelock JE (1979). *Gaia: A new look at life on earth*. Oxford University Press.

Lovelock JE (1984). Causes and effects of changes in stratospheric ozone: Update 1983: Review. *Environment*; **26**(10): 25–6.

Lovelock JE (2000). *The Ages of Gaia: A biography of our living earth*. Oxford University Press.

Lovelock JE, RJ Maggs, and RJ Wade (1973). Halogenated hydrocarbons in and over the Atlantic. *Nature*; **241**(5386): 194–196.

Lydon C (1971a). Experts assure house SST would not be harmful. *New York Times*, 4 March 1971.

Lydon C (1971b). Senate SST hopes fade: Ozone peril called grave. *New York Times*, 18 May 1971.

Lydon C (1971c). White House and Proxmire in dispute on SST hazard. *New York Times*, 18 March 1971.

MacIlwain C (1995). Climate critics claim access blocked to unpublished data. *Nature*; **378**(6555): 329.

Mann ME, RS Bradley, and MK Hughes (1998). Global-scale temperature patterns and climate forcing over the past six centuries. *Nature*; **392**(6678): 779–787.

Mason BJ (1974). A winter's tale – much ado about nothing? *Guardian*, 24 December 1974.

Masood E (1995). Climate panel confirms human role in warming, fights off oil states. *Nature*; **378**(6557): 524.

Masood E (1996a). Head of climate group rejects claims of political influence. *Nature*; **381**(6582): 455–455.

Masood E (1996b). Sparks fly over climate report. *Nature*; **381**(6584): 639.

Mastenbrook HJ (1971). The variability of water vapor in the stratosphere. *Journal of the Atmospheric Sciences*; **28**(8): 1495–1501.

Matthews S (1976). What's happening to our climate? *National Geographic*; **150**(5).

Matthews WH, WW Kellogg, and GD Robinson (eds) (1971). *Man's Impact on the Climate*. MIT Press.

Maugh T (1988). Ozone depletion far worse than expected. *Los Angeles Times*, 16 March 1988.

McElroy MB, RJ Salawitch, SC Wofsy, and JA Logan (1986). Reductions of Antarctic ozone due to synergistic interactions of chlorine and bromine. *Nature*; **321**(6072): 759–762.

Mead M and WW Kellogg (eds) (1977). *The Atmosphere: Endangered and endangering. Report of a conference at the National Institute of Environmental Health Sciences, North Carolina, October 1975*. US National Institutes of Health. NIH 77-1065.

Mellanby K (1992). *The DDT Story*. British Crop Protection Council.

Meyer A (1995). Economics of climate change. *Nature*; **378**(6556): 433.

Michaels P (1995). Climate policy of the 'anointed': The avoidance of peer review. *World Climate Report*; **1**(3). Online at: http://www.worldclimatereport.com/archive/previous_issues/vol1/v1n3/feature.htm. The authorship of the article is not indicated, but it is assumed to be Michaels.

Michaels P (1996). Santer springs forth. *World Climate Report (online)*; **2**(1). Michaels is the assumed author of the article.

Michaels P and P Knappenberger (1996). Human effect on global climate? *Nature*; **384**: 522–523.

Mitchell JM (1966). Climatic change: report of a working group of the Commission for Climatology. Report 79, World Meteorological Organization.

Mitchell JM (1971). The problem of climatic change and its causes. In: W. Matthews, W. Kellogg, and G. Robinson (eds), *Man's Impact on the Climate*, pp. 133–40. MIT Press.

Mitchell JM (1972). The natural breakdown of the present interglacial and its possible intervention by human activities. *Quaternary Research*; **2**(3): 436–445.

Möller F (1963). On the influence of changes in the CO_2 concentration in air on the radiation balance of the Earth's surface and on the climate. *Journal of Geophysical Research*; **68**(13): 3877–3886.

Monbiot G (1995). Last warning on earth. *Guardian*, 14 December 1995.

Montford A (2010). *The Hockey Stick Illusion*. Stacey International.

Montford A (2012). Nullius in verba: the Royal Society and climate change. Report 6, The Global Warming Policy Foundation.

Mormino J, D Sola, and C Patten (1975). Climatic Impact Assessment Program: development and accomplishments, 1971–1975. Report, United States Department of Transportation.

NASA (2013). Consensus: 97% of climate scientists agree. http://climate.nasa.gov/ scientific-consensus.

Nature (1995). Global warming rows. *Nature*; **378**(6555): 322.

Nature (1996). Climate debate must not overheat. *Nature*; **381**(6583): 539.

Nature–Times News Service (1975). Atmosphere: Controversy over fluorocarbon. *The Times*, 19 July 1975.

New Scientist (1975). Hysteria ousts science from halocarbon controversy. *New Scientist*; **66**(954): 643.

New Scientist (1976). US row over aerosol ban. *New Scientist*; **72**(1025): 262.

New Scientist (1978). EEC to study climatic change. *New Scientist*; **79**(1121): 831.

New York Daily News (2010). Nixon warned of global warning over 30 years ago. *New York Daily News*, 3 July 2010.

Nixon R (1973). Address to the nation about policies to deal with the energy shortages. Televised live from the Oval Office, 7 November 1973.

Nordhaus WD (1977). Strategies for the control of carbon dioxide. Report 443, Cowles Foundation for Research in Economics, Yale University.

NRC Committee on Causes and Effects of Changes in Stratospheric Ozone (1984). Causes and effects of changes in stratospheric ozone: update 1983. Report, National Academies Press.

NRC Committee on Chemistry and Physics of Ozone Depletion and NRC Committee on Biological Effects of Increased Solar Ultraviolet Radiation (1982). Causes and effects of stratospheric ozone reduction, an update. Report, National Academies Press.

Olaguer E (2012). Interview regarding Olaguer's participation at the Working Group 1 session, Madrid November 1995, as a representative of Dow Chemicals. 9 September 2012.

Oreskes N and EM Conway (2010). *Merchants of Doubt: How a handful of scientists obscured the truth on issues from tobacco smoke to global warming*. Bloomsbury Press.

Osmundsen JA (1963). Weather scientists optimistic that new findings are near. *New York Times*, 23 September 1963.

Ottawa Citizen (1988). Dramatic move of ozone hole to Australia 'shakes' experts. *Ottawa Citizen*, 7 December 1988.

Pearce F (1994). Frankenstein syndrome hits climate treaty. *New Scientist*; **142**(1929): 5.

Pearce F (1995). Global warming 'jury' delivers guilty verdict. *New Scientist*; **148**(2007): 6.

Pearce F (1996). Sit tight for 30 years, argues climate guru. *New Scientist*; **149**(2013): 7.

Pearce F and I Anderson (1989). Is there an ozone hole over the north pole? *New Scientist*; **1653**: 32.

Pearman G, N Quinn, and J Zillman (1989). The changing atmosphere. *Search*; **20**(2): 59–65.

Peterson JT and RA Bryson (1968, October). Atmospheric aerosols: Increased concentrations during the last decade. *Science*; **162**(3849): 120–121.

Petit J, J Jouzel, D Raynaud, N Barkov, J Barnola, I Basile, and M Bender (2001). Vostok ice core data for 420,000 years. *IGBP PAGES/World Data Center for Paleoclimatology Data Contribution Series*; **76**.

Plass GN (1956). Carbon dioxide and the climate. *American Scientist*; **44**(3): 302–316.

Priestly C (1972). Atmospheric effects of supersonic aircraft. Report 15, Australian Academy of Science.

Ramanathan V (1975). Greenhouse effect due to chlorofluorocarbons: Climatic implications. *Science*; **190**(4209): 50–52.

Ramanathan V, RJ Cicerone, HB Singh, and JT Kiehl (1985). Trace gas trends and their potential role in climate change. *Journal of Geophysical Research: Atmospheres*; **90**(D3): 5547–5566.

Rasool S and SH Schneider (1971). Atmospheric carbon dioxide and aerosols: Effects of large increases on global climate. *Science*; **173**(3992): 138–141.

Reeves R and D Gemmill (2004). *Climate Prediction Center: Reflections on 25 years of analysis diagnosis and prediction*. NOAA.

Reuters (1971). Blindness seen from SST change in ozone. *Los Angeles Times*, 22 September 1971.

Reuters (1989). Scientists to reveal results of ozone-depletion study. *Sun Sentinel*, 17 February 1989.

Reuters (1995). Expert panel poised to link man to global warming despite conflicts. *Orlando Sentinel*, 27 November 1995.

Reuters (1995). Scientists confirm that man affects climate. *Orange County Register*, 3 December 1995.

Reuters (1996). Australia disappointed at US greenhouse gas switch. Reuters News, 18 July 1996.

Revelle R, W Broecker, H Craig, C Keeling, and J Smagorinsky (1965). Atmospheric carbon dioxide. In: *Restoring the Quality of our Environment: A report of the Environmental Pollution Panel*, pp. 111–133. President's Science Advisory Committee.

Revelle R and HE Suess (1957). Carbon dioxide exchange between atmosphere and ocean and the question of an increase of atmospheric CO_2, during the past decades. *Tellus*; **9**(1): 18–27.

Revkin AC (1988). Endless summer: living with the greenhouse effect. *Discover Magazine*; **9**(10): 50–61.

Ridley BA, HI Schiff, AW Shaw, L Bates, C Howlett, H Levaux, LR Megill, and TE Ashenfelter (1973). Measurements in situ of nitric oxide in the stratosphere between 17.4 and 22.9 km. *Nature*; **245**(5424): 310–311.

Roan S (1989). *Ozone Crisis: The 15-year evolution of a sudden global emergency*. Wiley.

Rowlands IH (1995). *The Politics of Global Atmospheric Change*. Manchester University Press.

Royal Swedish Academy of Science (1995, October). Press release: The 1995 Nobel Prize in Chemistry. http://www.nobelprize.org/nobel_prizes/chemistry/laureates/1995/press.html.

Rubin CT (1994). *The Green Crusade: Rethinking the roots of environmentalism*. Free Press.

Ruckelshaus W (1979). To Mr Allan Grant, President, American Farm Bureau Federation. Letter, 26 April 1979.

Sanderson M (2002). *The History of the University of East Anglia, Norwich*. A&C Black.

San Francisco Examiner (1995). Panel warns global warming is a human-induced reality 2 oil-producing nations dissent. *San Francisco Examiner*, 1 December 1995.

Santer BA (2016). Response to request for assistance with research on the IPCC SAR. Email, 16 April 2016.

Santer BD (1996). Subject: Seitz editorial in *Wall Street Journal*. Email, 12 June 1996. The email had a wide circulation list.

Santer BD, K Taylor, T Wigley, T Johns, P Jones, D Karoly, J Mitchell, A Oort, J Penner, V Ramaswamy, and others (1996). A search for human influences on the thermal structure of the atmosphere. *Nature*; **382**(6586): 39–46.

Santer BD, TML Wigley, TP Barnett, and E Anyamba (1995). Chapter 8: Detection of climate change and attribution of causes. 18 April 1995 draft of Chapter 8.

Sawyer K (1995). Experts agree humans have 'discernible' effect on climate. *Washington Post*, 1 December 1995.

SBSTA (1996). Report of the subsidiary body for scientific and technological advice on the work of its second session, held at Geneva from 27 February to 4 March 1996. Report FCCC/SBSTA/1996/8, Framework Convention on Climate Change.

SCEP (ed.) (1970). *Man's impact on the global environment: assessment and recommendations for action*. MIT Press.

Schmeck Jr HM (1976). FDA urges a curb on fluorocarbons. *New York Times*, 16 October 1976.

Schmidt H (1979). Opening address at the European Nuclear Conference, Hamburg, 7 May 1979. In: *Der Kurs Heisst Frieden*. Econ Verlag.

Schmitt LE (ed.) (1980). *Carbon Dioxide Effects Research and Assessment Program: Proceedings of the Carbon Dioxide and Climate Research Program Conference, Washington, DC, April 24–25, 1980*. United States Department of Energy.

Schneider SH (1989). The greenhouse effect: Science and policy. *Science*; **243**(4892): 771–781.

Schoeberl MR and AJ Krueger (1986, November). Overview of the Antarctic ozone depletion issue. *Geophysical Research Letters*; **13**(12): 1191–1192.

Schoon N (1995). Global warming is here, experts agree. *Independent*, 30 November 1995.

Scorer R (1974). Freon in the stratosphere. *New Scientist*; **64**(918): 140.

Scorer R (1975). The danger of environmental jitters. *New Scientist*; **66**(955): 702–3.

Scotto J, G Cotton, F Urbach, D Berger, and T Fears (1988). Biologically effective ultraviolet radiation: surface measurements in the United States, 1974 to 1985. *Science*; **239**(4841): 762–764.

Seidel S (1983). *Can We Delay a Greenhouse Warming?: The effectiveness and feasibility of options to slow a build-up of carbon dioxide in the atmosphere*. United States Environmental Protection Agency.

Seitz F (1996). A major deception on 'global warming'. *Wall Street Journal*, 12 June 1996.

Shabecoff P (1988a). Global warming has begun, expert tells Senate. *New York Times*, 24 June 1988.

Shabecoff P (1988b). Study shows significant decline in ozone layer. *New York Times*, 16 March 1988.

Singer F (1996). Climate debate. *Nature*; **382**(6590): 392.

Singer SF (1989). My adventures in the ozone layer. *National Review*; **41**(12): 34–38.

Slade DH (1980). The US Department of Energy and the carbon dioxide issue. In: W. Bach, J. Pankrath, and J. Williams (eds), *Interactions of Energy and Climate*. Springer.

Sladen WJL, CM Menzie, and WL Reichel (1966). DDT residues in Adelie penguins and a crabeater seal from Antarctica. *Nature*; **210**(5037): 670–673.

SMIC (1971). *Inadvertent climate modification*. MIT Press.

Solomon S, RR Garcia, FS Rowland, and DJ Wuebbles (1986). On the depletion of Antarctic ozone. *Nature*; **321**(6072): 755–758.

Solomon S, GH Mount, RW Sanders, RO Jakoubek, and AL Schmeltekopf (1988). Observations of the nighttime abundance of OClO in the winter stratosphere above Thule, Greenland. *Science*; **242**(4878): 550–555.

Stehr N and H von Storch (eds) (2000). *Eduard Brückner: The sources and consequences of climate change and climate variability in historical times*. Kluwer.

Stevens WK (1995). Global warming experts call human role likely. *New York Times*, 10 September 1995.

Strong M (1973). Introductory statement of the executive director to the first session of the UNEP Governing Council. Report UNEP/GC/L.10, United Nations Environment Programme.

Strong M (1992). Opening statement to the Rio Earth Summit. Speech, 3 June 1992.

Subcommittee on Energy and Environment (1996). *Scientific integrity and public trust: The science behind federal policies and mandates: Case study 2: climate models and projections of potential impacts of global climate change: Hearing before the Subcommittee on Energy and Environment of the Committee on Science, U.S. House of Representatives, One Hundred Fourth Congress, first session, November 16, 1995*. United States Congress House Committee on Science, US GPO.

Sullivan W (1965). Jet trails' effect on climate studied. *New York Times*, 1 May 1965.

Sullivan W (1970). Navy tests show humidity in stratosphere rose 50% in the last six years. *New York Times*, 28 May 1970.

Sullivan W (1971). Ozone: Sorry but there's still more to say on the SST. *New York Times*, 30 May 1971.

Bibliography

Sullivan W (1974). Tests show aerosol gases may pose threat to earth. *New York Times*, 26 September 1974.

Sullivan W (1975). Ozone depletion seen as a war tool. *New York Times*, 28 February 1975.

Sullivan W (1976). Two climate experts decry predictions of disasters. *New York Times*, 22 February 1976.

Sullivan W (1977). Scientists fear heavy use of coal may bring adverse shift in climate. *New York Times*, 25 July 1977.

Sullivan W (1978). International team of specialists finds no end in sight to 30-year cooling trend in northern hemisphere. *New York Times*, 5 January 1978.

Sullivan W (1981). Study finds warming trend that could raise sea levels. *New York Times*, 22 August 1981.

Sullivan W (1985). Low ozone level found above Antarctica. *New York Times*, 7 November 1985.

Sumner B (1962). Space taint could twist weather. *Boston Globe*, 21 January 1962.

Sweeney EM (1972). Consolidated DDT hearing: Hearing Examiner's recommended findings, conclusions and orders (40 CFR 164.32). Report, United States Environmental Protection Agency.

Swihart J (1971). Cancer charge refuted (letter). *Aviation Week and Space Technology*, 12 April 1971.

Tendler S (1977). Legacy of Dickens's snowy childhood. *The Times*, 28 December 1977.

Thatcher M (1990). Speech at Second World Climate Conference. 6 November 1990.

Tickell C (1986). *Climatic Change and World Affairs* (Revised edn). Center for International Affairs, Harvard University.

Time Magazine (1968). The age of effluence. *Time*; **91**(19): 58.

Times(1976). World's temperature likely to rise. *The Times*, 22 June 1976.

Tinker J (1976). Environmental imperialism. *New Scientist*; **69**(982): 50.

Tolba MK and I Rummel-Bulska (1998). *Global Environmental Diplomacy: Negotiating environment agreements for the world, 1973–1992*. MIT Press.

Toronto Star(1989). Scientists defend Arctic ozone study. *Toronto Star*, 18 February 1989.

Trenberth K (1988). Report on reports: Executive Summary of the Ozone Trends Panel report. *Environment: Science and Policy for Sustainable Development*; **30**(6): 25–26.

Tully MB, AR Klekociuk, and SK Rhodes (2015). Trends and variability in total ozone from a mid-latitude southern hemisphere site: The Melbourne Dobson Record 1978–2012. *Atmosphere-Ocean*; **53**(1): 58–65.

UN (1973). Report of the United Nations Conference on the Human Environment, Stockholm, 5–16 June 1972. Report, United Nations.

UN (1978). United Nations conference on desertification, 29 August–9 September 1977, Round-up, plan of action and resolutions. United Nations. Available from: http://www.ciesin.org/docs/002-478/002-478.html.

UN (1992). Framework Convention on Climate Change. United Nations.

UN General Assembly (1988). Resolution on protection of global climate for present and future generations of mankind (A/RES/43/53). United Nations.

UN General Assembly (1989). Resolution on protection of global climate for present and future generations of mankind (RES/44/207). United Nations.

UN General Assembly (1990). Resolution on protection of global climate for present and future generations of mankind (RES/45/212). United Nations.

UNEP Governing Council (1974). Report of the Governing Council on the work of its second session, Nairobi, 11–22 March 1974. Report, United Nations.

UNEP Governing Council (1979). Report of the Governing Council on the work of its seventh session, Nairobi, 18 April–4 May 1979. Report, United Nations.

UNEP Governing Council (1980). Report of the Governing Council on the work of its eighth session, Nairobi, 16–29 April 1980. Report, United Nations.

UNEP Governing Council (1990). Report of the Governing Council on the work of its second special session, Nairobi, 1–3 August 1990. Report, United Nations.

UNEP Insurance Industry Initiative (1996). Position paper on climate. 9 July 1996.

UNESCO (1963). *Changes of Climate: Proceedings of the Rome symposium organized by UNESCO and the World Meteorological Organization, 2–7 October 1961.* UNESCO.

UNFCCC (1995). Report of the Conference of the Parties on its first session, held at Berlin from 28 March to 7 April 1995, Addendum, Part two: Action taken by the Conference of the Parties at its first session. FCCC/CP/1995/7/Add.1. Report, United Nations.

UNFCCC (1996a). The Geneva Ministerial Declaration FCCC/CP/1996/Add.1.

Bibliography

UNFCCC (1996b). Report of the Conference of the Parties on its second session, held at Geneva from 8 to 19 July 1996, Part One: Proceedings. Report FCCC/CP/1996/15/, United Nations Framework Convention on Climate Change.

US Bureau of International Organization Affairs (1974). Cable: Climate change: follow-up on secretary Kissinger's General Assembly proposal from 'Geneva for Cartwright'. 8 May 1974.

US Committee for GARP (1975). Understanding climatic change: a programme for action. Report, National Academy of Sciences.

US Congress House Appropriations (1971). Civil supersonic aircraft development (SST): Hearings ... 92d Congress, 1st session, Continuing appropriations.

US Congress House Committee on Interstate and Foreign Commerce, House of Representatives (1974). Fluorocarbons – impact on health and environment: Hearings before the Subcommittee on Public Health and Environment of the Committee on Interstate and Foreign Commerce, House of Representatives, Ninety-third Congress, second session, on H.R. 17577 ... and H.R. 17545 ... December 11 and 12, 1974.

US Delegation to the IPCC (1995). US Government specific comments on the draft IPCC WG I Summary for PolicyMakers prepared by the US delegation for consideration at the IPCC WG I plenary for submission on November 15, 1995.

US DOE (1980). Carbon dioxide research progress report: Fiscal year 1979. Report, US Department of Energy.

USCDCP (2010). Elimination of malaria in the United States (1947–1951). United States Centers for Disease Control and Prevention. http://www.cdc.gov/malaria/about/history/elimination_us.html.

Valery N (1975). SSTs are clean – in small numbers. *New Scientist*; **68**(969): 19–21.

von Neumann J (1955). Can we survive technology? *Fortune*; **91**(6): 106–108, 151–152.

Walgate R (1985). The aerosol and Antarctica. *Guardian*, 30 May 1985.

Walker M (2012). *History of the Meteorological Office*. Cambridge University Press.

Wall Street Journal (1984). Heads in the ozone (Editorial). *Wall Street Journal*, 5 March 1984.

Watson RT (1988). Atmospheric ozone. In: *Conference Proceedings. The Changing Atmosphere: Implications for Global Security, Toronto, Canada, 27–30 June 1988*, pp. 70–91.

Watson RT, MJ Prather, and MJ Kurylo (1988). Present state of knowledge of the upper atmosphere 1988: An assessment report. Report RP-1208, NASA.

Weart SR (2003a). *The Discovery of Global Warming*. Harvard University Press.

Weart SR (2003b). Government: The view from Washington, DC. In: *The Discovery of Global Warming*. American Institute of Physics. https://www.aip.org/history/climate/Govt.htm.

Webster B (1970). Scientists ask SST delay pending study of pollution. *New York Times*, 2 August 1970.

Weinberg AM (1972). Social institutions and nuclear energy. *Science*; **177**(4043): 27–34.

Weinberg AM (1974). Global effects of man's production of energy. *Science*; **186**(4160): 205.

White RM (1989). Greenhouse policy and climate uncertainty. Speech given at the Annual Meeting of the National Academy of Sciences, April 1989, Washington, DC. *Bulletin of the American Meteorological Society*; **70**(9): 1123–1127.

Whittle CE, EL Allen, CL Cooper, HG MacPherson, DL Phung, AD Poole, WG Pollard, RM Rotty, NL Treat, and AM Weinberg (1976). Economic and environmental implications of a US nuclear moratorium, 1985–2010. Report ORAU/IEA-76-4, Institute for Energy Analysis.

Wigley TML (1995). A successful prediction? *Nature*; **376**: 463–4.

Wigley TML (2014). Emails to the author. 19 and 20 June 2014.

Wiin-Nielsen A (1999). The greenhouse effect yes or no? A scientific evaluation. *Water Resources Management*; **13**(1): 59–72.

Wirth T (1996). Making the international climate change process work. Speech delivered to the second session of the conference of parties to the Framework Convention on Climate Change, 17 July, 1996.

WMO (1975). Seventh World Meteorological Congress (Cg-VII): abridged final report with resolutions. Report WMO 416, World Meteorological Organization.

WMO (1976). WMO statement on climatic change. *WMO Bulletin*; **25**(3): 211–2.

WMO (1977). Twenty-ninth session of the Executive Committee, Geneva, 26 May–15 June 1977: abridged report with resolutions. Report, WMO Executive Committee.

WMO (1981). Joint WMO/ICSU/UNEP meeting of experts on the assessment of the role of CO_2 on climate variations and their impact (Villach, Austria, November 1980). Report, World Meteorological Organization.

Bibliography

WMO (1986). Report of the International Conference on the Assessment of the Role of Carbon Dioxide and of Other Greenhouse Gases in Climate Variations and Associated Impacts, Villach, Austria, 9–15 October 1985. Report 661, World Meteorological Organization.

WMO (1988). Report of the International Ozone Trends Panel, 1988. Report, World Meteorological Organization.

WMO (1989). *Conference Proceedings. The Changing Atmosphere: Implications for Global Security, Toronto, Canada, 27–30 June 1988.* World Meteorological Organization.

WMO Committee (1979). Thirty-first session of the Executive Committee: Geneva, 28 May–1 June 1979: abridged report with resolutions. Report, World Meteorological Organization.

WMO Conference (1979). Proceedings of the World Climate Conference: a conference of experts on climate and mankind, Geneva, 12–23 February 1979. Report, World Meteorological Organization.

WMO Executive Committee (1970). Twenty-second session of the Executive Committee: Geneva, 8-16 October 1970: abridged report with resolutions. Report, World Meteorological Organization.

WMO Executive Committee (1975). Twenty-seventh Session of the Executive Committee, Geneva, 26–30 May 1975: abridged report with resolutions. Report, World Meteorological Organization.

WMO Executive Committee (1980). Thirty-second session of the Executive Committee: Geneva, 8–28 May 1980: abridged report with resolutions. Report, World Meteorological Organization.

WMO Executive Committee (1981). Thirty-third session of the Executive Committee: Geneva, 1–17 June 1981: abridged report with resolutions. Report, World Meteorological Organization.

WMO Executive Committee (1982). Thirty-fourth session of the Executive Committee, Geneva, 7–24 June 1982: abridged report with resolutions. Report, World Meteorological Organization.

WMO Executive Council (1984). Thirty-sixth session of the Executive Council, Geneva, 6-23 June 1984: abridged report with resolutions. Report, World Meteorological Organization.

WMO Executive Council (1985). Thirty-seventh session of the Executive Council, Geneva, 5-22 June 1985: abridged report with resolutions. Report, World Meteorological Organization.

WMO Executive Council (1990). Forty-second session Geneva, 11–22 June 1990: Abridged report with resolutions. Report 739, World Meteorological Organization.

World Commission on Environment and Development (1987). *Our Common Future*. Oxford University Press.

World Commission on Environment and Development and Australia Commission for the Future (1990). *Our Common Future* (Australian edn). Oxford University Press. With a foreword by R.J.L. Hawke.

Wright P (1976). Cold but no ice age is longer-range forecast. *The Times*, 18 March 1976.

Wurster CF (1968). DDT reduces photosynthesis by Madne phytoplankton. *Science*; **159**(3822): 1474–1475.

Zillman J (1991). Science and the greenhouse debate. Presentation to 'Workshop on the use and misuse of science in the environmental debate' organised by Sustainable Development Australia in association with AAS and AATSE, 12 August 1991.

Zillman J (1995). Third drafting session for the IPCC Second Scientific Assessment of Climate Change, Asheville, USA, 25–28 July 1995. Report, Australian Delegation to the IPCC.

Zillman J (2007). Some observations on the IPCC assessment process 1988–2007. *Energy & Environment*; **18**(7): 869–892.

Zillman J (2009). A history of climate activities. *WMO Bulletin*; **58**(3): 141–150.

Zillman J (2015). Climate science and greenhouse policy: some observations from early years at the science-policy interface. *Bulletin of the Australian Meteorological and Oceanographic Society*; **28**(4 (suppl.)): 106–126.

Zillman J (2016). Email correspondence regarding the history of atmospheric scares. 6–18 April 2016.

Zubrin R (2012). The truth about DDTs and *Silent Spring*. *The New Atlantis*, 27 September 2012. Online at: http://www.thenewatlantis.com/publications/the-truth-about-ddt-and-silent-spring.

Index

Index

European Economic Community (EEC), 90, 118, 168
Evans, Wayne, 115–117
Ewing, Maurice, 159

Farman, Joe, 99–103, 114, 115, 118, 125, 289–290
Federal Aviation Administration (FAA), 54, 56
Ferguson, Howard, 216, 219, 243
Fertiliser, 86
Fingerprinting, *see* Pattern analysis
Fleming, James, 14, 258
Flohn, Hermann, 167, 179
Food and Drug Administration (FDA), 88–89
Ford, Gerald, 127
France, 90, 91, 115, 227
Frankenstein syndrome, 268, 278
Friends of the Earth, 25, 27, 129

G77, 246, 299
Gaia hypothesis, 24, 73–75, 97
Gaia: A New Look at Life on Earth (book), 24
Gerasimov, IP, 185
Germany, 12, 63, 115, 134, 167, 227, 228
Gibbs, WJ, 179n
Gille, John, 109
Global Atmospheric Research Programme (GARP), 13, 140, 173, 177, 179, 188, 191, 199, 202, 237, 240–241
Global Change Research Program, Office of the, 244, 274
Goldburg, Arnold, 40, 44, 45, 57–58
Goldemberg, Jose, 247n
Goldwater, Barry, 57
Gonzalez, Felipe, 227
Goodman, Gordon, 215–232
Gorbachev, Mikhail, 228
Gore, Al, 224, 232, 299, 310
Gori, Gio, 43
Greenpeace, 283, 297

Grubb, Michael, 286n
Gummer, John, 288–289, 299, 310
Gutowsky, Herb, 88

Hague Declaration (1989), 227, 246
Hallgren, Richard, 233
Hansen, James, 193–197, 211, 223–224, 234, 261–262
Hare, Ken, 212, 216
Harrison, Halstead, 39–40, 45, 57–58, 60–61, 78, 87, 125
Heath, Donald, 103
Heffner, Hubert, 153
High Speed Dreams (book), 53
Hirschfelder, Joe, 45–46
Hockey Stick graph, 158, 309, 310
Holocene epoch, 148, 157, 158, 173, 179
Houghton, John, 2–3, 191, 222, 236–241, 245, 246, 253n, 262, 268–269, 272, 274–275, 279, 281–283, 294–296, 301–310
Howard, John, 301
Human Environment Conference (1972), 22, 29, 247

Ice ages, *see also* Little Ice Age, 151–174, 179, 193, 256, 259
Ice sheet, 169, 172, 194, 195
Idso, Sherwood, 253n
Inadvertent Modification of the Stratosphere (IMOS), 84, 88, 94
India, 161, 226, 246, 271
Insecticides, 18, 22, 23, 76
Institute for Energy Analysis, 63, 130–134, 144, 208, 216
Insurance industry, 298
Intergovernmental Negotiating Committee (INC), 250, 252, 264–271, 287
International Geophysical Year, 12, 35, 99, 111, 138, 217
International Ozone Trends Panel, 107, 109–111
Iranian revolution, 127, 184

About the author

Bernie Lewin became involved in the Australian environment movement in the early 1980s, where he was involved in the campaigns to save old growth forests in Gippsland and the Franklin River in Tasmania. However, he was always sceptical of the science behind the global warming scare. Amazed at how this scare had come to overwhelm all other environmental campaigns, in 2009 he started to investigate how there came to be such a widespread belief that it had a firm foundation in the authority of science. Approaching this question through a separate interest in the history of science, his findings were posted on his blog *Enthusiasm, Scepticism and Science*, and also on the *Watts Up With That?* and *Bishop Hill* blogs. These explorations led to his being invited to contribute a chapter in *Climate Change: The Facts 2014* and, for the Global Warming Policy Foundation, a report entitled *Hubert Lamb and the transformation of climate science* and now this comprehensive history of the IPCC.

371

About the GWPF

The Global Warming Policy Foundation is an all-party and non-party think tank and a registered educational charity which, while openminded on the contested science of global warming, is deeply concerned about the costs and other implications of many of the policies currently being advocated.

Our main focus is to analyse global warming policies and their economic and other implications. Our aim is to provide the most robust and reliable economic analysis and advice. Above all we seek to inform the media, politicians and the public, in a newsworthy way, on the subject in general and on the misinformation to which they are all too frequently being subjected at the present time.

The key to the success of the GWPF is the trust and credibility that we have earned in the eyes of a growing number of policy makers, journalists and the interested public. The GWPF is funded overwhelmingly by voluntary donations from a number of private individuals and charitable trusts. In order to make clear its complete independence, it does not accept gifts from either energy companies or anyone with a significant interest in an energy company.

Views expressed in the publications of the Global Warming Policy Foundation are those of the authors, not those of the GWPF, its trustees, its Academic Advisory Council members or its directors.